MOLECULAR AND CELLULAR MECHANISMS OF NEURONAL PLASTICITY

Basic and Clinical Implications

ADVANCES IN EXPERIMENTAL MEDICINE AND BIOLOGY

Recent Volumes in this Series

Volume 443
ADVANCES IN LACTOFERRIN RESEARCH
Edited by Geneviève Spik, Dominique Legrand, Joël Mazurier, Annick Pierce, and Jean-Paul Perraudin

Volume 444
REPRODUCTIVE TOXICOLOGY: *In Vitro* Germ Cell Developmental Toxicology, from Science to Social and Industrial Demand
Edited by Jesús del Mazo

Volume 445
MATHEMATICAL MODELING IN EXPERIMENTAL NUTRITION
Edited by Andrew J. Clifford and Hans-Georg Müller

Volume 446
MOLECULAR AND CELLULAR MECHANISMS OF NEURONAL PLASTICITY: Basic and Clinical Implications
Edited by Yigal H. Ehrlich

Volume 447
LIPOXYGENASES AND THEIR METABOLITES: Biological Functions
Edited by Santosh Nigam and Cecil R. Pace-Asciak

Volume 448
COPPER TRANSPORT AND ITS DISORDERS: Molecular and Cellular Aspects
Edited by Arturo Leone and Julian F. B. Mercer

Volume 449
VASOPRESSIN AND OXYTOCIN: Molecular, Cellular, and Clinical Advances
Edited by Hans H. Zingg, Charles W. Bourque, and Daniel G. Bichet

Volume 450
ADVANCES IN MODELING AND CONTROL OF VENTILATION
Edited by Richard L. Hughson, David A. Cunningham, and James Duffin

Volume 451
GENE THERAPY OF CANCER
Edited by Peter Walden, Uwe Trefzer, Wolfram Sterry, and Farzin Farzaneh

Volume 452
MECHANISMS OF LYMPHOCYTE ACTIVATION AND IMMUNE REGULATION VII: Molecular Determinants of Microbial Immunity
Edited by Sudhir Gupta, Alan Sher, and Rafi Ahmed

MOLECULAR AND CELLULAR MECHANISMS OF NEURONAL PLASTICITY

Basic and Clinical Implications

Edited by

Yigal H. Ehrlich

The College of Staten Island of The City University of New York, and
The CSI/IBR Center for Developmental Neuroscience
Staten Island, New York, and
The City University Graduate School
New York, New York

SPRINGER SCIENCE+BUSINESS MEDIA, LLC

Library of Congress Cataloging-in-Publication Data

Molecular and cellular mechanisms of neuronal plasticity : basic and clinical implications / edited by Yigal H. Ehrlich.
p. cm. -- (Advances in experimental medicine and biology ; v. 446)
"Proceedings of the Neuroscience Symposium on Molecular and Cellular Mechanisms of Neuronal Plasticity, held December 4, 1996, in Staten Island, New York"--T.p. verso.
Includes bibliographical references and index.
ISBN 978-0-306-46040-1 ISBN 978-1-4615-4869-0 (eBook)
DOI 10.1007/978-1-4615-4869-0
1. Neuroplasticity--Congresses. 2. Molecular neurobiology--Congresses. 3. Neurons--Congresses. I. Ehrlich, Yigal H. II. Neuroscience Symposium on Molecular and Cellular Mechanisms of Neuronal Plasticity (1996 : Staten Island, New York, N.Y.) III. Series.
[DNLM: 1. Neuronal Plasticity--physiology congresses. 2. Neurons--physiology congresses. W1 AD559 v.446 1998]
QP363.3.M65 1998
611'.0188--dc21
DNLM/DLC
for Library of Congress 98-31763
CIP

Proceedings of the Neuroscience Symposium on Molecular and Cellular Mechanisms of Neuronal Plasticity, held December 4, 1996, in Staten Island, New York, with chapters written by symposium speakers and additional invited authors during 1997 and updated through March 1998.

ISBN 978-0-306-46040-1

Originally published by Kluwer Academic / Plenum Publishers in 1998

PREFACE

Cells in the nervous system have the capability of undergoing extremely long-lasting alterations in response to hormonal, pharmacological and environmental stimulations. The mechanisms underlying this neuronal plasticity are activated by experiential inputs and operate in the process of learning and the formation of memories in the brain. Numerous studies have proven the biological basis of memory formation and have begun to identify the biochemical traces and cellular circuits that are formed by experience, and which participate in the storage of information in the brain, its retention for long durations, and its retrieval upon demand. Furthermore, the elucidation of molecular and cellular mechanisms operating in the adaptive processes which underlie neuronal plasticity have begun to yield novel strategies in the diagnosis and treatment of neurological diseases and neurophychiatric disorders. A major practical outcome of these investigations has been the emergence of a new generation of therapeutic agents; memory enhancing drugs developed on the basis of understanding the biochemistry, cellular biology and molecular genetics of processes underlying neuronal plasticity. These studies are at the focus of the present Volume, which has emerged initially from a Symposium held at the College of Staten Island of The City University of New York on December 4, 1996. This Volume includes contributions submitted during 1997 and early 1998 by presenting speakers of the Symposium, as well as additional invited authors who conduct unique studies in this area. As in my previous editorial work (Ehrlich et al., Volume 116 and Volume 221 of Advances in Experimental Medicine and Biology, Plenum Press, New York), I hope that the readers will find here a useful source of information and ideas for stimulating studies which will serve to further narrow the gap between basic neuroscience research and its clinical implications.

One of the guidelines in selecting the contributions invited as chapters for this Volume has been the presentation of research areas that have not been highlighted in previous books on this topic. Thus, in addition to studies on the involvement of functional proteins (enzymes, receptors) in neuronal adaptation, this Volume presents recent developments on the critical roles of bioactive lipids and nucleotides in these processes. In addition to the widely studied role of second messengers-regulated protein kinases in mechanisms underlying neuronal plasticity, a review of studies on extracellular protein phosphorylation systems operating on the surface of brain neurons is presented in this Volume. Together with studies of mammalian brain, the advantages of investigating avian hippocampus as a model to study synaptic plasticity is presented here. While the NMDA sub-type of glutamate receptors has been a major focus of studies on hippocampal synaptic plasticity, a chapter on the clinical impact of studies on the role of metabotropic glutamate receptors in the plasticity of excitatory responses in the hippocampus is presented here as a bridge between the two sections of the Volume. The first section presents studies of basic mechanisms operating in a wide-range of adaptive processes. These include synaptic changes responsible for associative memory, long-term habituation, long-term potentiation, neuronal development and programmed neuronal cell-death (apoptosis). The second section of this Volume, beginning with Chapter 9, presents recent advances in investigations that have demonstrated the clinical implications of this research. These include: state-of-the-art use of transgenic models in studies of molecular and cellular mechanisms implicated in familial Alzheimer's disease and Amyotropic Lateral Sclerosis (ALS); studies of specific proteins implicated in Alzheimer's disease—including an adapter that binds to the β-amyloid precursor protein (β-APP) and the microtubular protein Tau and its membrane-bound counterpart. The advantages of using cell culture models for elucidating the causes of neuronal degeneration and for identifying mechanisms of neuroprotection are also presented among the chapters in the section on Clinical Implications.

The Editor wishes to express here his thanks to a few of the many individuals who contributed to the organization of the Symposium: the President of the College of Staten Island (CSI) Dr. Marlene Springer and the Vice-President for Academic Affairs and Provost Dr. Mirella Affron, the Director of the New York State Institute of Basic Research in Developmental Disabilities (IBR) Dr. Henryk M. Wisniewski, the Deputy Director of the CSI/IBR Center for Developmental Neuroscience Dr. E. Trenkner, and the Dean of Science at CSI Dr. Martin Zeldin. Special thanks are due to the staff of the Center for Performing Arts of CSI, and in particular to the secretary of the Program in Neuroscience of CSI, Rosa Shirippa for her tireless efforts. Finally, the support and help of my colleague and friend Dr. Elizabeth Kornecki in the preparation of this Volume is acknowledged with affection and admiration.

Yigal H. Ehrlich

CONTENTS

1

MOLECULAR SPECIFICITY OF SYNAPTIC CHANGES RESPONSIBLE FOR ASSOCIATIVE MEMORY

Daniel L. Alkon

Laboratory of Adaptive Systems
NINDS, NIH
Bethesda, Maryland

INTRODUCTION

Specificity of molecular mechanisms for synaptic weight regulation is critical for understanding the cellular basis of learning and memory. Thus, while we have known for decades that protein synthesis is involved in long-term memory (Flexner et al., 1963; Hyden et al., 1965; Agranoff et al., 1966; Matthies, 1989; Van der Zee et al., 1992; Nelson and Alkon, 1992; see Kandel, this volume), protein synthesis is also necessary for so many cellular functions that its direct contribution to memory function remains obscure. Here, however, we describe a new signalling protein, *Calexcitin*, which powerfully and specifically regulates synaptic weight as well as synaptic sign and is activated by a molecular cascade more directly implicated in associative learning of diverse molluscan and mammalian species . This cascade begins when temporally related training stimuli (e.g. an auditory tone and a touch stimulus to the cornea) elicit temporally related synaptic signals that in turn elicit temporally related second messengers such as Ca^{++}, DAG (diacylglycerol), and AA (arachidonic acid). Temporally associated second messengers activate protein kinase C (PKC) through its translocation to the inner surface of neuronal membranes (and possibly membranes of subcellular organelles such as the ER). Membrane-associated PKC, now sensitive to low levels of Ca^{++} (0.1–1.0 uM), phosphorylates critical signaling proteins such as the recently sequenced cp20 (now called "*Calexcitin*"), the first known protein to bind both Ca^{++} and GTP. *Calexcitin*, shown to be a high affinity substrate of Ca^{++}-dependent isozymes of PKC (particularly the α-isozyme), when phosphorylated during learning, also becomes membrane associated and inactivates voltage-dependent K^+ channels (such as I_A, I_C etc.). Recent experiments (Weh et al., in press) also implicate Calexcitin in the regulation of Ca^{++} release from the endoplasmic reticulum and thus further Ca^{++}-mediated transformations of synaptic weight and perhaps structure. Still other recent experiments (Cavallero et al., Soc.for Neuroscience, Abstract, in press) showed increased

Molecular and Cellular Mechanisms of Neuronal Plasticity, edited by
Ehrlich. Plenum Press, New York, 1998.

activation of the gene responsible for ryanodine receptor expression after spatial maze learning.

Through this molecular cascade, stimuli associated during training increase dendritic excitability and thus increase synaptic weight through enhanced post-synaptic responsiveness, as demonstrated for associative memory within the molluscan (*Hermissenda*) visual-vestibular network, the hippocampus, and the H6 cerebellar cortex. In more recent studies, *Calexcitin* was found to markedly enhance synaptic EPSP's and to inactivate IPSP's. These mechanisms of memory storage induced *in vivo* show some convergence with *in vitro* models of cellular memory (e.g LTP, LTD, LTT Long-term synaptic transformation-see below) such as involvement of PKC activation. One divergence, however, was confirmed when antisense oligonucleotides to mRNA that codes for mkv1.1 *in vivo* eliminated memory retention (Meiri et al., in press) along with specific voltage-dependent K^+ channels, but leaving LTP unimpaired. These conserved molecular and biophysical mechanisms that can regulate synaptic weight during associative memory have also recently been implicated as targets of pathophysiology induced by nM levels of soluble β-amyloid in the clinical entity Alzheimer's Disease, recognized by its memory deficits.

SYNAPTIC TRANSFORMATIONS DURING ASSOCIATIVE LEARNING

In vitro changes of synaptic efficacy in fully differentiated nervous systems have been known and analyzed since Feng first identified post-tetanic potentiation at the neuromuscular junction in the late 1930's. Since that time, many additional examples of synaptic efficacy modification have been identified in diverse species (Katz, 1966; Kuffler et al., 1984). These include synaptic facilitation at the lobster and crab neuromuscular junctions, heterosynaptic facilitation in the mollusc abdominal ganglion, long-term potentiation (LTP) of glutamatergic EPSP's recorded from pyramidal cells in the rat hippocampus, and long-term depression (LTD) of glutamatergic EPSP's recorded from Purkinje cells of the rat cerebellum. All of these changes of synaptic efficacy are candidates for synaptic mechanisms postulated by such pioneers as Cajal, Sherrington, and Pavlov to provide a physiologic basis for memory. Yet, these examples of synaptic modification have been difficult to observe during associative learning and memory behavior of living animals.

To assess *in vitro* synaptic change(s) in the context of *in vivo* associative learning and memory, LAS scientists set out many years ago to correlate synaptic changes of efficacy with behavioral acquisition and retention of associative memory. Pavlovian conditioning of the mollusc *Hermissenda* (Alkon, 1983*)*, for example, produced long-term enhancement of type B cell excitability (and thus responsiveness) and thereby enhanced input to and output from downstream motor neurons. Pavlovian conditioning of the rabbit (Thompson, 1986) was shown in LAS studies to enhance post-synaptic dendritic excitability and the amplitude of summated EPSP responses of CA1 pyramidal cells (Disterhoft et al., 1986; Lo Turco et al., 1985). Still other LAS studies (Schreurs et al., 1997) demonstrated intradendritic excitability increases in cerebellar microzones after rabbit Pavlovian conditioning (Figure 1). All of these examples correlated with and/or could be accounted for by long-term inactivation of voltage-dependent K^+ channels (as explained below) and are consistent with current understanding of ion channel distribution within the pyramidal cell dendritic tree (Migliore et al., 1995). A crucial part of this past research was accomplished when LAS investigators pioneered the analysis of cellular changes within brain slices taken from trained and control animals exposed to a variety of stimulus paradigms.

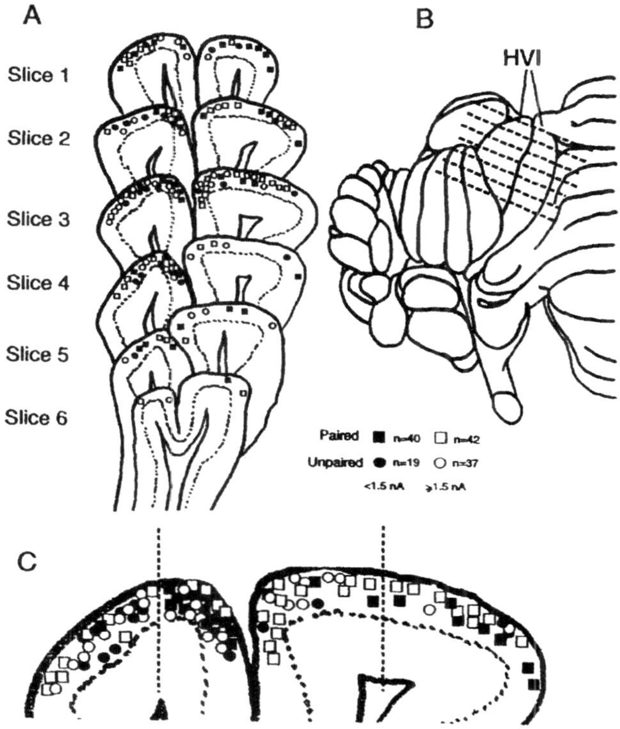

Figure 1. Membrane excitability is localized to a specific microzone. A:recording sites for individual Purkinje cells located on sequential parasaggital slices from Larsell's lobule HV1. B: an anterior view of the right cerebellum indicating the area from which the slices were cut (dashed lines). C: a composite of slices 2,3, and 4 showing electrode locations to the left and right of the midline of each folium. The slices are depicted so that the rabbit's left folium of lobule HV1 is shown on the left and the rabbit's right folium is shown on the right. The cell locations for Paired (□ and ■) and Unpaired (● and O) animals are based on visual inspection (x12 magnification) of the electrode position in the slice at the time or recording. Locations are divided between those below (■ and ●) or at and above (□ and O) the mean dendritic spike threshold for cells from Paired animals (1.5 nA). The individual slices were obtained by tracing cresyl violet-stained parasaggital sections cut through the right cerebellar hemisphere. Dotted line within each slice indicates the border between the molecular and granular layers.

All of the major changes of brain network functions measured in slices (synaptic weight, voltage-dependent K+ channels, PKC activation, RNA turnover etc.) were always closely correlated with the acquisition and storage of memory in living animals (Alkon, 1989).

Once these correlates were obtained we could then develop *in vitro* model systems to analyze synaptic and molecular events necessary and sufficient to produce the *in vivo* synaptic changes. In the mollusc *Hermissenda*, for example, we began with the complete visual and vestibular networks (Alkon, 1983) and all of their synaptic contacts. Within the synaptic connectivity, mapped by thousands of simultaneous pre-and post-synaptic intracellular recordings, are many of the neurochemical messenger systems (cholinergic, GABAergic etc.) now known to characterize hippocampal and cortical networks in mammalian brains. Stimulus paradigms for isolated *Hermissenda*, hippocampal slice, and cerebellar slice networks, could then be designed to simulate the paradigms of stimuli used to train living animals.

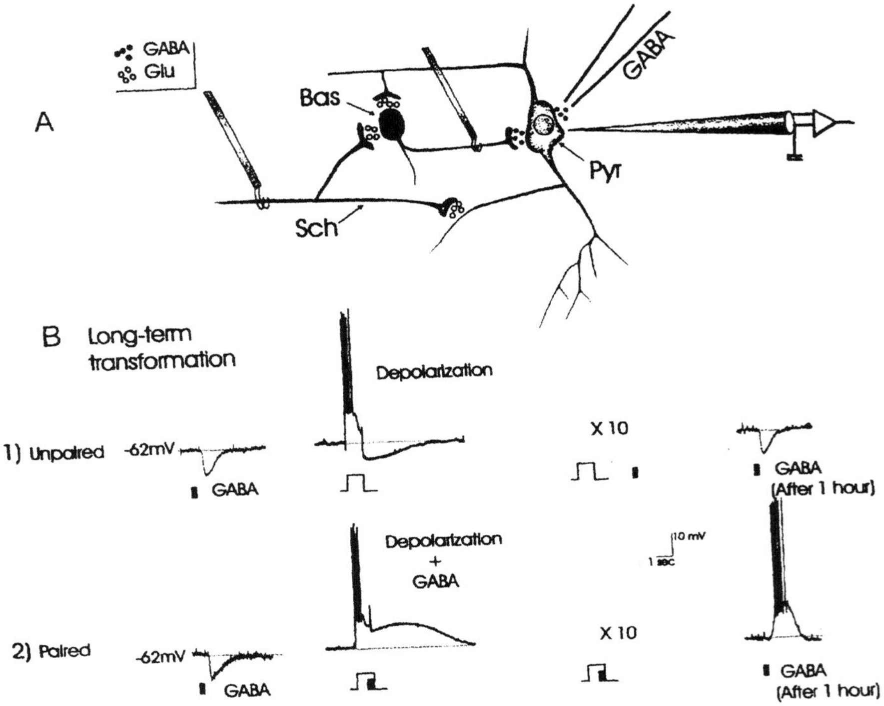

Figure 2. Long-term transformation of somatic hyperpolarization in response to exogenous GABA application. (A) Excitatory and inhibitory inputs to the CA1 pyramidal (Pyr) neurons in rat hippocampus. Excitatory input from Schaffer collaterals (Sch) to CA1 dendrites is mediated by glutamate (Glu). Inhibitory inputs are mediated by GABA released from three spatially segregated types of basket cell (Bas) interneurons (9), each specifically innervating dendrites, soma or axon of the postsynaptic CA1 neuron. Only basket-to-soma interneurons are shown here to illustrate recording conditions. (B) GABA (indicated by black bar) was applied with a 30-msec 20-psi (1 psi = 6.9 kPa) pressure pulse through an ejection pipette (containing 1 mM GABA dissolved in artificial cerebrospinal fluid) near the CA1 pyramidal soma and elicited a pure hyperpolarizing response (left trace) with repeated applications at 2-min intervals. (1) After test applications of GABA alone that elicited hyperpolarizing responses (Left trace) depolarizing steps alone elicited a train of action potentials followed by an afterhyper-polarization (Middle trace). Repeated alteration (10 times at 5-sec intervals) of depolarizing steps with GABA application alone (unpaired stimulation) did not change the responses to GABA puffs (Right trace). (2) Similar GABA puffs were delivered immediately before termination of a 1-sec depolarizing current step, as indicated by an open bar (paired stimulation). Paired stimulation was repeated 10 times at 5-sec intervals. The middle trace shows a typical voltage response during one such pairing. Current intensity was sufficient (1.0–2.0 nA) to elicit at least five action potentials that were typical voltage response during one such pairing. Current intensity was sufficient (1.0–2.0 nA) to elicit at least five action potentials that were interrupted by the onset of the GABA puff. Typically, paired stimulation produced a late, prolonged depolarization after the train of action potentials (Middle trace). After 10 pairings, GABA alone elicited a transient and sometimes reversible depression of the fast hyperpolarizing response together with a depolarizing wave (data not shown). Within 10 min. of such a pairing, the responses to GABA alone had become transformed into depolarizations, which often elicited action potentials (Right trace).

An *in vitro* associative memory model, for example, was designed to mimic the actual stimuli received by the molluscan visual-vestibular networks during the *in vivo* Pavlovian conditioning. Paired visual-vestibular stimuli were found to cause a novel form of synaptic change which LAS scientists named "Long-term Transformation" or LTT. LTT, as first identified in the mollusc (Alkon, et al., 1992; Usherwood, 1993) and then, with remarkable similarity, in the rat hippocampus (Collin et al., 1995), was shown to cause GABAergic inhibition to be transformed into GABAergic excitation (Figure 2). In the hippocampus, release of GABA from pre-synaptic basket cell terminals temporally associated with depolarization of post-synaptic CA1 pyramidal cells caused LTT lasting more than one hour. After LTT, the GABAergic excitatory potential had a new reversal potential (-45 vs -75 mv for the GABAergic IPSP), involved an increased membrane conductance, and was eliminated by the $GABA_A$- Cl^-channel blocker, bicuculline, but not by the glutametergic blockers, CNQX and/or APV. LTT, which persisted with no further paired stimulation for 60 minutes or longer and was shown to contribute to pairing-induced long-term potentiation but not to long-term potentiation induced by presynaptic stumulation alone..

Subsequent LAS experiments then showed that release of ACh (acetylcholine) from *S. oriens* terminals presynaptic to the pyramidal cells, when associated with GABA release from presynaptic basket cell terminals, caused the same LTT (Dahl et al., in press). Temporal association of the cholinergic agonist carbachol (20μM) with tetanization of GABAergic basket cell terminals induced LTT in hippocampal CA1 pyramidal neurons. Furthermore, either pharmacologic enhancement (with physostigmine, 20 μM) or electrical activation of endogenous cholinergic transmission also induced LTT. Other experiments with dantrolene (a blocker of Ca^{++} release from intracellular stores), and acetazolamide (a blocker of the pump responsible for the HCO_3-gradient across the cell membrane) suggested that LTT arises, following elevation of Ca_i^{2+}, when HCO_3- flows out of the pyramidal cell through GABA-activated Cl-channels (with reduced Cl^-influx) to cause depolarization. Remarkably, very recent experiments showed that injection of *Calexcitin* (see below) into hippocampal pyramidal cells, induces similar long-term transformation (LTT) of GABAergic inhibition in excitation.

Thus, temporally associated stimulation of two or more pre-synaptic inputs (cholinergic, GABAergic and glutamatergic) causes long-term changes of synaptic sign that could have *in vivo* relevance. Classical LTP typically requires tetanization of only a single glutamatergic pathway, whereas LTT involves the induction of long-term synaptic change by the synergistic interaction of multiple neurotransmitter pathways. These findings demonstrate that the sign of hippocampal synapses can undergo long-lasting reversal when GABA release is associated with post-synaptic ACh receptor activation.

INTRACELLULAR SIGNALING DURING ASSOCIATIVE LEARNING AND MEMORY

Molecular signals that regulate synaptic transmission and ion channels within neuronal membranes, like examples of synaptic change, constitute an embarrassment of riches. Second messengers such as Ca^{++}, DAG, AA, and cyclic-AMP and their kinases (PKC, adenylate cyclase, Ca^{++}/CAM-II kinase, and tyrosine kinases) have all been implicated in one or more of the many *in vitro* and *in vivo* strategies adopted by the LAS and other laboratories.

The LAS strategy of correlating *in vivo* associative learning and memory behavior with electrophysiologic events, however, has proven useful for identifying particularly those biochemical cascades that underlie associatively-induced synaptic change. The physiclogic correlates, in fact, were used to provide a cellular context for crucial molecular steps during synaptic change. LAS studies showed, for example, that prolonged elevation of intracellular Ca^{++} and activation of protein kinase C produced the same persistent inactivation of voltage-dependent K^+ currents and increased membrane excitability that had been correlated with associative learning in identified snail neurons and in hippocampal pyramidal cells.

Many years ago we undertook to define calcium-dependent phosphorylating pathways that might be responsible for the persistent changes of voltage-dependent K+ currents and enhanced synaptic weight we had correlatied with associative learning and memory. The first major pathway we uncovered was mediated by CA^{++}/CAM-II kinase which, together with Ca^{++} loading conditions, induced persistent K^+ current reduction (Acosta-Urquido et al., 1984; Sakakibara et al., 1993). Activation of protein kinase C dependent on a combination of Ca^{++}, DAG, and AA, was then demonstrated for *Hermissenda* learning-specific changes of voltage dependant K^+ currents (see Alkon et al., 1986; 1990) and imaged by LAS scientists in the rabbit hippocampus for classical conditioning of the rabbit nictitating membrane (Bank et al., 1988; Olds et al., 1989). Since then, we have found patterns of PKC activation in hippocampal and cortical regions specific for one- and two-platform spatial maze learning (Olds et al., 1993) olfactory discrimination learning, and in single identified cells for *Hermissenda* Pavlovian conditioning (McPhie et al., 1992). Many other laboratories (Van der Zee, 1992) around the world have now confirmed

Figure 3 . (A) Single-channel current traces of K^+ channel activity in an excised inside-out membrane patch of human skin cultured fibroblast recorded in the absence of CE 7 min after excision of the membrane patch (middle trace), at 10 min after addition of heat-inactivated CE (bottom trace), and at 3 min after addition of intact CE (top trace). The channel shown was typical of those recorded at -40 mV. Pipette solution was 140 mM KCl/2 mM $CaCl_2$/1 mM $MgCl_2$/10 mM Hepes (NaOH)/0.001 mM tetrodotoxin, pH 7.4; bathing solution was 150 mM NaCl/5 mM KCl/2 mM $CaCl_2$/1 mM Mg Cl_2/10 MM Hepes (NaOH, pH 7.4. Addition of protein was made by a 20-μl plastic pipette near the tip of the patch micropipette as 10-μl aliquots of CE in 10 nM imidazole HCl. Petri dish contain 2 ml of bath solution. CE markedly reduced the mean open time (^{T}open) and probability of openings. (B) Amplitude distributions of unitary K^+ current records were fitted with a Gaussian function. Unitary current peak amplitudes were -10.6, -11.0, and -9.9 pA after addition of CE and heat-inactivated CE, and no addition, respectively. There were no significant differences in distribution among the three groups. (C) Open time distributions for the data collected during 30-s periods after excision of the membrane (as mentioned in A) were fitted to a single exponential. Exponential decay constants for opern time distribution were 1.4, 12.0, and 12.5 ms after addition of CE and heat-inactivated CE, and no addition, respectively. CE application markedly recuced the open times of the K+ channel. (D) Membrane excitability increases due to CE. (Left) Passively propagated somatic spikes (small amplitude) occurred spontaneously in most cells, including the one illustrated here, before CE injection (pre). Spontaineous local dendritic calcium spikes (large amplitude) did not occur in some cells (upper trace, pre-calexcitin) but then did occur after injection of CE (post) through the recording electrode. In Purkinje cells hyperpolarized 20 mV below the membrane potential for somatic spikes, calcium spikes could be elicited by lower levels of injected positive current pulses after injection of CE. (Right) Mean percent change in the threshold current required to elicit local calcium spkes was lower for CE-injected cells vs. controls injected with heat-inactivated CE or 3 M potassium acetate ($P < 0.001$, Student's *t* test). (E) Intracellular recordings of the *Hermissenda* type B photoreceptor response to a 1-s flash of light (10^3 erg/cm 2sec) before (Upper) and after (Lower) injection of purified cloned CE. The *Hermissenda* photoreceptors were isolated and submerged in artificial sea water (430 mM NaCl/10 mM KCl/10 mM $CaCl_2$/50 mM $MgCl_2$/10 MM Hepes Na. pH 7.4). The CE (intraelectrode concentration 364 nM) was brought to 1 M in KAc (pH 7.4) and injected (3 min, 2 nA) into the photoreceptor with the recording electrode. Recordings were obtained using intracellular amplifier (Axopatch 2A), digitized at 50 HZ (Digidata 1200), and analyzed by computer. The normal light response returned to the original resting potential within a minute (Upper) Injection of CE (n=5 cells) enhanced the light response returned to the original resting potential within a minute (Upper). Injection of CE (n=5 cells) enhanced the light response and the cell remained depolarized for more than 5 min (Lower, only 1 min is represented). Injection of heat-inactivated CE (n=6 cells) using the above protocol did not alter the light response. Time scale at right is compressed by 3X.

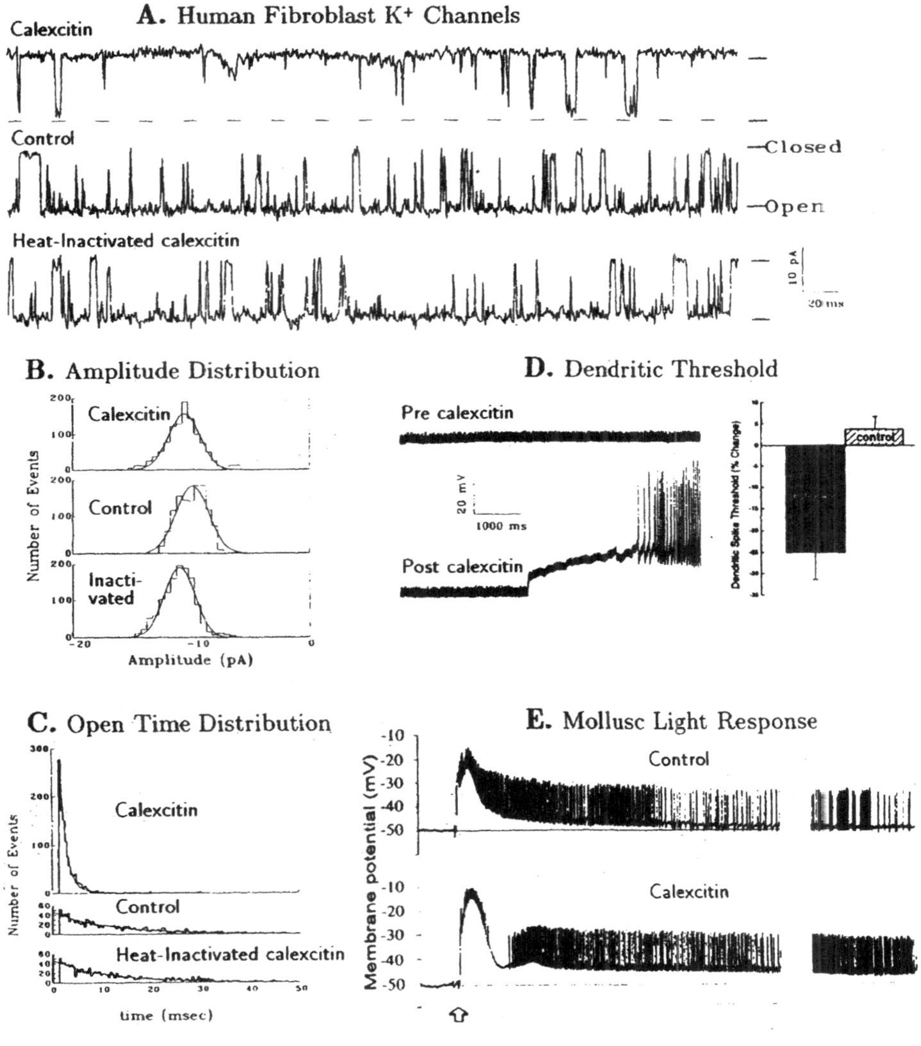

and extended these findings with other learning paradigms and other methodologies including monoclonal antibodies to specific PKC isozymes within particular neuronal compartments such as dendritic branches (Olds et al., 1993). Activation of PKC is often manifest as "translocation", i.e. a movement of PKC from cytosolic domain(s) to the inner surface of the neuronal membrane (Schacter et al., 1996). Membrane-associated PKC is responsive to low levels of Ca^{++} for phosphorylation activity.

LAS studies have also identified a new Ca^{++} and GTP-binding protein (Nelson et al., 1996) that is phosphorylated by the Ca^{++}-activated PKC isozymes, α, β, and γ. This recently sequenced 22 kd protein, called *Calexcitin*, was previously identified with chromatographic techniques and shown to change during Pavlovian conditioning (Nelson et al., 1990). *Calexcitin* injected into molluscan and mammalian neurons or applied to the inner surface of inside-out membrane patches inactivates voltage-dependent K^+ channels (Figure 3). Although

Calexcitin appears to act directly on these K^+ channels, much more research will be necessary to determine the role of possible intermediate(s) on conductance regulation (Wickman and Clapham, 1995). In any case, a Ca^{++}-activated PKC substrate produces changes of membrane excitability similar to those observed during associative learning and memory. *Calexcitin* also increases mRNA turnover as was also observed with associative memory consolidation.

For purified *Calexcitin*, a single peak containing protein, GTPase activity, and GTP-binding activity was detected, indicative of a G protein. *Calexcitin* also bound azidoanilido GTP, a photoaffinity label for G-proteins. The GTPγ^{35}S binding of *Calexcitin* corresponded to a stoichiometry of about 0.9 mol GTPγ^{35}S/mol protein. The turnover rate of GTPase under these conditions was estimated to be about 0.1 min^{-1}, a value comparable to that of other G proteins. No GTP-binding activity comigrated with the other conditioning-associated proteins (cp16, cp18, or cp27) (Nelson et. al., 1990).

In addition to two EF-hand consensus domains (Kretsinger, 1987) the sequence contained a possible GTP-binding consensus motif (Walker et al., 1982), a PKC phosphorylation site (Turner et al., 1985), two possible N-myristoylation sites (Musil et al., 1988), and a polyisoprenylation consensus (CAAX) domain at the C-terminal (Nelson et al., 1996). CAAX is a domain shared among GTP-binding proteins used for attachment of an isoprenoid tail which facilitates membrane binding (Glomset et al., 1990; Lowy and Willumsen, 1989). The theoretical pI (4.97) of cloned *Calexcitin* agreed with the pI measured by isoelectic focusing (5.3) and with that observed for purified native squid and *Hermissenda Calexcitin* (pI=5.2) on 2D gel electrophoresis (Nelson et al., 1994).

In still other experiments, *Calexcitin* was found to regulate axonal transport. At 8,000–10,000 x magnification, 0.1–0.3 μ m particles in crab axons were found to undergo reduced transport in the retrograde (but not orthograde) direction. Reduced transport was found not to involve a change in the velocity of individual particle movements, but in the total number of particles moving per unit time. Thus, *Calexcitin* appears to be increasing the intraxonal binding or docking of particles, a function consistent with the homology of the C terminal end of its sequence with the ARF (ADP ribosylation factor) proteins. These proteins have been independently implicated in the transport of membrane particles between the Golgi and the endoplasmic reticulum. This transport regulating property of *Calexcitin* may be related to its ability to induce rearrangement of terminal synaptic branches when injected into neurons. A similar rearrangement of terminal synaptic branches, called "focusing", was also observed in single identified Type B neurons many days after Pavlovian conditioning of the mollusk *Hermissenda*.

Finally, a major new function of *Calexcitin*, has been identified in the regulation of intracellular Ca^{++} release from the endoplasmic reticulum (Nelson, et al., in press; Weh et al., in press). This function not only represents a potentially new molecular cascade for Ca^{2+} control, but also a possible upstream event for some of the other functions mentioned above.

RELEVANCE OF *IN VITRO* TO *IN VIVO* MECHANISMS OF SYNAPTIC TRANSFORMATION

The relevance of *in vitro* learning models to the *in vivo* mechanistic constructs for memory storage outlined above is suggested by specific convergences. LTT, for example, is induced by the same temporally associated pathway stimuli required for *in vivo* memory. LTT, in addition, apparently involves PKC activation and K^+ channel inactivation.

Both LTP and LTD also show a requirement for prolonged elevation of Ca_i^{++} and PKC activation. However, no differences of voltage-dependent K^+ channels have ever been observed for either LTP or LTD. Conversely conditioning-specific enhancement of EPSP summation that correlated with reduction of I_{AHP} of hippocampal pyramidal cells occurred in the absence of LTP or LTD (LoTurco et al., 1988). Similarly, Schreurs et al. found enhanced excitability of Purkinje cell dendrites in the absence of LTD. These studies suggested that while LTP and LTD (Linden and Connor, 1995) could share common molecular and biophysical steps with *in vivo* learning and memory, there may be important differences (such as involvement of changes in voltage-dependent K^+ channels).

To further examine this possibility of mechanistic divergence between *in vitro* and *in vivo* learning models, we used antisense to voltage-dependent K^+ channel (mkv1.1) mRNA (Meiri et al., in press). In contrast to *Drosophila* memory mutant and mouse transgenic manipulations that can impact on the development of brain networks as well as network function, antisense ODN has the potential of reversibly disrupting the expression of proteins in a fully differentiated brain without deranging its genetically specified neuronal or synaptic architecture. *In vivo* antisense ODN administration has in the past been compromised by (1) the instability of the short single stranded oligonucleotides involved, and (2) poor uptake of the oligonucleotides into cells. Biochemical modification of the phosphodiester backbone to better resist nucleases and co-incubation with cationic lipids to facilitate intracellular uptake have improved *in vivo* oligo efficacy. These methods were applied here to block specific K^+ channels with anatomically specific localization to induce memory loss without affecting sensory, motor, and state-dependent behavior. A daily antisense injection into the rat lateral ventricle did cause a marked decrease (for 2–4 days) of the mkv1.1 m-RNA levels in the hippocampus. Spatial-maze learning during the days in which mkv1.1 mRNA was found to be markedly reduced showed no differences. as compared to scrambled anti-sense injected or non-injected controls. Retention of spatial maze memory, however, was virtually eliminated for the anti-sense but not control animals (Figure 4). In a related study with collaborators at the University of Florence using the same antisense for mkv1.1 K^+ channels, memory of a passive avoidance conditioning task for mice was markedly impaired. Intracellular recordings from the rat dentate gyrus granular cells demonstrated, furthermore, that late voltage-dependent K^+ currents (which are encoded for by the mkv1.1 gene) were markedly reduced as indexed by absent after-hyperpolarizations of granule cell action potentials and a doubling of maximum action potential duration for antisense but not control animals. Thus, antisense injection clearly reduced m-RNA and the late K+ currents while eliminating all memory of the spatial maze.

In contrast to these consistent behavioral and electrophysiologic differences, there were no measurable differences in LTP recorded from both the CA1 and dentate gyrus regions of the hippocampus. This is the first reported demonstration, therefore, of a molecular biologic manipulation that eliminates associative memory without effects on LTP. This experiment does not rule out, however, contributory roles for LTP in learning and memory. LTP may be one of several alternative means for long-term synaptic modification. It may also be involved in attentional and/or arousal mechanisms that gate learning of new information. Other laboratories have reported findings complementary to those just described. Knock-outs of the γ isozyme of PKC or mGluR1 receptors minimally affected spatial maze learning and memory, but did drastically impair LTP (Tonegawa et al., 1994). Knockout of the cell adhesion molecule Thy-1, similarly abolished CA3 LTP, but not spatial maze learning and memory (Nosten-Bertrand et al., 1996). Other more recent "knock-out" studies (recently presented in symposia, but not yet published) for mkv1.1 and

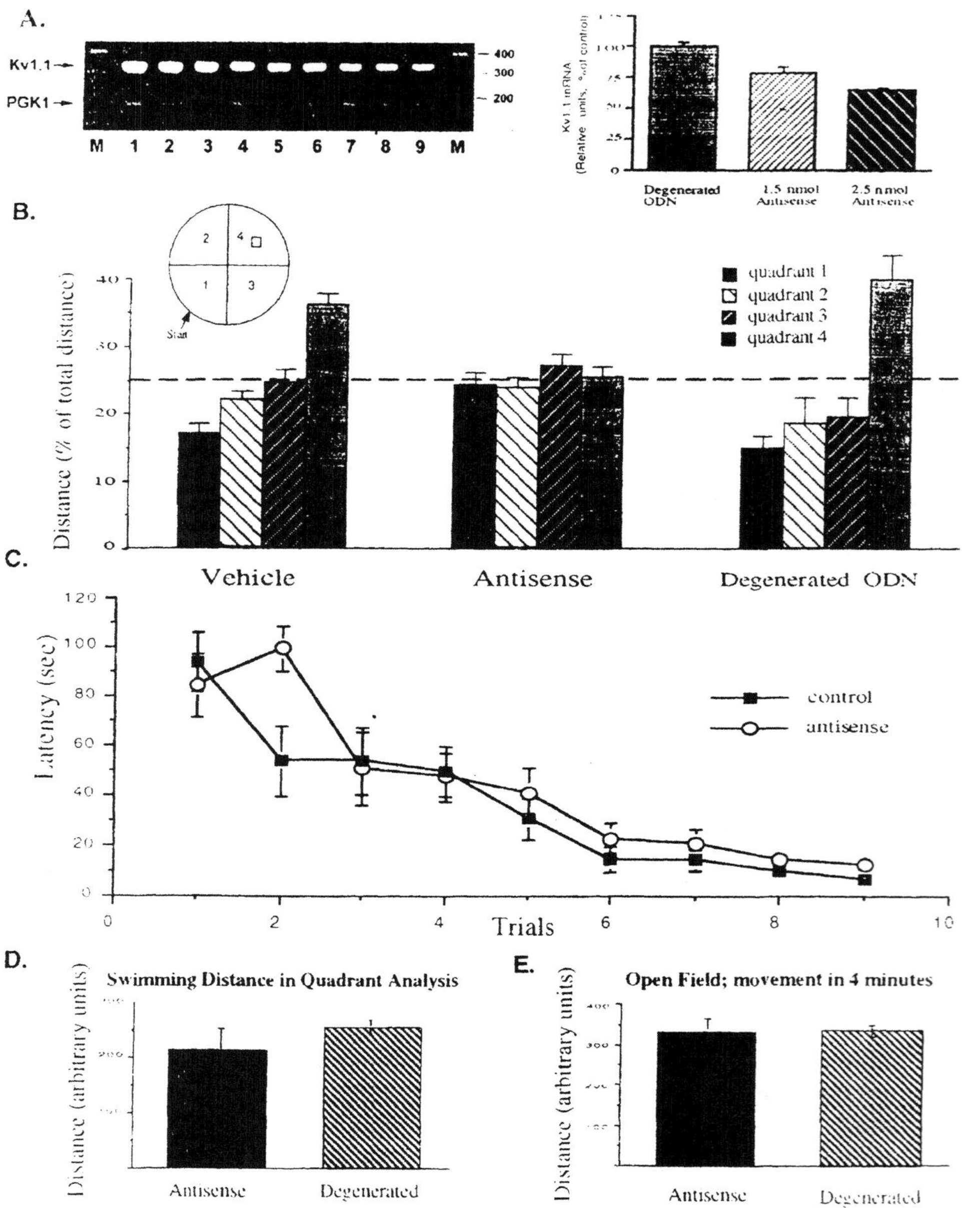

Figure 4. Analysis of Kv1.1 mRNA by reverse transcription polymerase chain reaction (RT-PCR) and the effect or antisense on rat spatial memory (A) Agarose gel of amplification products arising from RT-PCR with PGK1 and Kv1.1 primers from rats injected with either degenerated ODN (lanes 1–3), 1.5 nmol antisense ODN (lanes 4–6) or 2.5 nmol antisense ODN (lanes 7–9) that underwent water maze testing (see below). (B). Quadrant analyses were performed on the fourth day, after 3 training days in the water maze, in the absence of the island. The square in quadrant 4 indicates the previous position of the submerged platform. (antisense N=18; Degenerated ODN's N=19; Vehicle N=10) (C) Water maze Rats were trained 3 consecutive training trails each day for 3 consecutive days to locate a submerged island (N=20 in each group) (D). Distance of swimming in the first encounter with the pool. (N=9 in each group) (E) Open field exploratory behaviour was monitored in the same poole as the water maze but without the water, a day after the quadrant analysis was monitored (N=9 in each group).

mkv1.4 with rat spatial maze learning and LTP show remarkable consistency with the LAS "knockdown" studies just described.

LAS results, together with the above results of other labs (as well as from other studies), therefore, do suggest substantial mechanistic divergence for the *in vitro* LTP from *in vivo* associative learning and memory behavior. Many more comparative studies will be needed and useful, to establish when *in vitro* models are applicable to *in vivo* learning as they are for PKC (particularly the α-isozyme) activation and Ca_i^{++} elevation, and when they are not applicable (for LTP) as for K^+ channel inactivation.

ALZHEIMER'S DISEASE

Implication of voltage-dependent K^+ channels (and other conserved mechanisms) for *in vivo* learning and memory suggested their potential relevance for human function and dysfunction as occurs in dementias (e.g. Alzheimer's Disease-AD). To test this relevance, the LAS began several years ago, to analyze cells from AD patients and controls for evidence of AD-specific abnormalities within the memory cascade(s) implicated in our animal studies. We began with the assumption that AD pathophysiology would have systemic phenotypic expression, particularly if that expression involved β-amyloid metabolism. β-amyloid has been shown to be ubiquitous in mammalian tissues that range from CNS to skin fibroblasts and thus could express and/or cause cellular dysfunction in many tissues although clinical symptoms might only reflect CNS involvement. Of all the dementias, AD is the most specific for memory loss, particularly early in it's course.

Skin fibroblasts, did show a lack of or reduced activation of specific K^+ channels (113pS but not 166pS) in AD (Figure 5) as compared to control individuals (also Bondy et al., 1996 for a confirmatory study). These differences were independently confirmed with patch-clamp and Ca^{++} imaging techniques. Furthermore, incubation of control fibroblasts in 10nM soluble β-amyloid (a concentration comparable to endogenous levels) produced similar dysfunction of the 113 pS but not 166 pS K^+ channels (Etcheberrigarry et al., 1994). Other laboratories have subsequently confirmed the selectivity of soluble β-amyloid for K^+ channel inactivation in central neurons such as the CA1 pyramidals of rat hippocampus (Good et al., 1996; Furukawa et al., 1996).

Measures of other elements of the previously implicated memory cascade also uncovered additional AD-specific differences. Release, for example, of intracellular Ca^{++} via activation of the phospholipase C (PLC) - IP_3 receptor cascade was shown to be enhanced in AD as compared to non-AD dementia and age-matched control individuals. Bradykinin at picomole levels released much more Ca^{2++} and was associated with increased receptor binding in AD patients. This latter observation has also been confirmed by others (Gibson et al., 1995). Other LAS studies suggested AD-specific differences in levels of the PKC α-isozyme substrate, *Calexcitin*. Furthermore, soluble β-amyloid (10nM) induced a reduction of *Calexcitin* in normal fibroblasts. Other laboratories have also observed lower levels of the α–isozyme of PKC in AD fibroblasts.

These and other results suggest that early AD may involve soluble β-amyloid effects on K^+ channels and on their molecular regulation. AD-specific enhanced IP3-mediated calcium release is not induced by soluble β-amyloid and thus could be more closely related to early pathophysiology of AD. In any case, these results suggest some potential generality to humans of the molecular and biophysical events in memory storage uncovered by the LAS in animal models. The animal models themselves, such as anti-sense reduction of K^+ channel expression (see below), could prove useful, therefore, for developing therapeutic and diagnostic strategies for Alzheimer's disease.

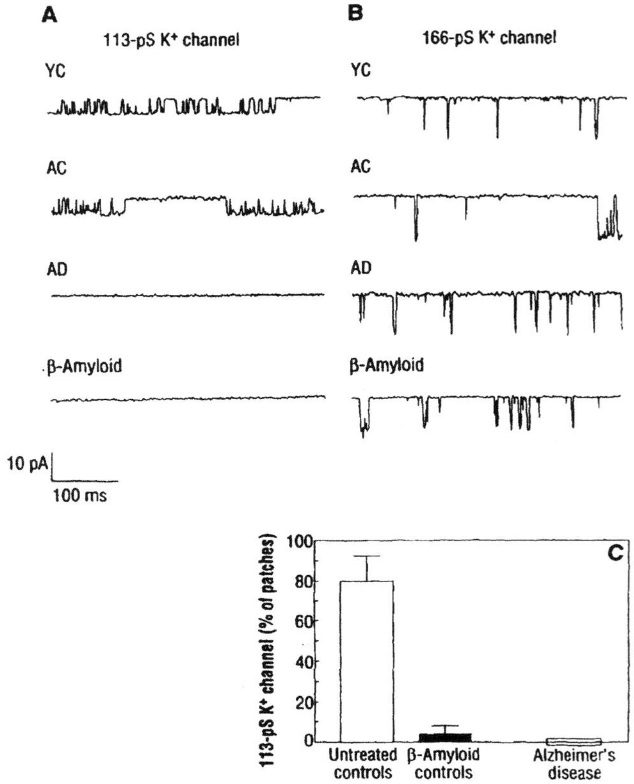

Figure 5. Cell-attached recordings in human fibroblasts. (A) Sample traces of the 113-PS K^- channel. This channel was not detected when AC fibroblasts were treated with βAP(1–40) (10 nM) (bottom trace). Upward deflections represent channel closures. (B) The K^+ channel with larger conductance (166 pS) in AC cell lines was unaffected by βAP treatment (bottom trace). The 113-pS channel was open about 60% of the time in control cell lines compared with ~10% for the 166-pS channel present in control (AC and YC), AD, and βAP-treated (10 nM) AC cell lines. (C) the 113-pS channel was present in only 1 of 21 cells (five cell lines of treated controls, middle column). This channel, however, was present in 94% of the previously recorded untreated controls (five different cell lines, 17 cells examined). Untreated control observations were also repeated and confirmed. This is almost identical to the situation in AD fibroblasts (right symbol), where no 113-pS channel was observed (27 cells from nine cell lines) Much of the data for untreated control (YC and AC) and AD cell lines had been collected fore this study.

CONSERVATION, MODEL IMPLICATIONS, AND HUMAN RELEVANCE

LAS correlation of quantitatively assessed associative learning and memory behaviors with synaptic, channel, molecular and structural transformations in a variety of animal models are all consistent with an emerging view that memory mechanisms have been conserved across diverse species. Briefly summarized, associated stimuli first interact within post-synaptic compartments of dendritic branches. Mammalian memory appears to involve clusters (Alkon, 1984; Alkon et al., 1991; Olds et al., 1989) of such dendritic interaction sites that are distributed across arrays of neurons such as the CA1 pyramidal cell neurons within the hippocampus. Intracellular messengers released at dendritic sites then trigger molecular cascades that change ion channel properties, membrane excitability, synaptic weight, and most likely synaptic morphology. These same dendritic convergence cascades may also release cell signaling molecules that affect the expression of specific genes (Nelson et al., 1993 & Cavallero et al., 1997, in press) and possibly pre-synaptic terminals to make synaptic weight changes more permanent.

While this emerging view of conserved memory storage is still hypothetical, a number of its critical elements have received important corroboration both from LAS studies and the studies of other labs. These critical elements include elevated Ca_i^{++}, translocated and thus activated PKC, inactivation of voltage-dependent K^+ channels, and phosphorylated target molecules such as the Ca^{++} and GTP-binding protein, *Calexcitin*, and enhanced expresssion of specific genes such as the ryanodine receptor. As discussed above, the relevance of *in vitro* learning models to the *in vivo* mechanistic constructs for memory storage outlined above is suggested by specific convergences as well as significant divergences. While many future studies will be required to validate the generality of conserved memory mechanisms, their relevance for humans has been supported by the observed Alzheimer's-specific deficits in cellular and subcellular functions.

Undoubtedly involved in far more complex cascades than those we know at present, these molecular elements, nevertheless, have already begun to reveal how memory is stored in different spatial and temporal domains. They extend from brief (msec) intervals of action potential propagation (as modified by I_A), to synaptic transmitter release (msec to seconds), to release of second messengers (seconds), to kinase activation and ion channel inactivation (minutes to days), to enhanced gene expression (hours to days), and to changes of neuronal architecture (days to weeks).

Memory storage involves networks distributed across the space of diverse brain regions from the hippocampus to cerebellar cortex to olfactory cortex. Yet critical molecular signaling occurs within the subcellular spaces of compartments that include pre-synaptic terminals, post-synaptic dendritic branches, and nuclei within cell bodies. Within subcellular spatial domains, synaptic-transmitter receptor- G-protein-mediated events begin at the cell membrane, signal through cytosolic second messengers, and cause kinases such as PKC and its target *Calexcitin* to translocate to the inner surface of the cell wall membrane, the ER, and to the nucleus. At the outer membrane, ion channel regulation changes electrical signals (and thus synaptic weight) received by other neurons within brain networks. In the nucleus, regulation of gene expression can in turn participate in longer term changes affecting the levels of other proteins as well as in the adjustment of neuronal and synaptic architectures.

REFERENCES

Acosta-Urquidi J., Alkon D. L., Neary J. T. (1984) Ca^{2+}-calmodulin-dependent protein kinase injection in a photoreceptor mimics biophysical effects of associative learning in *Hermissenda*. *Science* ***224***, 1254–1257.

Agranoff B.W., Davis R. E., and Brink J. J. (1966) Chemical studies on memory fixation in goldfish. *Brain Res.* **1**, 303–309.

Alkon D. L. (1983) Learning in a marine snail. *Sci. Amer.* **249**, 70–84.

Alkon, D. L. (1984) Calcium-mediated reduction of ionic currents: A biophysical memory trace. *Science* **226**, 1037–1045.

Alkon D. L. (1989) Memory storage and neural systems. *Sc. Amer.* **260**, 42–50.

Alkon D. L., Blackwell K. T., Barbour G. S., Werness S. A., and Vogl T. P. (1994) Biological plausibility of synaptic associative memory models. Neural Networks **7**, 1005–17.

Alkon D. L., Sanchez-Andres J. V., Ito E., Oka K., Yoshioka T. and Collin C. (1992) Long-term transformation of an inhibitory into an excitatory GABA-ergic synaptic response. *Proc. Nat. Acad. of Sci. USA* **89**, 11862–6.

Bank B., DeWeer A., Kuzirian A. M., Rasmussen H. and Alkon D. L. (1988) Classical conditioning induces long-term translocation of protein kinase C in rabbit hippocampal CA1 cells. *Proc. Natl. Acad. Sci. USA* **85**, 1988–92.

Bondy B., Hofmann M., Muller-Spahn F., Witzko M., and Hock C. (1996) The PHA-induced calcium signal in lymphocytes is altered after blockade of K^+-channels in Alzheimer's disease. *J. Psych. Res.* **30**, 217–27.

Colin C., Devan, W. A., Dahl D., Lee C. J., Axelrod J. and Alkon D. L. (1995) Long-term synaptic transformation of hippocampal Ca1γ-aminobutyric acid synapses and the effect of anadamide. *Proc. Nat. Acad. of Sci. USA* **92**,10167–71.

Dahl D., Arrowood J. L., Olds J. L., McPhie D. L. and Alkon D. L. (1997) Cholinergic transformation of GABAergic inhibition into excitation: heterosynaptic receptor interaction in the hipocampus. *J. Neuropyhysiology*, (in press).

Davies H. R. and Squire L. R. Protein synthesis and memory: a review. (1984) Psychol. Bull. **96**, 518–559.

Disterhoft J. F., Coulter D. A. and Alkon D. L. (1986) Conditioning-specific membrane changes of rabbit hippocampal neurons measured *in vitro*. *Proc. Natl. Acad. Sci. USA* **83**, 2733–7.

Etcheberrigaray R., Ito E., Kim S., and Alkon D. L. (1994) Soluble β-amyloid induction of Alzheimer's phenotype for human fibroblast K+ channels. *Science* **264**, 276–9.

Flexner J. B., Flexner L. B., Stellar E. (1963) Science **141**, 57.

Furukawa K., Barger S. W., Blalock E. M., and Mattson M. P. (1996) Activation of K^- channels and suppression of neuronal activity by secreted β-amyloid-precursor protein. *Nature* **379**, 74–8.

Glomset J.A., Gelb M.H., and Farnsworth C.C. (1990) Prenyl proteins in eukaryotic cells: a new type of membrane anchor. *Trends in Biochem. Sci.* **15**, 139–42

Good T.A., Smith D.O. and Murphy R.M. (1996) β-amyloid peptide blocks the fast-inactivating K^- current in rat hippocampal neurons. *Biophysical Journal* **70**, 296–304.

Hyden H., Lange P.W. (1965) A differential RNA response in neurons early and late during learning. Proc. Natl. Acad. Sci. USA **53**, 946–952.

Katz B. (1966) *Nerve, Muscle, and Synapse*. New York: McGraw Hill.

Kretsinger R.H. (1987) Calcium coordination and the calmodulin fold: divergent versus convergent evolution. Cold Spring Harbor Symp. Quant. Biol. **52**, 499–510

Kuffler S.W., Nicholls J.G. and Martin A.R. (1984) *From Neuron to Brain* (2nd edition) Sunderland, MA, Sinauver.

Linden D.J. and Connor J.A. (1995) Long-term synaptic depression. Ann. Rev. Neurosci. **18**, 319–357.

LoTurco J. L., Coulter D. A. and Alkon D. L. (1988) Enhancement of synaptic potentials in rabbit CA1 pyramidal neurons following classical conditioning. *Proc. Natl. Acad. Sci. USA* ***85***, 1672–76.

Lowy D., and Willumsen B. (1989) Protein modification: new clue to Ras lipid glue. *Nature* **341**, 384–5.

Matthies H. (1989) In search of cellular mechanisms of memory. Prog. Neurobiol **32**, 277–349.

McPhie DL, Matzel LD, Olds JL, Lester DS, Kuzirian AM, Alkon DL. (1993) Cell specficity of molecular changes during memory storage. *J Neurochem* **60**, 646–51

Meiri N., Ghelardini C., Tesco G., Galeotti N., Dahl D., Tomsic D., Cavallaro S., Quattrone A., Capaccioli S., Bartolini A. and Alkon D. (1997) Antisense inhibition of the rodent Shaker-like potassium channel Kv1.1: disruption of memory without effects on LTP. Proc. Natl. Acad. Sci. USA (in press).

Migliore M., Coo E. P., Jaffe D. B., Turner D. A., and Johnston D. (1995) Computer simulations of morphologically reconstructed CA3 hippocampal neurons. J. Neurophysiol. **73**, 1157–68.

Musil L., Carr C., Cohe, J., and Merlie J. (1988) Acetylcholine receptor-associated 43K protein contains covalently bound myristate. *J. Cell Biol.*, **107**,: 1113–21.

Nelson T. and Alkon D.L. (1990) Specific high-molecular weight mRNAs induced by associative learning in *Hermissenda*. Proc. Natl. Acad. Sci. USA **87**, 195–200.

Nelson T. J., Cavalaro S., Yi C.-L., McPhie D., Schreurs B. G., Gusev P. A., Favit A., Zohar O., Kim J., Beuschausen S., Ascoli G., Olds J., Neve R., and Alkon D. L. (1996) *Calexcitin*, a signaling protein that binds calcium, inhibits potassium channels, and enhances membrane excitability. *Proc. Nat. Acad. Sci. USA* **93**, 13808–13813

Nelson T., CollinC. and Alkon D. L. (1990) Isolation of a G protein that is modified by learning and reduces potassium currents in *Hermissenda*. *Science* **247**, 1479–83.

Nelson T. J., Yoshioka T., Toyoshima S., Han Y.-F., and Alkon D.L. (1994) Characterization of a GTP-binding protein implicated in both memory storage and interorganelle vesicle transport. *Proc. Nat. Acad. Sci. USA* **91**, 9287–91.

Nosten-Bertrand M., Errington M.L., Murphy K.P., Tokugawa Y, Barboni E., Kozlova E., Michalovich D., Morris R.G., Silver J., Stewart C.L., Bliss T.V., and Morris R.J. (1996) Normal spatial learning despite regional inhibition of LTP in mice lacking Thy-1. Nature **379**, 826–9.

Olds J. L., Anderson M., McPhie D. L., Staten L., Alkon D. L. (1989) Imaging memory-specific changes in the distribution of protein kinase c within the hippocampus. *Science*; **245**, 866–869.

Olds JL, Bahalla US, McPhie DL, Lester DS, Bower JM, Alkon DL. (1994) Lateralization of membrane-associated protein kinase C in rat piriform cortex: specific to operant training cues in the olfactory modality. *Beh Brain Res* **61**: 37–46.

Olds JL, Golski S, McPhie DL, Olton D, Mishkin M and Alkon DL. (1990) Discrimination learning alters the distribution of protein kinase C in the hippocampus of rats. J. Neurosci. **10**, 3707–13.

Sakakibara M., Alkon D. L., Neary J. T., Heldman E., and Gould R.: (1986) Inositol trisphosphate regulation of photoreceptor membrane currents. *Biophys. J.* **50**, 797–803.

Sakakibara M., Ikeno H., Usui, S., Collin C., Alkon, D. L. (1993) Reconstruction of ionic currents in a molluscan photoreceptor. *Biophys J* **65**, 519–27.

Schachter, J., Lester, D., and Alkon, D.L. (1996) Synergistic activation of protein kinase C by arachidonic acid and diacylglycerols *in vitro*: generation of a stable membrane. *Biochemica Biophysica Acta.*

Schreurs B. G., Tomsic D., Gusev P. A., and Alkon D. L. (1997) Dendritic excitability microzones and occluded long-term depression after classical conditioning of the rabbit's nictating membrane response. *J. Neurophysiology*,**75**, 1051–1060.

Thomson R. F. (1986) The neurobiology of learning and memory. *Science Wash. DC* **233**, 941–7.

Tonegawa S. (1994) Gene Targeting: A new approach for the analysis of mammalian memory and learning. In: *Molecular Neurobiology: Mechanisms Common to Brain, Skin, and Immune System*, Wiley-Liss, Inc. pp. 5–18.

Turner R.S., Kemp B., Su H.-D., and Kuo J.F. (1985) Substrate specficity of phospholipoid/Ca^{2+} dependent protein kinase as probed with synthetic peptide fragments of the bovine myelin basic protein. *J. Biol. Chem.* **260**, 11503–7.

Walker J.E., Saraste M., Runswick M.J., and Gay N.J. (1982) Distantly related sequences of the alpha- and beta-subunits of ATP synthase, myosin, kinase, and other ATP-requiring enzymes and a common nucleotide binding fold. *EMBO J.* **1**, 945–951.

Wickman K. and Clapham D. E. (1995) Ion channel regulation by G proteins. Physiol. Rev. **75**: 865–85.

Usherwood P. N. R. (1993) Memories are made of this. *Trends in Neuroscience* **93**, 427–9.

Van der Zee E.A., Compaan J.C., de Boer M., Luiten P.G.M. (1992) Changes in PKC immunoreactivity in mouse hippocampus induced by spatial discrimination learning. J. Neurosci. **12**,: 4808–4815, 1992.

2

BEHAVIORAL AND MECHANISTIC BASES OF LONG-TERM HABITUATION IN THE CRAB *CHASMAGNATHUS*

Daniel Tomsic,* Arturo Romano, and Héctor Maldonado

Laboratorio de Fisiología del Comportamiento Animal
Facultad de Ciencias Exactas y Naturales
Depto Biología, Universidad de Buenos Aires
Buenos Aires, Argentina

1. INTRODUCTION

The exact relationship between findings at the cellular level and the behavioral phenomenon of learning remain complex and obscure in mammals. With invertebrate preparations, however, relationships between cellular and behavioral phenomena are much more readily clarified (Krasne and Glanzman, 1995).

Ten years ago, we reported that an opaque screen moving overhead provokes an escape response in the crab *Chasmagnathus granulatus* that declines after few presentations of the eliciting stimulus and meets most of the parametric criteria of habituation (Brunner and Maldonado, 1988).

Although habituation is a behavioral change due to a specific previous experience, it has often been considered not to be an appropriate candidate for mechanistic studies on long-term memory. Doubts came from two current views about how classical and instrumental conditioning differ from habituation (Marlin and Miller, 1981). According to the first one, habituation is an extremely simple process dependent solely upon the parameters of the eliciting stimulus (e.g. Groves and Thompson, 1973), that is, habituation only describes a tendency to drop out rather than incorporating new responses or complicating those ready present (Thorpe, 1963). Therefore, some authors concluded that habituation is a marginal case of learning whose study cannot help to understand a real mnemonic process i.e. an associative process (e.g. Hilgard, 1956) though other authors consider that as a simple process, habituation is worthy of being studied since it can be conceived as an irre-

* Present address: Laboratory of Adaptive Systems, NINDS, NIH. Building 36, Room B308. Bethesda, MD 20892. Fax: (301) 402-2281 e-mail: tomsic@codon.nih.gov

Molecular and Cellular Mechanisms of Neuronal Plasticity, edited by Ehrlich. Plenum Press, New York, 1998.

ducible module of higher order forms of learning (Hawkins and Kandel, 1984). The second objection considers that even though examples of habituation have been found at all levels of the phylogenetic scale, many of these instances are short-lived phenomena and therefore of little consequence for long-term changes in behavior. A short persistence of the habituated response is expected mainly when a defensive response is involved, such as the escape reaction of *Chasmagnathus*, since in such a case a relatively permanent waning of the response would compromise the survival of the species.

An increasing body of results from experiments performed at our laboratory are at variance with the foregoing views. The long-term habituation (LTH) of *Chasmagnathus*, built up after a single session of repeated presentation of a visual danger stimulus, lasts for five days at least. LTH is quickly acquired and during acquisition animals not only gain information on the visual danger stimulus (= the phasic stimulus) but also on the contextual cues and the circadian phase. LTH seems not to be an unique process but to include two memory components, one of them being determined by a context-unconditioned stimulus (US) association.

Therefore, *Chasmagnathus* LTH encompasses an elaborated mnemonic process that leads to an enduring change of behavior, providing a suitable model to study long-term memory. In addition, the short duration of the training and some peculiarities of the crab's nervous system, mainly those regarding the blood-brain barrier, makes it an appropriate model for a mechanistic study of memory using a pharmacological approach.

Throughout this chapter, we present the main advances reached during this decade in the behavioral characterization of LTH in *Chasmagnathus* as well as disclosing the underlying mechanisms that subserve it.

2. MATERIALS AND METHODS

2.1. The Crab

The semiterrestrial crab *Chasmagnathus* is found along the coast of south Brazil, Uruguay and Argentina, occupying mud flats of the mesolittoral and supralittoral zones of fresh-seawater transition. For our study, adult male crabs of similar size are easily captured during the whole year from the *rias* (narrow coastal inlets) of San Clemente del Tuyú, Argentina, and transported to the laboratory. They are kept in plastic tanks filled to 2 cm depth with artificial sea water, fed rabbit pellets every 3 days and show healthy conditions up to fifteen days after arrival.

2.2. An Automatic Device to Study the Crab's Long-Term Habituation

Upon sudden presentation of a rectangular screen passing overhead, the crab responds with a running reaction in an attempt to escape. The flight reaction declines over trials and the decrement persists for at least five days (Lozada et al., 1990; Pedreira et al., 1995).

To study this process a wholly automated device that allows simultaneous work with 40 animals is used. Briefly, the experimental unit is the *actometer* (Fig 1a), consisting of a bowl-shaped plastic container (C) with a steep concave wall and a circular central flat, covered to a depth of 0.5 cm with water. The crab is placed in the container which is suspended by three strings from an upper wooden framework and illuminated by a 10 W lamp placed 30 cm above the animal. An opaque rectangle screen (R) is moved horizontally by

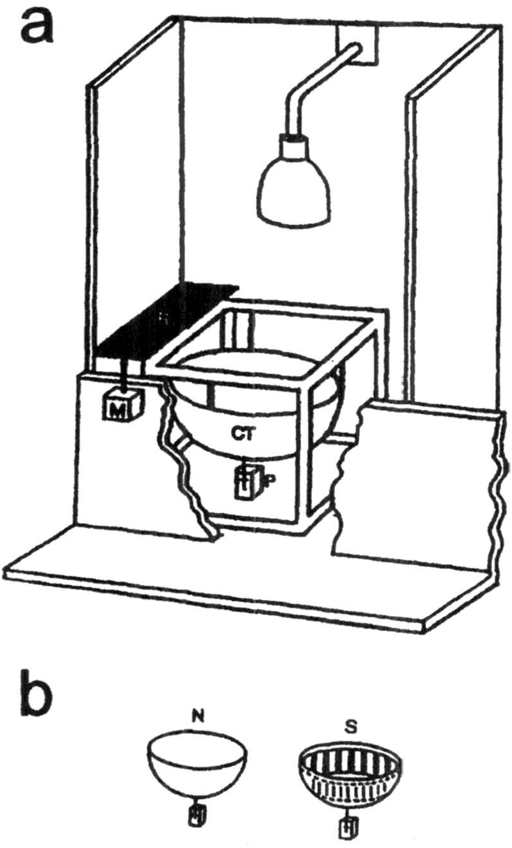

Figure 1. a) *The actometer*, one of the 40 units of the apparatus. C: plastic container; R: rectangular screen; M: motor; P: piezoelectric transducer. b) *Training-to-testing change in context experiments*: a ring with black and white vertical strips (S) is located inside the usual actometer container (N).

a motor (M) over the animal and across the upper surface of the framework, thus provoking a crab's running response and consequent container oscillations. A stylus is centrally cemented to the bottom of the container and connected to a piezoelectric transducer (P). Container oscillations induce, through the transducer, electrical signals proportional to the velocity of the oscillations. Such signals are amplified, integrated during the recording time (9 s) and translated into numerical units ranging from zero to 1530 before being processed by computer. The experimental room has 40 actometers, isolated from each other by partitions. A computer is employed to program trial sequences, trial duration and intertrial intervals, as well as to monitor experimental events.

During an experiment, crabs are usually distributed in pairs of groups, each pair consisting of a *trained group* (TR) and a *control group* (CT). The experimental procedure for a TR is sketched as follows. A stimulation session has a fixed number of trials given with a fixed intertrial interval (ITI) and preceded by 15 min of adaptation in the actometer. Each trial lasts 9 s and consists of passing the screen four times over the actometer, recording the crab's activity during the entire trial time. Two sessions per experiment are run, i.e. the *training session* (5, 10,15, 30 or 300 trials) and the *testing session* (6 trials), separated by a 24 h or more. During the intersession interval, crabs are individually

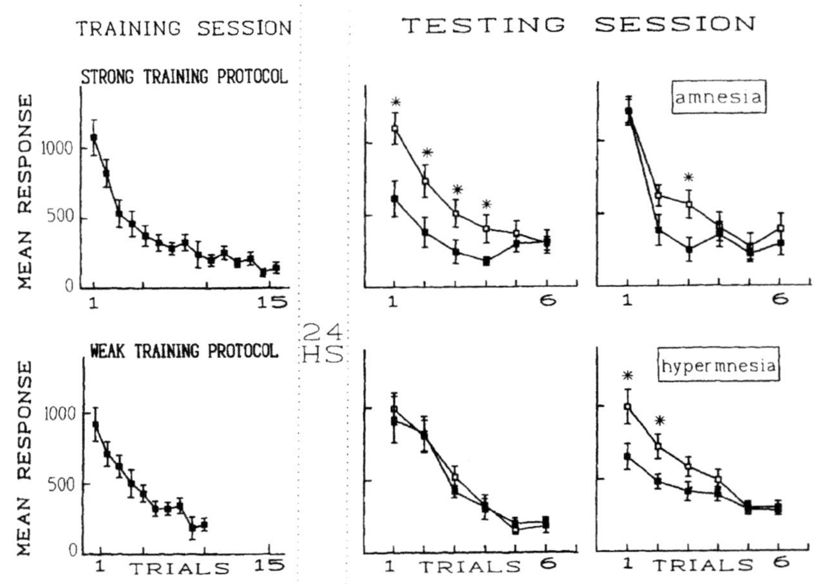

Figure 2. Strong vs weak training protocol. Ordinate: mean score per trial ± SEM. *Left panels*: *top*: the trained group receives 15 trials (ITI=171 s) (the control group stays in the actometer without training); *bottom* idem but the trained group receives 10 trials. *Middle panels*: testing of untreated groups. *Right panels*: *top*: testing of groups treated with an amnestic agent; *bottom*: idem with a hypermnesic agent. Open squares curves correspond to control groups; closed squares curves to trained groups. * stands for a significance difference ($p < 0.01$) between control and trained group at each trial. Intersession interval: 24 h.

lodged in rest containers covered to a depth of 0.5 cm with water and kept inside drawers dimly lit.

The experimental protocol for a CT is as that of TR in all respects except that animals of this group stay in the actometers during the entire training session without being trained, that is, without been presented the habituating stimulus. Retention for long-term habituation (LTH) is operationally defined as significant difference between CT and TR at testing session.

2.3. Strong vs Weak Training Protocol

The difference between these two modalities of training is illustrated in Fig 2. A trained group (TR) undergoes 15 or more training trials, ITI=171 s, when a strong training protocol is used. The sharp fall of responding shown by TR during the training session corresponds to the so-called short term habituation, STH (left panels, Fig 2). After the 24-h intersession interval, most of the testing trials show a significant difference for TR vs CT, namely, the strong protocol yields a robust LTH (middle upper panel, Fig 2). A similar pattern of results was invariably found in all experiments at our laboratory provided than 20 or more crabs per groups were used. In many instances the CT-TR difference is exhibited at any testing trial but the most constant difference was found at the first or two first testing trials. In contrast, ten or less training trials (i.e. weak training protocol) invariably fails to induce LTH (middle lower panel, Fig 2). The advantage of having such contrasting and reliable results with these two protocols is apparent, mainly for *in vivo* pharmacological experiments. Thus, a strong training protocol is used when an agent presumably has an amnesic effect, while a weak training protocol is adopted when the agent is assumed to have a hypermnesic (= facilitatory) effect. In Fig 2, testing results exhibited in the middle panels correspond to groups post-training injected with saline solution, while those shown at right panels to groups injected with either an amnesic agent (top right panel) or a facilitatory agent (bottom right panel).

3. BEHAVIORAL EXPERIMENTS

3.1. The Adaptive Value of LTH

Since the very beginning of our studies with *Chasmagnathus*, the finding that LTH was built up by a single session of 15 training trials (= 45 min of training) and exhibited for more than 24 h latter, was considered a rather surprising result. In fact, the crab's escape response induced by an object moving overhead suggests that this visual stimulus stands for an impending threat (gulls are common predators for this crabs), so that a persistent reduction of a defensive response would seem to be nonadaptive for the species. Therefore, an extensive study was performed to ascertain the possible biological meaning of the robust LTH (Tomsic et al., 1993). With such purpose, the capacity for yielding LTH in *Chasmagnathus* and in *Pachygrapsus*, two closely related grapsid species that diverge widely in ecology, was compared, using the same experimental set up and design. Clear-cut results were obtained: *Chasmagnathus* showed a far greater capacity for LTH than *Pachygrapsus* and such difference appears to correlate well with ecological disparity. The former lives immersed in an environment featuring wind-oscillations of cord grasses while the latter inhabits a bare biotope of rocky promontories, so that the probability that a passing object were signaling an actual predator is lower for *Chasmagnathus* and consequently larger its capacity for acquiring LTH. However, this aptitude of *Chasmagnathus* to easily suppress its escape response proved to be highly stimulus- and context-specific, assuring the animal the possibility of fully reacting when a real threat became present.

3.2. LTH Includes a Component of Associative Memory

Decrease in crab's reactivity within the training session (short-term habituation, STH), was shown to meet most of the parametric criteria of habituation (Brunner and Maldonado, 1988). Consistently, LTH showed stimulus specificity, that is, no retention after 24 h is revealed when the passing screen at testing differs from that presented at training (Lozada et al., 1990), a result in keeping with the view that habituation is dependent solely upon the parameters of US presentation (e.g. Groves and Thompson, 1973). However, several lines of evidence suggest that *Chasmagnathus* LTH can be appropriately categorized as an instance of associative memory linking the phasic stimulus and the context , a view in keeping with Wagner's associative theory of habituation (Wagner et al., 1981; Whitlow and Wagner, 1984). Specifically: 1) *Chasmagnathus* LTH is impaired when the context is changed between sessions (Tomsic et al., 1997) by placing a ring with black and white vertical strips inside the actometer (Fig 1b); 2) pre-training exposure to the contextual cues in the absence of the phasic stimulus, namely the classical conditioning procedure of latent inhibition (Lobow and Moore, 1951), reduces LTH (Tomsic et al., 1997); 3) post-training exposure to the context without phasic stimulation, namely the classical conditioning procedure of extinction, produces a decrease in LTH (Tomsic et al., 1997); 4) LTH tends to vanish when decreasing the intertrial interval (Pedreira et al., 1997), an outcome in keeping with a large body of evidence from experiments of classical conditioning (e.g. Hintzman, 1974) but at odds with a current view of the nonassociative theory of habituation (Thompson and Spencer, 1966).

A closer analysis of the foregoing results shows that the LTH impairment produced by context shift as well as by pre-training or post-training exposure is expressed without exception during the initial phase of testing (first trial), but not always during the final phase (the following five testing trials). Based on this disparity, we hypothesized that LTH would include both an associative and a nonassociative component: the former, evoked by the initial exposure to the context before the test stimulus presentation, would be purely expressed at the first testing trial (though occasionally also affecting the following ones), while the latter, evoked by the first trial at testing, is expressed at the subsequent test trials. Such an interpretation is similar to the dual-process habituation model proposed for explaining drugs tolerance in humans (Baker and Tiffany, 1985; Dafterg et al., 1988) and is strongly supported by recent results presented in the following section 3.4.

3.3. Chasmagnathus Learns the Phase of the Day in Which LTH Is Acquired

The response level of *Chasmagnathus* to the phasic stimulus during the *dark phase* (defined for these experiments as a time interval between 2230 and 0130) is higher than that observed during the *light phase* (between 1030 and 1330) and coincidentally, the level of locomotor activity in the actometer shows a circadian rhythm with a peak during the subjective night. However, LTH is impaired in crabs tested at a time of the day that differs from that of their original training, regardless of either the daily phase of the training or testing session or the extent of the intersession interval. That is, retention was found after 24, 48 or 72 h after training but not after 12 or 36 h (Pereyra et al., 1996).

The foregoing finding is reminiscent of the so-called state-dependent retention, an effect commonly observed with drug-induced states (Overton, 1978) but also induced non-pharmacologically (Peck and Ader, 1974; Santucci and Riccio, 1986). On an interpretative level, it appears that the recovery of the habituated response depends on the congruency between the crab internal state at the time of training and testing, a notion consistent with Spear's view of memory retrieval (Spear, 1973, 1978). This type of interpretation assumes that there is an oscillation of the internal state, namely, a differential set of interoceptive cues across time of day; and on the other hand, that memory for the internal state at training time is stored together with the memory of the context-stimulus association. In brief, LTH is circadian-phase specific.

3.4. Massed and Spaced Training Build Up Different Memory Components of LTH

Several results from our laboratory suggest that the expression of LTH at the first or at the two first testing trials (i.e. during the initial testing phase) is different to that at the following trials (retraining phase). This imply that different mechanisms may subserve memory at different testing phases. The following results support such distinction. 1) When animals undergo a strong training protocol, the largest difference between CT and

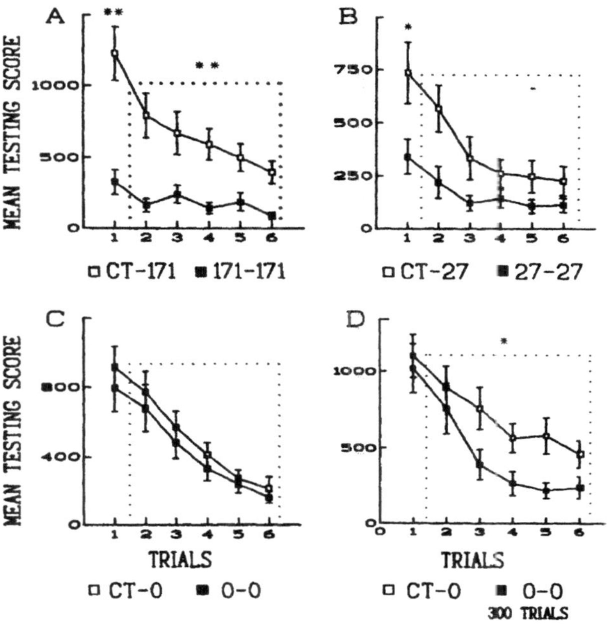

Figure 3. Massed and spaced training. Performances at testing (6 trials). Ordinate: mean score per trial ± SEM. Open squares stands for a control group (CT); closed squares for a trained group (TR). *A*: TR had received 15 training trials, ITI=171 s. *B*: idem but ITI=27 s. *C*: idem but ITI=0 s. *D*: TR had received 300 trials, ITI=0 s. * : statistical significance between TR and CT at first testing trial (initial phase) or at retraining (dotted box), $p < 0.01$; ** : idem but $p < 0.005$. Intersession interval: 24 h.

TR groups takes generally place at the initial part of the testing curve, then decreases over trials. It is worth emphasizing that an opposite picture of results was obtained for *Pachygrapsus*: CT and TR were invariably indistinguishable at the initial phase of testing but significant differences emerged over trials (Tomsic et al., 1993). 2) When animals undergo a strong training protocol followed by an injection of an amnesic agent, any CT-TR difference at the initial phase of testing vanishes (e.g. Pedreira et al., 1995; Pedreira et al., 1996), though sporadic CT-TR differences appear at some few trials of the retraining phase. 3) When a weak training is carried out and a facilitatory agent is administered, a CT-TR difference is disclosed at the initial testing phase but not necessarily at the following trials (Delorenzi et al., 1996). 4) As described above, a context shift as well as a pre- or post-training exposure to the actometer in the absence of the phasic stimulus, invariably impair LTH at the first testing trial but not necessarily at the following ones, suggesting that LTH would be constituted by an associative and a nonassociative component, respectively (Tomsic et al., 1997).

A recent study (Pedreira et al., 1997) was aimed at exploring how the magnitude of these two putative memory components of LTH are influenced by the training stimulus frequency, that is, the length of the training intertrial interval (ITI). With such purpose, experimental data were analyzed distinguishing between initial testing phase (confined in this case to the first testing trial) and retraining (the block of the last five testing trials). From a long series of experiments four instances are exhibited in Fig 3. In Figs 3 a, b and c all groups received 15 trials and stayed in the actometers during the 45 minutes, but one group was given a 171-s ITI (Fig 3a), other group had a 27-s ITI (Fig 3b) and a third group had no rest interval between trials (Fig 3c). With an ITI longer than 27 s a significant difference at both initial phase and retraining was always found; with 27-s ITI (a boundary case), LTH was expressed at the initial phase but not at retraining; and with

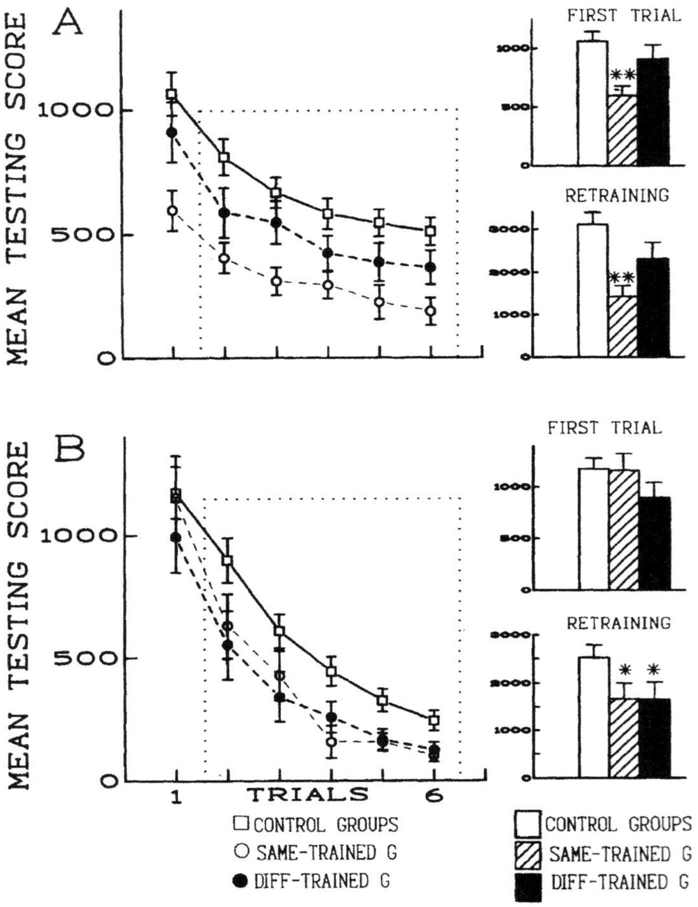

Figure 4. Training to testing context shift impairs LTH induced by spaced-training but not by massive training. Performances at testing (6 trials). Ordinate: mean score per trial ± SEM. Intersession interval: 24 h. *A*: Groups trained with spaced-training: 15 trials; ITI=171 s. Open circles and stripped bars: SAME-TRAINED GROUP: a group trained and tested with the usual context (n=40 crabs); closed circles and black bars: DIFF-TRAINED GROUP: a group trained with the stripped context and tested with the usual one (n=40 crabs); open squares and white bars: pooled control groups exposed to the same or different context during training and testing (n=80 crabs); *: statistical significance between a group and pooled control groups, $p < 0.05$; **: $p < 0.01$. Dotted box: retraining. *B*: Same protocol as in A but with massed training: 300 trials without intertrial interval. Symbols as in A.

massive training no retention was expressed at any testing phase. However, when the number of trials of the massive training was largely increased up to reaching 300 trials (i.e. 45 min of continuous stimulation) (Fig 3 d), LTH was expressed at retraining but not at the initial phase of testing.

The main conclusion from this behavioral analysis is that a) spaced training produces a memory component of LTH that is expressed at the initial testing phase, though also at retraining when longer ITIs are used; b) massed training yields another component that is expressed only at retraining.

In order to test if such distinction between two memory components of LTH is correlated with the distinction between associative and nonassociative components (section 3.2), two experiments were performed. In each experiment, crabs from a CT-TR pair of groups underwent the training session in the usual actometers while those of other pair were trained in actometers with a wall of black and white vertical strips (Fig 1 c). In one experiment, TR groups received a spaced training protocol (15 trials spaced by 171 s) while in the other, TR groups underwent a massive training (300 training trials separated by 0 s). All groups were tested in the usual actometers with the same intertrial interval used at training. Testing results corresponding to the experiment with groups trained with the spaced protocol (Fig 4 A) showed that context shift impaired LTH at both testing phases. However, the effect seems to be larger at the first trial that at retraining, as shown by the fact that sporadic CT-TR differences appeared at some trials of the retraining phase. Results of each trained group were compared with those of pooled control groups since no significant difference was found between untrained groups. The result of the second experiment (Fig 4 B) showed that when crabs underwent a massive training, context shift had no effect on LTH at retraining (the only testing phase where LTH could be expressed with this protocol). consequently, LTH at first testing trial, but not at retraining (purely evaluated with a massed training), is context specific. In other words, the memory expressed at the initial testing phase is associative and that expressed at the retraining phase is nonassociative. At this juncture, it is pertinent to recall that early experiments at our laboratory (Lozada et al., 1990) demonstrated that, unlike concerning context, both memory components are stimulus-specific since no retention of LTH is shown at both testing phases if the passing screen presented at testing differed markedly from that shown at training.

4. PHARMACOLOGICAL AND BIOCHEMICAL EXPERIMENTS

4.1. The Lack of an Endothelial Brain-Blood Barrier in *Chasmagnathus* Makes More Effective the Action of Compounds Given by Systemic Administration

The injection procedure in pharmacological experiments with *Chasmagnathus* is as follows. Fifty μl of vehicle or drug solution are given through the right side of the dorsal cephalothoracic-abdominal membrane by means of a syringe fitted with a sleeve to control depth of penetration to 4 mm, thus ensuring that the injected solution is released roughly at the center of the pericardial sac (e.g. Godoy and Maldonado, 1995; Maldonado et al., 1989). A remarkable feature shared by all our experiments is that the effective drug doses, given by systemic administration, were manifestly low, that is, equivalent to or even lower than doses administered by intracranial injections in vertebrates e.g. cycloheximide (Pedreira et al., 1995), actinomycin-D (Pedreira et al., 1996), angiotensin II (Delorenzi et al., 1995), enkephalin(Godoy and Maldonado, 1995), serotonin(Aggio et al., 1996). The relative simplicity of brain organization and the lack of an endothelial brain-blood barrier in crabs (Abbott, 1970), together with the fact that blood is distributed throughout an extensive capillary system in various neuropil areas of the brain (Abbott, 1992; Sandeman, 1986), could explain the low threshold for drug action. The possibility of working with such low doses given by systemic injection makes this a suitable approach to study mechanistic aspects of the mnemonic processes. Pharmacological manipulations in behaving organisms allows to test for the relevance of the purported molecular processes on long-term memory, a necessary step to bring the search for the molecular basis of neuronal plasticity back to the behavior that it was designed to explain (Silva and Giese, 1994).

4.2. LTH Is Blocked by Inhibitors of Protein and RNA Synthesis

A first step in a study aimed at elucidating the cellular mechanisms that subserve *Chasmagnathus* LTH was to inquire in what extent this mnemonic process is protein synthesis dependent. With such a purpose a series of experiments was performed to test the effect of cycloheximide (CYX) and actinomycin-D (ATY) on acquisition and retention of LTH (Pedreira et al., 1995; Pedreira et al., 1996).

An injection of 10 μg of CYX inhibited incorporation of [^{14}C]-aminoacid into the brain protein by 88% for 2 h and 20 μg inhibited by 92 %, while 0.6 μg of ATY inhibited 59.7% for 1 h, but no inhibition was found at 24 h postinjection. To test the effect of blocking protein synthesis on LTH, the strong training protocol (15 or 30 trials separated by 171 s) was used and the impact was assessed considering only the performance at the first testing trial or at the first 2-trial block, that is during the initial phase of testing. Both CYX (10–20 μg) and ATY (0.6 μg) administered immediately after training impaired LTH tested at 24 h. The amnesic effect of CYX was also shown at 72 or 120 h of intersession interval when 10–20 μg CYX but not 5 μg were injected. The amnesic window of these drugs proved to be very narrow since no memory impairment was found with doses equal to or less than 5 μg CYX or 0.5 μg ATY and on the other hand, overt symptoms of sickness appeared 24 h after injection with doses hardly higher than 20 μg CYX or 0.7 μg ATY. Concerning the time window of the CYX effect, experiments carried out with 20 μg CYX showed amnesia when the drug was injected 2 h after training but not after 6 h. To address the possibility that the above effect of CYX might be an instance of state dependence (e.g. Lee et al., 1989; Patterson et al., 1984), an experiment was conducted in which vehicle or drugs were pre-training and pre-testing injected. LTH impairment was only shown in those combinations that included a injection of CYX before training.

Contrasting with the effect shown on LTH, neither CYX nor ATY had effect on STH, a result consistent with that obtained with other species in different laboratories (e.g. Gibbs and Ng, 1977).

Apart from their amnesic effect, CYX and ATY have in common only their inhibitory effect on protein synthesis, that is, CYX acts directly on the translation step while ATY acts indirectly by blocking total RNA synthesis. Therefore, present results support the view that *Chasmagnathus* LTH, unlike STH, requires *de novo* protein synthesis.

When in some cases the analysis of results from experiments with CYX was extended to includ the six testing trials, it was apparent that the amnesic effect was conspicuous at the initial phase of testing but poorer or absent at retraining. Therefore, we recently conducted two experiments to test a probable differential effect of CYX on both memory components of LTH. In each experiment, crabs of a CT-TR pair of groups was post-training injected with saline solution and those of other pair with CYX (15 μg/crab). In one experiment, TR groups received a spaced training protocol (15 trials spaced by 171 s) while in the other, TR groups underwent a massive training (300 training trials separated by 0 s). Crabs were tested with the same intertrial interval used at training. Testing results corresponding to the experiment with groups trained with the spaced protocol (Fig 5 A) confirmed previous findings, namely, CYX impaired LTH at the initial phase of testing but

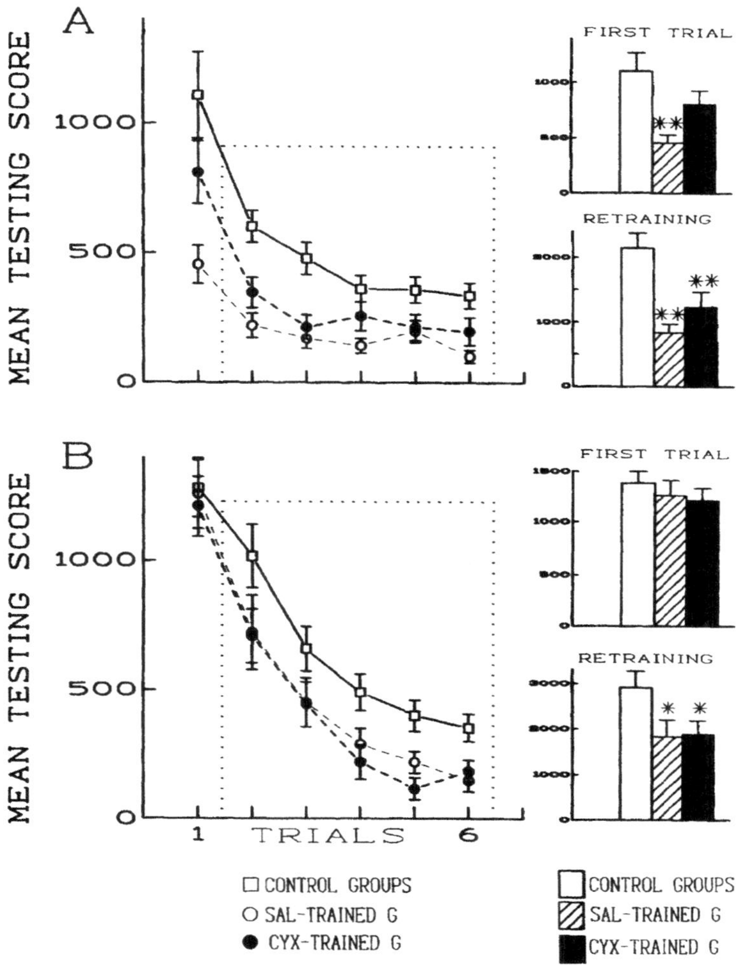

Figure 5. Cycloheximide (CYX) impairs LTH induced by spaced-training but not by massive training. Performances at testing (6 trials). Ordinate: mean score per trial ± SEM. Intersession interval: 24 h. *A*: Injections after spaced-training: 15 trials; ITI=171 s). Open circles and stripped bars: trained group injected with saline (n=40 crabs); closed circles and black bars: trained group injected with 15 μg/crab CYX (n=40 crabs). Open squares and white bars: pooled control groups injected with saline or CYX (n=80 crabs). *: statistical significance between a group and pooled control groups, $p < 0.05$; **: $p < 0.01$. *B*: Injections after massed training: 300 trials without intertrial interval. Symbols as in *A*.

not at retraining. Results of each trained group were compared with those of pooled control groups since no significant difference was found between untrained groups. The results of the second experiment also shown that CYX had no effect on LTH at retraining if crabs had received a massive training (Fig 5 B).

This differential effect of CYX is in keeping with the differential effect of context shift and with the massed and spaced training-induced memory components previously discussed. Therefore, it seems permissible to draw a comprehensive conclusion, namely, that the associative memory (determined by a context-stimulus association and mainly expressed at the first testing trial) is protein synthesis dependent, while the nonassociative memory (expressed at the retraining phase after a massive training) is protein synthesis independent. Incidentally, it is worth noticing the likeness between the hypothesis proposed for *Chasmagnathus* and that advanced for olfactory learning in *Drosophila* (Tully et al., 1994). Namely: associative and nonassociative LTH of *Chasmagnathus* appear to parallel long-term memory and anesthesia-resistant memory of *Drosophila*, respectively.

4.3. cAMP-Dependent Protein Kinase A (PKA) Is Mediating *Chasmagnathus* LTH

Initially, it was reported that systemic administration of the cAMP membrane permeable analog 8-(4-chlorophenylthio)-cAMP (CPT-cAMP), plus the phosphodiesterase inhibitor isobutyl methylxanthine (IBMX) induced LTH when injected before or immediately after a 5-trial training session (Romano et al., 1996a). Although such a result supports the view that augmentation of cAMP level is one of the key steps in the consolidation of crab's LTH, it is not possible to conclude straightforwardly that cAMP-dependent protein kinase A (PKA) is mediating the mnemonic process. Actually, it is known that CPT-cAMP also activates cGMP-dependent protein kinase (Sandberg et al., 1991) and that either IBMX or CPT-cAMP increases cGMP levels (Connoly et al., 1992) so that an explanation in terms other than activation of PKA could account for the facilitatory effect of CPT-cAMP + IBMX. Therefore, further experiments (Romano et al., 1996b) were aimed at studying the effect of the PKA activator Sp-5,6-DCl-cBIMPS (ACT) and the inhibitor Rp-8-Cl-cAMPS (INH) on retention level of the habituated response, tested 24 h after a weak training protocol (5-trials) or after a strong training protocol (15 or 30- trials), respectively. ACT and INH are membrane permeant, metabolically stable, analogs of cAMP whose action are highly specific of PKA, both interacting with the regulatory unit of this kinase (Sandberg et al., 1991; Yokosaki, 1992). If PKA activation were a necessary step for memory storage in this learning paradigm, then it should be expected a facilitatory effect of ACT on retention after a weak protocol and an amnestic effect of INH after a strong protocol. Results were conclusive. A CT-TR difference was shown at the initial phase of testing when crabs were injected immediately before or after a weak training . In contrast, a strong training failed to yield LTH when INH was pre- or post-training injected. Fig 6 summarizes the most relevant results from these experiments.

In all the experiments with pre-training injection, no difference at training between saline and drug (ACT or INH) groups was disclosed.

The foregoing results demonstrated that the two cAMP analogues employed were able to alter LTH in the crab. However, in order to conclude that PKA is actually implicated in the formation of *Chasmagnathus* LTH, a biochemical characterization was performed of the effects of both Sp-5,6-cDCl-cBIMPS and Rp-8-Cl-cAMPS on the crab cAMP-dependent protein kinase. *In vitro* kinase assays using extracts of the *Chasmagnathus* brain showed a dose dependent increase of phosphotransferase activity by Sp-5,6-DCl-cBIMPS and an antagonist effect of Rp-8-Cl-cAMPS. Addition of PKi clearly reduced the level of activity induced by ACT, thus indicating that the effect of the cAMP analogue was essentially due to the crab PKA activation. To our knowledge, this was the first report in which cAMP analogues' action have been studied by means of drug administration in intact animals.

In brief, results showed that PKA activation is a necessary step for LTH in *Chasmagnathus*, thus confirming that the cAMP-dependent signal transduction pathway seems

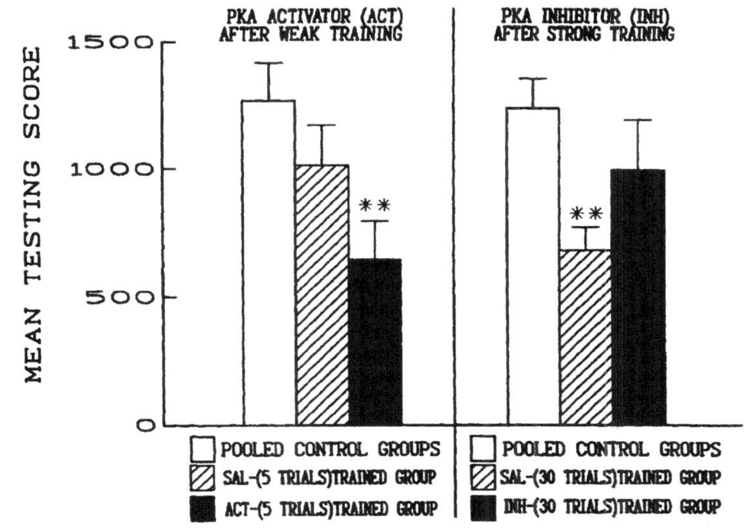

Figure 6. LTH is mediated by PKA. Performances at test (one trial). Ordinate: mean score per trial ± SEM. Intersession interval: 24 h. *Left panel*: effect of 75µM Sp-5,6-DCl-cBIMPS (ACT) pre-training injection on LTH. Weak training protocol (5 trials). White bar: pooled control groups injected with saline or ACT; stripped bar: trained group with saline; black bar: trained group with ACT. *Right panel*: effect of 25 µM Rp-8-Cl-cAMPS (INH) pre-training injection on LTH. Strong training protocol (30 trials). White bar: pooled control groups injected with saline or INH; stripped bar: trained group with saline; black bar: trained group with INH. Other symbols as in left panel.

to be a molecular conserved process implicated in long lasting storage of memory. Ongoing experiments are aimed at determining if PKA is differentially mediating the two memory components of LTH.

4.4. Components of the Renin-Angiotensin System (RAS) Are Involved in LTH Modulation

Manipulation of brain angiotensin levels appears to influence acquisition and recall of a newly learned task in vertebrates but results are still controversial (Morgan and Routtemberg, 1977; Braszko, et al., 1991; Wright et al., 1993). However, the research on this subject had been confined to this phylum, despite the fact that components of RAS have been described in invertebrates: instances are angiotensin II and I in leech (Laurent et al., 1995; Salzet et al., 1995), angiotensinogen-like epitopes in nervous system of *Aplysia* (Gonzalez et al., 1995) and angiotensin-converting enzyme-like activity in the blue crab (Smiley et al., 1994), reflecting the high conservation of angiotensins in the course of evolution.

Human angiotensin II (ANGII) (50 pm) injected in *Chasmagnathus* immediately after a weak training protocol (10 trials) facilitates LTH. Such facilitatory effect of ANGII proved to be dose-dependent and reversible by saralasin (5 pm), that is, by an AII analogous peptide with a well known inhibitory effect on the whole current family of AII receptors (Timmermans et al., 1993). Moreover, the effect of ANGII vanished either when the weak training protocol was reduced to only 5 trials or when the peptide was given before training or 1 h after. On the other hand, saralasin impaired LTH when injected before a strong training protocol, though no effect on STH was shown. An amnesic effect of saralasin was also obtained when given immediately after the strong training but not 1h later.

Biochemical experiments provide additional support to the existence of RAS components in *Chasmagnathus*. Brain showed ANGII-like immunoreactivity, estimated by RIA, and activity corresponding to angiotensin-converting enzyme, estimated by a significant hydrolytic activity of Hippuril-Hys-Leu that was inhibited by 5mM EDTA and 1 μM captopril (Delorenzi et al., 1996). In addition, angiotensin-like immunoreactivity was detected by immunohistochemical techniques in cell bodies of neurons corresponding to the two main divisions of the brain of *Chasmagnathus*, namely, the supraesophageal ganglia and in the terminal medulla of the lateral optic lobes (Florio, personal communication).

Recent results (Delorenzi et al., 1997) showed that a weak training protocol of 5 trials suffices to induce LTH when 5 pmol of angiotensin II (3–8) fragment, termed angiotensin IV (ANGIV), was post-training injected, suggesting that ANGIV has a facilitatory effect stronger than that obtained by ANGII. This effect proved to be dose dependent with a minimal effective dose coming near 0.5 pmol and reversible by the analog saralasin (5 pmol) (Fig 7), a result in conflict with reports indicating that saralasin fails to revert the effect of ANGIV (Roberts et al., 1995). However, this disparity could be accounted for by assuming either that in this case a cleavage in the analogue molecule produces a fragment that competes with the AT4 receptor or that the amnesic effect of saralasin masks the facilitatory effect of ANGIV. Fifty pmol of the nonpeptide angiotensin receptor antagonists DUP and PD, selective for AT1 and AT2 receptor subtypes respectively, failed to revert the ANGIV effect when coinjected after weak training (5 trials) and, unlike saralasin, showed no amnestic effect when given after a strong training protocol (15 trials).

Two hypothesis were offered (Delorenzi et al., 1997) to account for the precedent results and to assist as a theoretical framework for further research about the relation between ANGII and ANGIV relative to crab's memory. According to a first view, either ANGII or ANGIV would induce memory enhancement in crabs through different specific receptor subtypes. The action of ANGII would be mediated by receptors other than those described for mammals since neither DUP or PD showed amnesic effect after a strong training protocol. It should be noted that no angiotensin receptors have been yet described for invertebrates.The second hypothesis proposes that the effects obtained with ANGII are a consequence of its cleavage to ANGIV, that is, that ANGIV is the actual agent of the angiotensin facilitatory effect on long-term memory of *Chasmagnathus*. A basic assumption of this view is that the formation of the crab's ANGIV involves , like in mammals, successive cleavages of N-terminal from ANGII to ANGIV (Wright et al., 1994).

The analysis of the effect of angiotensin on LTH has been confined to the initial phase of testing, therefore, research is underway to explore whether or not angiotensins also modulate the retraining component of LTH.

5. CONCLUSIONS AND PERSPECTIVES

Main conclusions and perspectives from experiments on LTH in *Chasmagnathus* could be summed up in the following points.

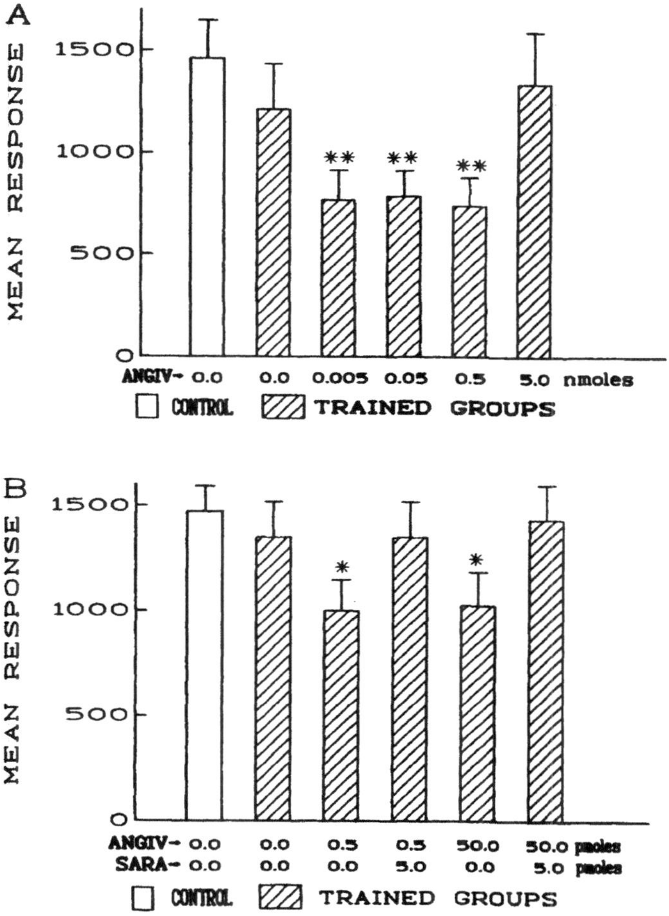

Figure 7. Angiotensin IV (ANGIV) has a facilitatory effect on LTH. Performances at test (one trial). Ordinate: mean score per trial ± SEM. Injections after a weak training protocol (5 trials). *A*: White bars: injection with vehicle: stripped bars: idem with diverse doses of ANGIV. *B*: Idem but stripped bars stand for ANGIV coinjected with diverse doses of saralasin. Intersession interval: 24 h. Other symbols as in Fig 4.

1. The long-lasting decrement of the *Chasmagnathus* escape response to a visual dangerous stimulus as a result of repeated stimulation entails a mnemonic process we call long-term habituation (LTH) since it meets most of the parametric conditions of habituation. Operationally, LTH is considered to be expressed when a statistical difference between crab's responsiveness of control and trained groups is disclosed, either at the initial phase of a 6-trial testing session (first trial or first two-trial block) or at the retraining phase (remaining trials), or at both phases.
2. LTH involves an enduring mnemonic process. During the acquisition period (the training session) the animal gathers information on diverse aspects of the experimental condition, as the parameters of the moving opaque screen presented for short temporal duration (the phasic stimulus), the unchanging or tonic stimuli

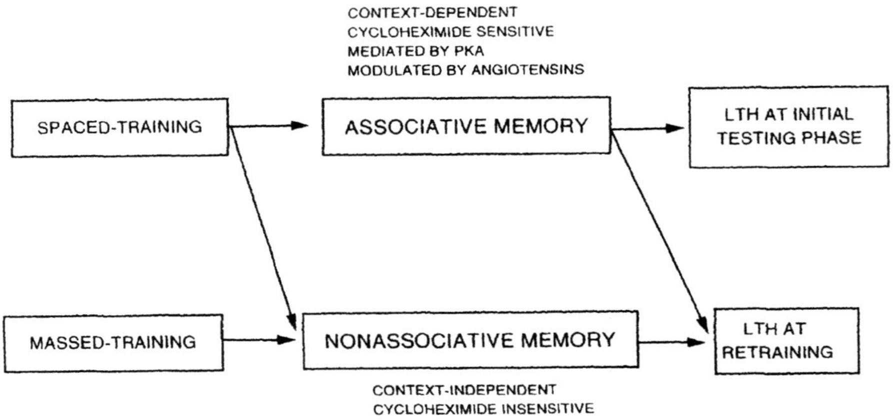

Figure 8. The dual-process hypothesis of LTH in Chasmagnathus. See text.

usually grouped as contextual or background cues (the context) and the phase of day or more precisely the corresponding internal state at training time (the circadian phase). A dual-process model is advanced to account for the mechanisms that would mediate some aspects of *Chasmagnathus* LTH (Fig 8). An association between context and phasic stimulus is established and stored as *associative memory* , provided that a sufficient number of training trials separated by a significant time interval were used (spaced-training). The associative memory is evoked by the exposure to context before the first test trial is presented, resulting in expression of LTH from the beginning of the testing session. A representation of the parameters of the phasic stimulus is stored as *nonassociative memory*, i.e. depending solely upon the iterative presentation of the stimulus (spaced or massive training). The nonassociative memory is retrieved by the first presentation of the phasic stimulus and LTH is expressed at the retraining phase only. The associative memory is context-dependent, cycloheximide sensitive, mediated by PKA and modulated by angiotensins. On the other hand, the nonassociative memory is context-independent and cycloheximide insensitive.

3. Research in our laboratory is underway to elucidate the possible role of PKA and angiotensins in the nonassociative memory as well as to explore how PKA activity, protein synthesis and angiotensin action are related each other within the process subserving LTH. Preliminary results suggest that changes of PKA activity in the crab's brain occurs after a strong training protocol (Locatelly, personal communication) and that CRE (cAMP response element) binding activity of nuclear extracts from the brain is enhanced between 4 and 18 h after 50 pm angiotensin II administration.
4. Recent preliminary results open a field to be studied within the dual-process hypothesis of habituation. A specific inhibitor of fucoglycoprotein synthesis, 2-deoxygalactose, diminishes retention of LTH in *Chasmagnathus*, but unlike cycloheximide, the impairing effect is here shown in both memory components (Saraco, personal communication). On an interpretative level, it appears that while the associative memory calls necessary for both *de novo* synthesis of pro-

tein and fucosilation of proteins, the nonassociative memory would only require the fucosilation of pre-existing proteins.

5. An item to be studied is the persistence of each memory component of LTH. Paralleling results obtained in other models (e.g. olfactory memory in *Drosophila*, Tully et al., 1994) it should be expected that the associative memory, dependent on *de novo* synthesis of protein, would last for much longer time than the nonassociative memory, insensitive to cycloheximide. However, no clear-cut differences have been reached yet, and further studies using longer training-to-testing intervals are required to draw a conclusion.
6. The behavioral studies performed in *Chasmagnathus* show that LTH is not necessarily a simple or marginal case of learning. Moreover, the clear-cut differences in the expression of LTH found between the two species closely related (Tomsic et al., 1993), indicate that the same operationally defined paradigm can be extremely different in nature. These conclusions are at odds with the hypothesis posed by Hawkins and Kandel (1984) that all the instances of complex memories could be explained by combining units of exceedingly simple memories (irreducible modules of memory) as habituation and sensitization.

REFERENCES

Abbott J. (1970). Absence of blood-brain barrier in a crustacean, *Carcinus maenas*. *Nature*, **225**:291–293.

Abbott N. J. (1992). Comparative physiology of the blood-brain barrier. In Bradbury MWB (Editor). *Physiology and Pharmacology of the Blood-Brain barrier*, Springer-Verlag, Berlin.

Aggio J., Rakitin A. and Maldonado H. (1996) Serotonin-induce short- and long-term sensitization in the crab *Chasmagnathus*, *Pharmacol. Biochem. Behav.*, **53**:441–448.

Baker T.B. and TiffanyS.T. (1985). Morphine tolerance as habituation. *Psychol Rev* **97**:78–108.

Braszko J.J., Wlasienko J., Koziolkiewicz W., Janecka A. and Wisniewski K. (1991). The 3–7 fragment of angiotensin II is probably responsible for its psychoactive properties, *Brain Res*, **542**: 49–54.

Brunner D. and Maldonado H. (1988). Habituation in the crab *Chasmagnathus granulatus*. Effect of morphine and naloxone. *J. Comp. Physiol. (A)*. **162**:687–694.

Connoly J.B., Willits P.B., Warrington B.H. and Murray J.K. (1992), 8-(4-chlorophenyl)thio-cyclic AMP is a potent inhibitor of the cyclic GMP-specific phosphodiesterase (PDE V_A), *Biochem. Pharmacol*, **44**:2303–2306.

Dafters R.I., Odber J., and Miller J. (1988) Associative and non associative tolerance to morphine: support for a dual process habituation model. *Life Science*, **42**:1987–1996.

Delorenzi A., Locatelli F., Romano A., Nahmod V. & Maldonado H. (1997). Angiotensin II(3–8) (ANG IV) induces long-term memory improvement in the crab *Chasmagnathus* (submitted).

Delorenzi A., Pedreira M. E., Romano A., Pirola C.J., Nahmod V.E. and Maldonado H. (1995). Acute administration of angiotensin II improves long-term habituation in the crab *Chasmagnathus*, *Neurosc. Letters*. **196**:193–196.

Delorenzi A., Pedreira M.E., Romano A., García S.I., Pirola C.J., Nahmod V.E. and Maldonado H. (1996). Angiotensin II enhances long-term memory in the crab *Chasmagnathus*. *Brain Res. Bull.* **41**: 211–220.

Gibbs M.E. and Ng K.T. (1977) Psychobiology of memory: Towards a model of memory. *Neurosci. Biobehav. Rev.*, **1**: 113–136.

Godoy A. and Maldonado H. (1995). Modulation of the scape response by D(Ala^2) Met- enkephalin in the crab *Chasmagnathus*. *Pharmacol. Biochem. and Behav.* **50**:445–451.

Gonzalez G.C., Roger I., Gonzalez E., Martinez-Padron M., Moore G.J. andLukowiak K (1995). Angiotensinogen-like epitopes are present in the CNS of *Aplysia californica* and co-localize with urotensin I- and urotensin-like immunoreactivities in the cerebral ganglia. *NeuroReport* **6**: 541–544.

Groves P.M. and Thompson R.F. (1973). A dual-process theory of habituation: Neural mechanisms. In: H.V.S. Peeke and M.J. Herz (Eds), *Habituation (vol 2)*. New York: Academic Press, pp 175–205.

Hawkins R.D. and Kandel E.R. (1984). Is there a cell-biological alphabet for simple forms of learning?. *Psychol. Rev.*, **22**: 375–391.

Hilgard E.R. (1956). *Theories of learning*. (2nd. edition). New York: Appleton-Century-Crofts.

Hintzman D.L. (1974). Theoretical implications of the spacing effect. In: Solso RL (Editor). *Theories in cognitive psychology: Loyola Simposium*. Hillsdale New Jersey: L. Erlbaum assoc.

Krasne F.B. and Glanzman D.L. (1995). What can we learn from invertebrate learning. *Annu. Rev. Psychol.*. **46**: 585–624.

Laurent V., Bulet P.and Salzet M. (1995). A comparison of the leech *Theromyzon tessulato* angiotensin I-like molecule with forms of vertebrate angiotensinogens: a hormonal system conserved in the course of evolution. *Neuroscience Lett.* **190**:175–178.

Lee D.W., Perlmuter A.M., Beniston D.S., Bennett E.L. and Rosenzweig M.R. (1989). Is anysomycin-induced amnesia for a passive avoidance task in chicks the result of state-dependent learning? *Dev. Brain Res.* **48**:179–184.

Lobow, R.E., & Moore, A.U. (1959). Latent inhibition: effect of non reinforced preexposure to the conditioned stimulus. *J Comp Physiol Psychol*, **52**:415–419.

Lozada M., Romano A. and Maldonado H. (1990). Long-term habituation to a danger stimulus in the crab *Chasmagnathus*, *Physiol. Behav.*, **47**:35–41.

Maldonado H., Romano A. and Lozada M. (1989). Opioid action on response level to a danger stimulus in the crab *Chasmagnathus granulatus*. *Behav. Neurosci.* **103**:1139–1143.

Marlin, N. A. & Miller, R. R. (1981). Association to contextual stimuli as a determinant of long term habituation. *J. Exp. Psychol: An. Behav. Proc.* ,7: 313–333.

Morgan T.M. and Routtenberg A. (1977). Angiotensin injected into the neostriatum after learning disrupts retention performance. *Science* 196: **87**–89.

Overton D.A. (1978). Major theories of state dependent learning. In: Ho B. T. (Editor). *Drug discrimination and state dependent learning*. New York: Academic Press.

Patterson T.A., Rose S.P. and Bradley P.M. (1989). Anysomicin and amnesia in the chick: state dependent-effects are not present with intracranial injections. *Dev. Brain Res.* **49**:173–178 .

Peck J.H. and Ader R. (1974). Illness-induced taste aversion under states of deprivation and satiation. *Anim. Learn. Behav.* **2**: 6–8.

Pedreira M.E., Dimant B. and Maldonado H. (1996). Inhibitors of protein and RNA synthesis block context memory and long-term habituation in the crab *Chasmagnathus*. *Pharmacol. Biochem. and Behav.* **54**:611–617 .

Pedreira M.E., Dimant B., Tomsic D., Quesada-Allue L.A. and Maldonado H. (1995). Cycloheximide inhibits context memory and long-term habituation in the crab *Chasmagnathus*. *Pharmacol. Biochem. and Behav.*, **52**:385–395.

Pedreira M.E., Romano A., Tomsic D., Lozada M. and Maldonado H. (1997). Massed and spaced training build up different components of long-term habituation in the crab *Chasmagnathus*. (submitted).

Pereyra P., De La Iglesia H. and Maldonado H. (1996). Training-to-testing intervals different from 24 h impair habituation in the crab *Chasmagnathus*. *Physiology & Behavior*. **59**:19–25.

Roberts K.A., Krebs L.T., Kramar E.A., Shaffer M.J., Harding J. W. and Wright J. W. (1995). Autoradiographic identification of brain angiotensin IV binding sites and differential c-Fos expression following intracerebroventricular injection of angiotensin II and IV in rats. *Brain Research*, **682**:13–21.

Romano A., Delorenzi A., Pedreira M.E., Tomsic D. and Maldonado H. (1996a). Acute administration of a permeant analog of cAMP and a phosphodiesterase inhibitor improve long-term habituation in the crab *Chasmagnathus*. *Behav. Brain Res.* **75**:119–125.

Romano A., Locatelli F., Delorenzi A., Pedreira M.E. and Maldonado H. (1996b). Effects of activation and inhibition of cAMP-dependent protein kinase on long-term habituation in the crab *Chasmagnathus*. *Brain Research*, **735**:131–140.

Salzet M., Bulet P., Wattez C. and Verger-Bocquet M. (1995). Structural characterization of a diuretic peptide from the central nervous system of the leech *Erpobdella octoculata*. *J. Biological Chem* **270**:1575–1582.

Sandberg M., Butt E., Nolte C., Fischer L., Halbrugge M., Beltman J., Jahnsen T., Genieser H.G., Jastorff B. and Walter U. (1991). Characterization of Sp-5,6-Dichloro-1-B-D-ribofuranosyl-benzimidazole-3',5'-monophosphorothioate as a potent and specific activator of cyclic-AMP-dependent protein kinase in cell extract and intact cells, *Biochem. J.*, **279**: 521–527

Sandeman D.C. (1986). Organization of the crustacean central nervous system. In Bliss DE (Editor), *The Biology of Crustacea*, Academic Press, New York.

Santucci A.C. and Riccio D.C. (1986). Hypothermia-induced anterograde amnesia and its reversal in rats trained on a T-maze task. *Physiol. Behav.* **36**: 1065–1069.

Silva A.J. and Giese K.P. (1994). Plastic gene are in! *Curr. Op. Neurobiol.* **4**: 413–420.

Smiley J.W. and Doig M.T. (1994). Distribution and characterization of angiotensin-converting enzyme-like activity in tissues of the blue crab *Callinectes sapidus*, *Comp. Biochem. Physiol.*, **4**: 491–496.

Spear N. E. (1978). The processing of memories: Forgetting and retention. Hillsdale, NJ: Erlbaum.

Spear N.E. (1973). Retrieval of memory in animals. Psychol. Rev. 80: 163–175.

Thompson R.F. and Spencer W.A. (1966). Habituation: A model phenomenon for the study of neural substrates of behavior. *Psychol. Rev.* **173**:16–43.

Thorpe W.H (1963). *Learning and instinct in animals* (2nd e). London: Methuen.

Timmermans, P.B.; Wong, P.C.; Chui, A.T.; Herblin, W.F.; Benfield, P.; Carini, D.J.; Lee, R.J.; Wexler, R.R.; Saye. J.A.M.; Smith, R.D. (1993) Angiotensin II receptors and angiotensin II receptor antagonists. Pharmacol.Rev. **45**:205–251; 1993 59

Tomsic D., Massoni V. and Maldonado H. (1993). Habituation to danger stimulus in two semiterrestrial crabs: ontogenic, ecological and opioids modulation correlates. *J. Comp. Physiol. (A).* **173**:621–633.

Tomsic D., Pedreira M.E., Romano A. and Maldonado H. (1997). Context- US association as a determinant of long-term habituation in the crab *Chasmagnathus*. (submitted).

Tully T., Preat T., Boyton S.C. and Del Vecchio M. (1994). Genectic dissection of consolidated memory in *Drosophila. Cell.* **79**:35–47 .

Wagner A.R. (1981). SOP as a model of automatic memory processing in animal behavior. In: Spear & RR Miller (Eds). *Information processing in animals: Memory mechanisms*. HillsdaleNJ. L Erlbaum associates.

Whitlow J.W. and Wagner A.R. (1984). Memory and habituation. In: Peeke HVS, Petrinovich L. (Eds). *Habituation sensitization and behavior*. Academic Press Inc. London.

Wright J.W. and Harding J.W. (1994). Brain angiotensin receptor subtypes in the control of physiological and behavioral responses. *Neuroscience and Behavioral Reviews*, **18**:21–53.

Wright J.W., Miller-Wing A.V., Shaffer M.J., Higginson C., Wright D.E., Hanesworth J.M. and Harding J.W. (1993). Angiotensin II(3–8) (ANG IV) hippocampal binding: potential role in the facilitation of memory. *Brain Res. Bull.* **32**:492–502.

Yokosaki H., Tortora G., Pepe S., Maronde E., Genieser H.G., Jastorff B. and Cho-Chung Y.S. (1992). Unhydrolyzable analogs of cyclic adenosine-3',5'-monophosphate demonstrating growth inhibition and differentiation in human cancer cells. *Cancer Res.*, **52**:2504–2508.

3

BIOACTIVE LIPIDS AND GENE EXPRESSION IN NEURONAL PLASTICITY

Nicolas G. Bazan*

Neuroscience Center of Excellence
Louisiana State University Medical Center
School of Medicine
New Orleans, Louisiana 70112-2272

INTRODUCTION

Excitable membranes are a reservoir of bioactive lipids that are contained as part of the structure of phospholipids. Signals such as those elicited by receptor occupancy trigger, via the activation of phospholipases, the release of phospholipid moieties. These enzymes are under control of receptors, calcium ions, protein kinase/phosphatase cascades, and of other signaling mechanisms which still need to be better characterized. The best defined regulatory events of these enzymes are those controlled by the seven transmembrane domain superfamily of receptors that, through G proteins, activate phospholipase C (PLC). This enzyme predominantly hydrolyzes phosphatidylinositol 4', 5' bisphosphate (PIP_2), resulting in the production of the second messengers, diacylglycerol (DAG) and inositol 1,4,5-trisphosphate (IP_3). These two messengers of inositol lipids represent a bifurcating, informational system inside the cell. The synaptic inositol lipid signaling system is activated by neurotransmitters, growth factors and drugs. IP_3 promotes intracellular Ca^{2+} mobilization by binding to a specific receptor in the endoplasmic reticulum, which in turn enhances calcium-dependent cellular processes.

The other product of inositol lipid degradation, DAG, is an activator of protein kinase C (PKC). PKC plays a modulatory role by catalyzing the phosphorylation of specific proteins that contribute to cell function. The modulation of cell function by PKC is complex because a) there are several isoforms of this enzyme, some are brain specific (Braquet et al., 1987); b) also, phospholipase A_2 (PLA_2) products such as free arachidonic and docosahexaenoic acids modulate some protein kinases C and c) there is a high degree

* Phone (504) 599-0832 FAX (504) 568-5801 E-mail nbazan@lsumc.edu

Molecular and Cellular Mechanisms of Neuronal Plasticity, edited by Ehrlich. Plenum Press, New York, 1998.

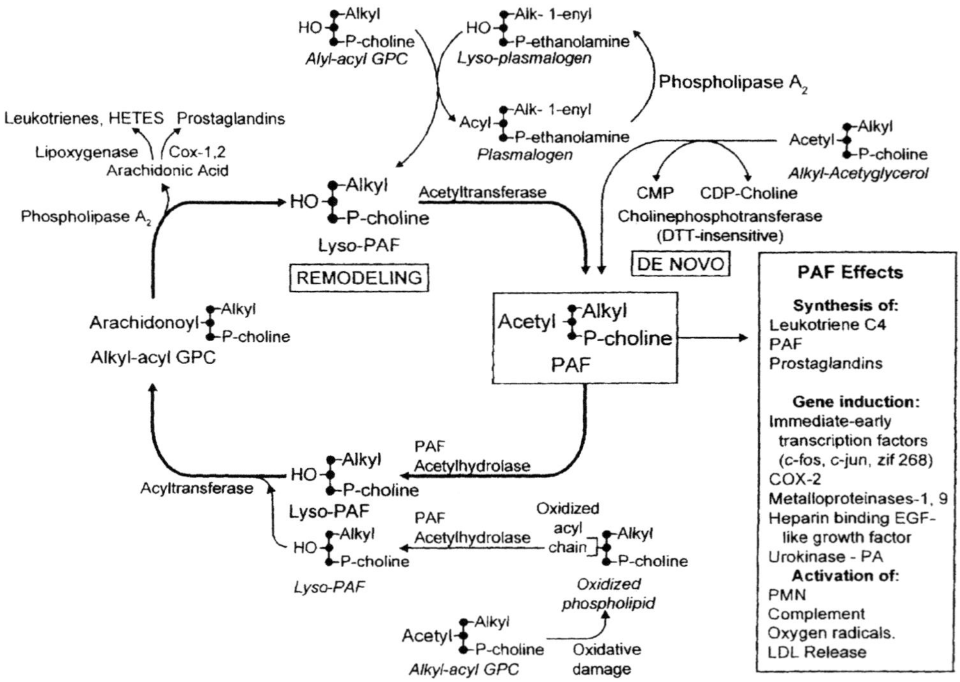

Figure 1. Summary of pathways of PAF synthesis and degradation. The central portion of this figure represents the "PAF cycle" between the membrane phospholipid PAF precursor alkyl-acyl-GPC and biologically active PAF. The remodeling synthesis involves the production of lyso-PAF, generated from the PAF precursor alkyl-acyl GPC either by the direct action of phospholipase A_2, or the transfer of the *sn*-2 acyl chain to a "donor" lyso-plasmalogen, which is itself mobilized from membrane plasmalogen by phospholipase A_2 action. The de novo pathway involves the direct transfer of a choline moiety to alkyl-acetylglycerol. Note that PAF acetylhydrolase inactivates all PAF molecules regardless of their biosynthetic route, and additionally inactivates oxidatively damaged phospholipids that possess biological activity at the PAF receptor.

of synergy and feedback modulation between various intracellular pathways of cell signal transduction, including also other protein kinases. Synaptic plasticity involves changes in PKC activity, hence phospholipases may play a role in regulating the formation of neuronal pathways.

In addition to the PLC-inositol signaling pathways, PLA_2 is an important modulator of the release from membrane phospholipids, of arachidonic acid and platelet-activating factor. PLA_2 cleaves the fatty acyl chain at the C2 position of a phospholipid. In excitable membranes, the C2 acyl chain of phospholipid is often polyunsaturated and, when cleaved, provides a substrate for oxidative metabolism (arachidonic acid) to biologically active lipid mediators. In addition, there is, in excitable membranes, a small, but metabolically active pool of 1-alkyl-2-acyl-glycero-3-phosphocholine. The 2-acyl is often arachidonic acid. The lyso-phospholipid PLA_2 product of this phospholipid is a precursor for the potent lipid mediator platelet-activating factor (PAF). Therefore, PLA_2 may release both arachidonic acid and the PAF precursor. Here the potential significance of PLA_2 and particularly of PAF in neurotransmission, gene expression, and neuronal plasticity are described.

PLATELET-ACTIVATING FACTOR

Platelet-activating factor (PAF, 1-O-alkyl-2-acetyl-glycero-3-phosphocholine), an acetylated alkyl ether phospholipid, is one of the most potent of all known bioactive lipids. PAF was initially described, as its name implies, as a leukocyte-derived mediator that induced the aggregation and activation of platelets (Henson, 1970; Benveniste et al., 1972). Today, it is known that there are there are wide-ranging functions for PAF, both as a mediator of inflammatory and immune responses, and as a second messenger involved in physiological and pathophysiological events of intracellular communication(Benveniste et al., 1972; Prescott et al., 1990; Snyder, 1995).

PAF is rapidly produced in brain in response to ischemia and seizures (Squinto et al., 1981; Panetta et al., 1987) and in neuronal cell cultures in response to neurotransmitters (Sogos et al., 1990) . Neurons and glial cells have the capacity for stimulus-evoked PAF synthesis.

There are at least three different routes for PAF biosynthesis (**Figure 1**). These involve either the modification of the membrane phospholipid precursor alkylacyl glycerophosphocholine (AAGPC), or *a de novo* synthesis in which the last step is the transfer of a choline moiety from CDP-choline to alkylacetyl glycerol. The remodeling pathway involves two alternative mechanisms, both using PLA_2, for the generation of lyso-PAF, and a common final step by the transacetylation of lyso-PAF to PAF. AAGPC may be directly converted to lyso-PAF by the PLA_2-catalyzed removal of the *sn-2* acyl group. Alternatively, the acyl group of AAGPC may be transferred, utilizing a CoA-independent acyltransferase, to a "donor" lyso-plasmalogen (1-alk-1'-enyl-2-lyso-*sn*-glycero-3-phosphoc- holine) that is itself generated by the action of PLA_2 on plasmalogen. All of the enzyme activities known to be involved in PAF biosynthesis are present in brain (Lee et al., 1988; Goracci and Francesangeli, 1991; Blank et al., 1995). The contribution of each of these pathways to the biosynthesis of PAF under physiological or pathological conditions in brain has not yet been fully assessed, but the *de novo* route regulates steady-state levels of PAF, whereas the remodeling route is activated under stimulatory conditions. There is also some evidence that PAF acetyltransferase activity is also inducible via covalent modification of the enzyme, thought to be phosphorylation (Holland et al., 1992). In addition, none of the enzyme activities involved in the *de novo* route are known to be inducible, and there is even a report that *in vitro*, activity of the brain choline phosphotransferease is inhibited by elevated levels of free calcium (Francescangeli and Goracci, 1989). These findings suggest that the remodeling pathway is an important contributor to PAF production in brain.

PAF is inactivated through the cleavage of the *sn*-2 acetyl group by PAF-acetylhydrolases (PAF-AH). PAF-AHs are serine esterases with a substrate specificity for ether phospholipids possessing a short carbon chain in the *sn*-2 position. PAF-AH are a phospholipase A_2, but they do not share significant protein sequence homology with the typical intracellular or secretory PLA_2 beyond the Gly-Xaa-Ser-Xaa-Gly (where Xaa is any amino acid residue) motif common to many lipases (Tjoelker et al., 1995) . There are two main classes of PAF-AH, plasma forms, consisting of a dimeric complex associated primarily with low-density lipoproteins (Stafforini et al., 1991) , and various intracellular cytosolic forms of the enzyme activity (Hattori et al., 1993; Stafforini et al., 1993). Various other forms of PLA_2 and phospholipases C and D are also able to hydrolyze PAF *in vitro* (Snyder, 1995), but their contribution to the degradation of PAF *in vivo* is unknown. The brain-specific form of intracellular PAF-AH is a complex enzyme, consisting of two different subunits possessing catalytic activity and a third, presumably regulatory subunit (Hattori et al., 1993; Hattori et al., 1994a; Hattori et al., 1994b). The gene encoding this regulatory subunit has been found to correspond to the LIS-1 gene, a gene that is mutated in Miller-Dieker lissencephaly (Hattori et al., 1994a).

PAF acetylhydrolase can also hydrolyze species of oxidatively-damaged phospholipids that possess PAF-like activity. Free radical damage and fragmentation of the *sn-2* polyunsaturated fatty acid residue of some phosphatidylcholines yields species that have a short residual fatty acyl chain in the *sn-2* position. These molecules not only structurally resemble PAF, but have biological activity at the PAF receptor (Smiley et al., 1991; Zimmerman et al., 1994; Heery et al., 1995). Although these PAF-like species have not yet been studied in brain, they may well have a significant role in oxidative stress.

PAF RECEPTOR

A PAF receptor has been cloned and characterized (Honda et al., 1991; Kunz et al., 1992; Sugimoto et al., 1992). Its predicted secondary amino structure identifies it as a member of the seven membrane-spanning domain, G protein-coupled receptor superfamily. The primary mode of signal transduction for this receptor is via a G protein-linked phospholipase C (Ali et al., 1994; Honda et al., 1995). PAF added to hippocampal neurons in culture elicits intracellular calcium fluxes, the dynamics of which are characteristic of release from intracellular stores rather than gating of extracellular calcium (Bito et al., 1992). However, additional signal transduction pathways have also been linked to the PAF receptor, most notably the mitogen-activated protein (MAP) kinase and associated protein kinase cascades (Honda et al., 1994; Kuruvilla et al., 1994; Mukerjee et al., 1998), and the down-regulation of cyclic AMP formation (Kester et al., 1992). Also, in some neuronal cell lines, PAF can raise intracellular calcium in a manner dependent upon the extracellular calcium concentration, implying the activation of cell-surface calcium channels (Kornecki and Ehrlich, 1988).

The PAF receptor gene structure displays two different transcription start sites contained in two distinct 5' non-coding exons (Mutoh et all, 1988). When alternatively spliced to a common exon containing the entire coding sequence, the result is the expression of two different transcripts (types 1 and 2) that contain the identical coding sequence, but differing in their 5' untranslated leader sequences. The leader sequences of types 1 and 2 both contain consensus binding sequences for the transcription factors NF-kappa-B and Sp-1, but type 2 alone contains additional binding sequences for AP-1 and AP-2. The transcripts show differential expression, with types 1 and 2 expressed in expressed in heart, spleen and kidney, and type 1 alone found in peripheral leukocytes and brain.

PAF receptors have also been detected in the brain using ligand binding assays on cortical subcellular fractions (Marcheselli et al., 1990). These studies showed two different types of binding sites, one in synaptosomal membranes and another in microsomal membranes. The two sites differ in the kinetics of [^{3}H]-PAF binding; the microsomal membranes in fact display cold-ligand displacement kinetics characteristic of two binding sites.

The synaptosomal and microsomal sites also differ in their sensitivity to different types of PAF receptor antagonist, that in turn appear to influence different aspects of PAF action in the CNS. BN 52021, a terpenoid extracted from the leaf of the *Ginkgo biloba* tree, binds preferentially to the synaptosomal site whereas the synthetic hetrazepine BN 50730 shows complete specificity for the microsomal sites (Marcheselli et al., 1990; Marcheselli et al., 1994). The BN 52021-sensitive site appears to correspond to the cloned PAF receptor and to mediate the effects of PAF at the synapse. This antagonist selectively inhibits glutamate release from hippocampal neurons (Clark et al., 1992), perforant path LTP in hippocampal slices (Kato et al., 1994), and PAF-induced enhancement of memory tasks related to structures in the limbic system (Izquierdo et al., 1995) and striatum (Bazan et al., 1996). The BN 50730-sensitive site has yet to be isolated and there is no indication whether this represents an isoform of the cell-surface receptor, or a novel receptor. There is also no indication as yet as to which cell type(s) in the CNS express this form of PAF receptor. Thus far, the BN 50730-sensitive receptor has been linked with PAF-mediated effects on gene expression in the CNS. BN-50730, but not BN 52021, inhibits PAF- and neurotrauma-induced expression of the immediate-early genes *zif-268*, *c-fos*, and COX-2 (Marcheselli et al., 1994; Marcheselli et al., 1995). The inhibition by this antagonist of COX-2 expression in a vasogenic cerebral edema model is accompanied by reduction of the blood-brain barrier breakdown (Marcheselli et al., 1995; Bazan et al., 1996). The selective effects of BN 50730 on PAF-induced gene expression are not limited to the brain or neuronal cells, as the drug also inhibits PAF-induced expression of immediate-early genes and selected metalloproteinase genes in a corneal epithelium organ culture system (HEP Bazan et al., 1993; Tao et al., 1994; Tao et al., 1995).

There are additional examples in neuronal and non-neuronal cells in which PAF selectively influences the expression of certain genes. PAF rapidly and transiently augments levels of the c-*fos* and c-*jun* transcription factor mRNAs in a neuronal cell line (Squinto et al., 1989). Phorbol esters and PAF synergistically stimulate c-*fos* expression suggesting that the transcriptional effects of PAF are not mediated by protein kinase C. PAF's effect is at the transcriptional level, as opposed to increasing the stability of the mRNA 5′ deletion mutagenesis studies of the c-*fos* promoter show that the calcium-response element is necessary for the PAF-induced response. There are other examples of PAF-induced gene expression in neuronal and non-neuronal cells including the heparin-binding epidermal growth factor in monocytes (Pan et al., 1995), *c-fos* and egr-2 in lymphoblastoid cell lines (Mazer et al., 1994), the transcription factor NF-kappa B and immunoglobulins in human B cell lines (Smith and Shearer, 1994), *c-fos* and zif/268 in rat astroglia (Dell'Albani et al., 1993), *c-fos* and TIS 1 in A-431 epidermoid carcinoma cells (Tripathi et al., 1991). In addition PAF is able to autoregulate gene expression of its own cell-surface receptor (Mutoh et al., 1994; Shirasaki et al., 1994).

PHYSIOLOGICAL SIGNIFICANCE OF PLATELET-ACTIVATING FACTOR IN GLUTAMATE NEUROTRANSMISSION, NEURONAL PLASTICITY AND MEMORY

PAF is an important modulator of glutamate neurotransmission, memory formation and neuronal plasticity. The following evidence supports this conclusion:

1. The PAF analog methylcarbamyl PAF (MC-PAF), but not biologically inactive lyso-PAF, when applied to synaptically connected hippocampal neurons, specifically augments endogenous excitatory synaptic responses, but not inhibitory GABA-mediated responses (Clark et al., 1992). It does not, however, augment the effects of exogenously-added glutamate, suggesting that this PAF effect is not at the level of glutamate receptors. The PAF receptor antagonist BN 52021 (at a concentration of 1μM) blocks the MC-PAF-enhanced neurotransmission. Since this antagonist is selective for the PAF receptor of synaptosomal membranes (Marcheselli et al., 1990), this implies that PAF under these conditions acts as a retrograde messenger on a presynaptic site. Incidentally, the neuroprotective effect of BN 52021 in ischemic-reperfusion in gerbil brain (Panetta et al.,

1987) may be due partly to an inhibition of the PAF-induced release of excessive amounts of glutamate which attenuates excitotoxic damage.

2. PAF is a retrograde messenger in LTP. When applied to the CA1 subfield of the hippocampus, as in the hippocampal cell cultures described above, the bioactive lipid acts presynaptically via the cell-surface PAF receptor. Bath application of BN 52021 to hippocampal slices 25 minutes before or 10 minutes after stimulation blocks the sustained enhancement of excitatory postsynaptic potentials that follows tetanic stimulation of the Schaffer collateral pathway (Kato et al., 1994). This is not due to interference with NMDA receptor activity because the NMDA component of synaptic currents remains unaltered under all conditions. Thus PAF must be acting downstream of the NMDA component of the LTP cascade. In the pretreated slices although an initial potentiation is seen, this declines to basal levels within 10 minutes, indicating that the effect of the antagonist is not on transmission of the initial stimulus. Indeed, there is a window of 5–10 minutes after commencement of stimulation in which BN52021 can still inhibit establishment of LTP. Very high concentrations of exogenous MC-PAF (40μM) can overcome this inhibition, indicating that the antagonist is acting directly on the presynaptic PAF receptor. Supporting this idea is the inability of the highly selective intracellular PAF receptor antagonist BN 50730 (Marcheselli and Bazan, 1994) to affect the establishment of LTP.
3. PAF enhances memory in rats performing an inhibitory avoidance task (Izquierdo et al., 1995) and in a water maze task (Bazan et al., 1996). In these instances specificity for the presynaptic PAF receptor was shown, as was neuroanatomical and temporal selectivity. For instance, in the inhibitory avoidance task PAF only had an effect when infused at specific times after training into defined areas of the limbic system, and in the water maze task PAF was only effective when infused into the striatum.

Arachidonic acid is another potential retograde messenger in hippocampal LTP (Williams et al., 1989) but slices require long-term exposure to high concentrations of the fatty acid. Furthermore inhibition of this form of LTP by the NMDA receptor antagonist D-2-amino-5-phosphopentanoic acid (APV) (O'Dell et al., 1991; Kato et al., 1991) indicates that arachidonic acid acts via a different mechanism than PAF, which acts independently of NMDA receptor. There is a report of PAF eliciting LTP which is sensitive to APV (Wieraszko et al, 1993). However, this was after prolonged exposure, and since arachidonic acid can potentiate NMDA responses (Miller et al., 1992) and PAF promotes the release of arachidonic acid (Kunievsky and Yavin, 1992), then this form of LTP could arise from the effects of arachidonic acid.

PAF ACTIVATES TRANSCRIPTION OF RESPONSE GENES

Several signals trigger the rapid and transient expression of immediate-early genes (primary genomic response genes, early response genes). In the brain physiological and pathological events (such as long-term potentiation, ischemia, seizures and NMDA receptor activation) initiate transcription of genes encoding transcription factors and thus have the potential to initiate cascades of gene expression. As a result of gene activation, long-term cellular responses such as neuronal plasticity occur.

PAF is a mediator of extracellular and intracellular signal/gene transcription coupling since it stimulates immediate-early gene expression in neuronal and other cells in culture. Moreover, selective PAF antagonists block gene transcription in transfected cells an *in vivo* in animal models of brain injury or seizures. PAF rapidly and transiently augments levels of the c-*fos* and c-*jun* transcription factor mRNAs in a neuronal cell line (Squinto et al., 1981). Phorbol esters and PAF synergistically stimulate c-*fos* expression suggesting that the transcriptional effects of PAF are not mediated by protein kinase C. The effect of PAF is at the transcriptional level, as opposed to increasing the stability of the mRNA 5′ deletion mutagenesis studies of the c-*fos* promoter show that the calcium-response element is necessary for the PAF-induced response. There are other examples of PAF-induced gene expression in neuronal and non-neuronal cells including the heparin-binding epidermal growth factor in monocytes (Pan et al., 1995), *c-fos* and egr-2 in lymphoblastoid cell lines (Mazer et al., 1994), the transcription factor NF-kappa B and immunoglobulins (Smith and Shearer, 1994) in human B cell lines, *c-fos* and zif/268 in rat astroglia (Dell'Albani et al., 1993), and *c-fos* and TIS 1 in A-431 epidermoid carcinoma cells (Tripathi et al., 1991). PAF is also able to up-regulate gene expression of its own cell-surface receptor (Mutoh et al., 1994; Shirasaki et al., 1994). The intracellular PAF binding site antagonist BN 50730 blocks *c-fos* and zif/268 transcription factor expression induced in rat hippocampus and cerebral cortex by a single electroconvulsive shock (Marcheselli and Bazan, 1994), or zif/268 and inducible cyclooxygenase (PGHS-2, COX-2, TIS 10) in cryogenically-induced brain edema (Bazan et al., 1996).

PAF is involved in the seizure-induced accumulation of PGS-2 in the brain, and is thus linked to the metabolism of free arachidonic acid generated during seizures. PGS-2 transcriptional activity and protein expression is up-regulated in a single electrically-induced seizure (ECS) and in multiple, kainic acid (KA)-induced seizures, and PAF is involved in this mechanism. Furthermore, the induction of PGS-2, but not the transcription factor immediate-early gene zif-268, is more sensitive to the severe KA-induced seizures than ECS in a brain region, the hippocampus, selectively vulnerable to damage in this model. Induction of both genes by KA shows neuroanatomical specificity in the order hippocampus > cerebral cortex > striatum > brain stem > cerebellum. Nuclear run-on and Western blotting shows that both genes are transcriptionally activated, and that kainic acid up-regulation of prostaglandin G/H synthase-2 mRNA expression in hippocampus matches increased protein levels. Whereas the magnitude of hippocampal *zif*-268 mRNA induction is similar in both seizure models, peak induction of prostaglandin G/H synthase-2 mRNA is seven-fold greater in the kainic acid model than in the electroconvulsive shock model, and is much more prolonged. Pretreatment of animals by intracerebroventricular injection with the intracellular platelet-activating factor receptor antagonist BN 50730 strongly attenuates kainic acid and electroconvulsive shock induction of prostaglandin G/H synthase-2 expression. The drug partially inhibits electroconvulsive shock induction of *zif*-268, but is relatively ineffective against kainic acid-induced *zif*-268 expression. Seizure-induced expression of both genes involves platelet-activating factor, but the mechanisms of induction must be otherwise distinct. The selectively elevated induction of hippocampal prostaglandin G/H synthase-2 by kainic acid correlates with a neuroanatomical region in which the agonist induces neuronal damage (Marcheselli and Bazan, 1991).

PAF ACTIVATES EXPRESSION OF THE INDUCIBLE PROSTAGLANDIN SYNTHASE GENE

The rate-limiting step of prostaglandin synthesis is conversion of arachidonic acid to prostaglandin $_{\mathrm{H_2}}$, catalyzed by prostaglandin G/H synthase. There are now known to be two forms of the enzyme, a "constitutive" enzyme (PGHS-1), and an "inducible" form (PGHS-2). The two enzymes have 61% amino acid sequence homology and share highly conserved domains essential for enzyme function, including glycosylation sites, heme binding sites, the aspirin acetylation sequence, and a region with homology to epidermal growth factor . They are, however, encoded by separate genes. PGHS-2 was first described as an immediate early gene induced by mitogens, cytokines, including by a number of mitogens and cytokines including serum, bFGF, TNFα, IL-1β, and pp60$^{v\text{-}src}$, and by phorbol esters, and down-regulated by glucocorticoids (Kujubu et al, 1991; O'Banion et al., 1991; Herschman, 1994). The rationale behind having two different forms of the enzyme is not entirely clear because, in cells which express both forms, up-regulation of PGHS-2 is not matched by an increase in prostaglandin synthesis. Nevertheless, recent findings suggest different subcellular localizations for the two enzymes: PGHS-1 is localized in the endoplasmic reticulum, while PGHS-2 is enriched in the nuclear envelope. One line of thought is that PGHS-2 is responsible for the synthesis of PGs which act as intracellular signals, while PGHS-1 activity leads to the generation of PGs which act as extracellular mediators (Goetzl et al., 1995). PGHS-2 has been found in rat brain, where it is expressed throughout the forebrain in discrete populations and is enriched in the cerebral cortex and hippocampus (Yamagata et al., 1993). Its expression is rapidly and transiently induced by seizure or NMDA-dependent synaptic activity, while basal expression in the developing and adult appears to be regulated by synaptic activity. For instance, there is detectable expression in the developing animal from post-natal day 5 onwards, a period during which the rat brain is undergoing massive synaptic plasticity and developmental changes. The inducibilty of PGHS-2 expression in response to synaptic stimuli implies a role in neuronal plasticity (Yamagata et al., 1993).

Using constructs of the promoter of the murine inducible prostaglandin synthase gene transfected to neural and non-neural cell it was found that PAF stimulates expression of the luciferase reporter gene (Bazan et al, 1993a). Retinoic acid modulates gene expression through nuclear receptors that are members of a superfamily of ligand-dependent transcription factors. Since PAF may also affect gene expression through an intracellular site, the effect of retinoic acid was studied. Retinoic acid induces the expression of constructs of the TIS-10/PGS-2 promoter with a luciferase reporter transfected into neuroblastoma cells. Using the calcium phosphate co-precipitation transfection procedure in the presence of retinoic acid, there is a PAF-dependent (from 1 to 50 nM) activation of luciferase reporter constructs driven by regulatory regions of the TIS-10/PGHS-2 gene. The effect of PAF in indicating that a preexisting latent transcription factor(s) is engaged in the effect. Deletion studies of the TIS-10/PGHS-2 promoter/luciferase reporter constructs showed the deletion of sequences between -371 and -300 reduced the PAF inductive effect from 31-fold to 4.1-fold. BN 50730 inhibited the PAF inductive effect when incubated for 1 h with cultured cells before transfection. This indicates that a PAF receptor is involved in the expression of the inducible prostaglandin synthase and that a major PAF-responsive element lies within this region. Further deletions showed no response to added PAF.

MEMBRANE-DERIVED LIPID SECOND MESSENGERS IN NEURONAL PLASTICITY

The membrane-derived bioactive lipids regulate normal synaptic function (Bazan et al., 1993b) and plasticity responses and repair/regenerative responses. PAF is one of the messengers that links synaptic activation with transcription early response genes. Al-

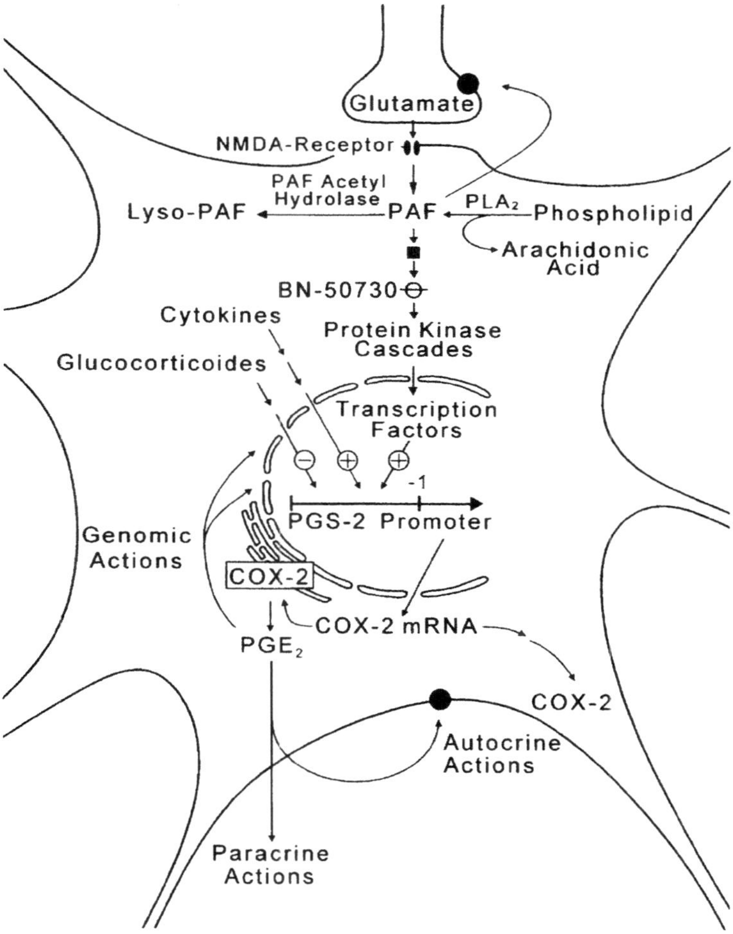

Figure 2. Excitatory synaptic transmission enhances the production of PAF, other messengers, and the expression of PGS-2. NMDA-receptor activation by glutamate leads to phospholipase A_2 activation and the generation of PAF and arachidonic acid. PAF is synthesized through other metabolic routes, as well. Other synaptic events, in addition to NMDA-receptor, may also be involved in the accumulation of PAF. PAF activates PGS-2 gene expression through a BN-50730-sensitive intracellular site, protein kinase cascades, and transcription factors. The PGS-2 promoter is also a target for cytokines (activation) and glucocorticoids (inhibition). PGS-2 protein (COX-2) then catalyzes the conversion of arachidonic acid into PGH_2, the precursor of eicosanoids. Constitutive PGS-1 also catalyzes this metabolic step (not shown). Prostaglandin E_2 is depicted as a product of PGS-2 (COX-2). The specific products of over expressed PGS-2 are not yet known. PGE_2 is shown to potentially elicit genomic, paracrine, and/or autocrine effects.

though PAF is most often referred to as a mediator of the inflammatory and immune responses as well as of cell injury, low PAF concentrations elicits sprouting in PC12 cells, whereas neuropathological changes occur when these cells are exposed to high PAF concentrations (Kornecki and Ehrlich, 1988). Therefore, PAF, when accumulated in large amounts, becomes a neuronal injury messenger (Bazan, 1997). This duality of PAF actions, both physiological and pathological, is also a feature of glutamate, an excitatory amino acid, that, by far, is the most abundant neurotransmitter of the mammalian brain. Glutamate plays a critical role in developmental plasticity and memory formation. However, in stroke, seizures, neurotrauma and some neurodegenerative diseases, there is an abnormally high synaptic accumulation of glutamate, making it a critical effector of excitotoxic damage. Other examples of neuronal mediators with this duality are interleukin-1 and, perhaps, amyloid peptide, that coexist with neural cells under physiological conditions but may, when overproduced, be engaged in pathophysiology.

The presynaptic PAF binding site is linked to glutamate release (Clark et al., 1992) and involves PAF as a retrograde messenger in long-term potentiation (Kato et al., 1994). There is also a very high affinity binding site, localized intracellularly and linked to early response gene expression (Bazan et al., 1991). It is not yet known if two or more of these actions are interlinked or if there are different effects among the diverse cellular responses stimulated by PAF. The significance of the effect of PAF on gene expression may be related to long-term adaptive responses. PAF's ability to increase intracellular Ca^{2+} (Bito et al, 1992; Kornecki and Ehrlich, 1988), likely by activating a Ca^{2+} channel, by activating phospholipase C with resultant intracellular Ca^{2+} mobilization by inositol trisphosphate and by other actions is also a significative set of mechanisms through which synaptic plasticity may be affected.

PAF stimulation of the inducible prostaglandin synthase (COX-2) gene establishes another link with synaptic plasticity. The gene encodes the inducible COX-2, one of the enzymes that catalyzes the cyclooxygenation of arachidonic acid. The inducible prostaglandin synthase utilizes only a small proportion of the free arachidonic acid available under these conditions. PGE_2 synthesis may comprise a feedback loop of synaptic responses (**Figure 2**). The signal transduction pathways enabled by enhanced PLA_2 activation at the synapse provide powerful modulators of plasticity that potentially regulate memory formation and long term potentiation of the nervous system.

ACKNOWLEDGMENT

Supported by NIH NS23002.

REFERENCES

Ali H., Richardson R. M., Tomhave E. D., DuBose R. A., Haribabu B. and Snyderman R. (1994) Regulation of stably transfected platelet activating factor receptor in RBL-2H3 cells. Role of multiple G proteins and receptor phosphorylation. *J Biol Chem.* **269**: 24557.

Bazan N.G., Squinto S.P., Braquet P., Panetta T., Marcheselli V.L. (1991) Platelet-activating factor and polyunsaturated fatty acids in cerebral ischemia or convulsions: Intracellular PAF-binding sites and activation of a Fos/Jun/Ap-1 transcriptional signaling system, *Lipids* **26**:1236–42.

Bazan H E P., Tao Y. and Bazan N G. (1993) Platelet activating factor induces collagenase expression in corneal epithelium cells. *Proc. Natl. Acad. Sci. U S A.* **90**: 8672.

Bazan N.B., Zorumski C.F., Clark G.D. (1993a) The activation of phospholipase A_2 and release of arachidonic acid and other lipid mediators at the synapse: The role of platelet-activating factor. *J Lipid Med* **6**:421–7.

Bazan N.G., Fletcher B.S., Herschman H.R., Mukherjee P.K. (1993b) Platelet-activating factor and retinoic acid synergistically activate the inducible prostaglandin synthase gene, *Proc. Natl Acad Sci U S A* **91**:5252–6.

Bazan N G., Packard M G., Teather L. and Allan G. (1997) Bioactive lipids in excitatory neurotransmission and neuronal plasticity. *Neurochem. Int.* **30**:225–231.

Bazan N G., Allan G. and Marcheselli V L. (1996) An inhibitor of injury-induced COX-2 transcriptional activation elicits protection of blood-brain barrier breakdown in a brain damage model, *Improved non-steroidal anti-inflammatory drugs: COX-2 enzyme inhibitors.* (Vane J R. and Botting R M., eds.) Kluwer Academic Publishers, Lancaster.

Bazan N.G. (1997): Lipid messengers and prostaglandin endoperoxide synthase-2 in neuronal cell death. In Primer Cerebrovascular Diseases. M Welsh, L Chaplan, D Reis, B Siesjö, B Weir (eds), Academic Press, pps. 1322–1326.

Benveniste J., Henson P M. and Cochrane C G. (1972) Leukocyte-dependent histamine release from rabbit platelets. The role of IgE, basophils, and platelet-activating factor. *J. Exp. Med.* **136**: 1356 .

Bito H., Nakamura M., Honda Z., Izumi T., Iwatsubo T., Seyama Y., Ogura A., Kudo Y. and Shimizu T. (1992) Platelet-activating factor (PAF) receptor in rat brain: PAF mobilizes intracellular Ca^{2+} in hippocampal neurons. *Neuron.* **9**: 285.

Blank M L., Smith Z L., Fitzgerald V. and Snyder F. (1995) The CoA-independent transacylase in PAF biosynthesis: tissue distribution and molecular species selectivity. *Biochim Biophys Acta* **1254**: 295.

Braquet P., Touqui L., Shen T Y. and Vargaftig B B. (1987) Perspectives in platelet-activating factor research. *Pharmacol. Rev.* **39**: 97.

Clark G D., Happel L.T., Zorumski C.F. and Bazan N.G. (1992) Enhancement of hippocampal excitatory synaptic transmission by platelet-activating factor. *Neuron* **9**: 1211.

Dell' Albani P., Condorelli D F., Mudo G., Amico C., Bindoni M. and Belluardo N. (1993) Platelet-activating factor and its methoxy analogue ET-18-OCH3 stimulate immediate-early gene expression in rat astroglial cultures. *Neurochem Int.* **22**: 567.

Francescangeli E. and Goracci G. (1989) The *de novo* biosynthesis of platelet-activating factor in rat brain. *Biochem Biophys Res Commun.* **161**: 107.

Goetzl E.J., An W.L., Smith S. (1995) Specificity of expression of eicosanoid mediators in normal physiology and human disease. *FASEB J.* **9**:1051–58.

Goracci G. and Francesangeli E. (1991) Properties of PAF-synthesizing phosphocholinetransferase and evidence for lysoPAF acetyltransferase activity in brain. *Lipids* **26**: 986.

Hattori M., Arai H. and Inoue K. (1993) Purification and characterization of bovine brain platelet-activating factor acetylhydrolase. *J Biol Chem* **268**: 18748.

Hattori M., Adachi H., Tsujimoto M., Arai H. and Inoue K. (1994a) Miller-Dieker lissencephaly gene encodes a subunit of brain platelet-activating factor acetylhydrolase *Nature* **370**: 216.

Hattori M., Adachi H., Tsujimoto M., Arai H. and Inoue K. (1994b) The catalytic subunit of bovine brain platelet-activating factor acetylhydrolase is a novel type of serine esterase. *J Biol Chem* **269**: 23150.

Heery J M., Kozak M., Stafforini D M., Jones D A., Zimmerman G A., McIntyre T M. and Prescott S M. (1995) Oxidatively modified LDL contains phospholipids with platelet-activating factor-like activity and stimulates the growth of smooth muscle cells. *J. Clin. Invest.* **96**: 2322 .

Henson P M. (1970) Release of vasoactive amines from rabbit platelets induced by antiplatelet antibody in the presence and absence of complement. *J Immunol.* **104**: 934.

Herschman H.R., (1994) Regulation of prostaglandin synthase-1 and prostaglandin synthase-2, *Cancer Metastasis Rev.* **13**:241–56.

Holland M R., Venable M E., Whatley R E., Zimmerman G A., McIntyre T M. and Prescott S M. (1992) Activation of the acetyl-coenzyme A:lysoplatelet-activating factor acetyltransferase regulates platelet-activating factor synthesis in human endothelial cells. *J. Biol. Chem.* **267**: 22883.

Honda Z., Nakamura M., Miki I., Minami M., Wantabe T., Seyama Y., Okado H., Toh H., Ito K., Miyamoto T. and Shimizu T. (1991) Cloning by functional expression of platelet-activating factor receptor from guinea-pig lung. *Nature* **394**: 342.

Honda Z., Takano T., Gotoh Y., Nishida E., Ito K. and Shimizu T. (1994) Transfected platelet-activating factor receptor activates mitogen-activated protein (MAP) kinase and MAP kinase kinase in Chinese hamster ovary cells. *J Biol Chem.* **269**: 2307.

Honda Z., Takano T., Hirose N., Suzuki T., Muto A., Kume S., Mikoshiba K., Itoh K. and Shimizu T. (1995) Gq pathway desensitizes chemotactic receptor-induced calcium signaling via inositol trisphosphate receptor down-regulation. *J Biol Chem.* **270**: 4840.

Izquierdo I., Fin C., Schmitz P K., Da Silva R C., Jerusalinsky O., Quillfeldt J A., Ferreira M B., Medina J H. and Bazan N G. (1995) Memory enhancement by intrahippocampal, intraamygdala, or intraentorhinal infusion of platelet-activating factor measured in an inhibitory avoidance task. *Proc Natl Acad Sci USA.* **92**: 5047.

Kato K., Uruno K., Saito K., Kato H. (1991) Both arachidonic acid and 1-oleoyl-2-acetyl

glycerol in low magnesium solution induce long-term potentiation in hippocampal CA1 neurons in vitro. *Brain Res* **563**: 94–100.

Kato K., Clark G D., Bazan N G. and Zorumski C F. (1994) Platelet-activating factor as a potential retrograde messenger in CA1 hippocampal long-term potentiation. *Nature.* **367**: 179.

Kester M., Thomas C P., Wang J. and Dunn M J. (1992) Platelet-activating factor stimulates multiple signaling pathways in cultured rat mesangial cells. *J. Cell. Physiol.* **153**: 244.

Kornecki E. and Ehrlich Y H. (1988) Neuroregulatory and neuropathological actions of the ether-phospholipid platelet-activating factor. Science. **240**: 1792.

Kujubu D.A., Fletcher B.S., Varnum B.C., Lim R.W., and Herschman H.R. (1991) TIS10, a phorbol ester tumor promoter-inducible mRNA from Swiss 3T3 cells, encodes a novel prostaglandin synthase/cyclooxygenase homologue. *J. Biol. Chem.* **266**:12866–12872.

Kunievsky B., Yavin E. (1992) Platelet-activating factor stimulates arachidonic acid release and enhances thromboxane B2 production in intact fetal rat brain ex vivo, *J Pharmacol Exp Ther* **263**:562–8.

Kunz D., Gerard N P. and Gerard C. (1992) The human leukocyte PAF receptor. cDNA cloning, cell surface expression and construction of a novel epitope-bearing analog. *J Biol Chem.* **267**: 9101 .

Kuruvilla A., Pielop C. and Shearer WT. (1994) Platelet-activating factor induces the tyrosine phosphorylation and activation of phospholipase C-γ1, fyn and lyn kinases, and phosphatidylinositol 3-kinase in a human B cell line. *J. Immunol.* **153**: 5433.

Lee T-C., Malone B. and Snyder F. (1988) Formation of 1-alkyl-2-acetyl-sn-glycerols via the *de novo* biosynthetic pathway for platelet-activating factor. *J Biol Chem.* **263**: 1755.

Marcheselli V L., Rossowska M., Domingo M T., Braquet P. and Bazan N G. (1990) Distinct platelet-activating factor binding sites in synaptic endings and in intracellular membranes of rat cerebral cortex. *J Biol Chem.* **265**: 9140.

Marcheselli V L. and Bazan N G. (1994) Platelet-activating factor is a messenger in the electroconvulsive shock-induced transcriptional activation of c-*fos* and *zif*-268 in hippocampus. *J. Neurosci. Res.* **37**: 54.

Marcheselli V L. and Bazan N G. (1995) Upregulation of expression of inducible prostaglandin synthase-2 in vasogenic cerebral edema: inhibition by BN 50730, a PAF antagonist. *Soc. Neurosci. Abs.* **21**: 1867.

Marcheselli VL, Bazan NG. (1996) Sustained induction of prostaglandin endoperoxide synthase-2 by seizures in hippocampus: Inhibition by a platelet-activating factor antagonist. *J Biol Chem* **271**:24794–24799.

Mazer B., Domenico J., Sawami H. and Gelfand E W. (1994) Platelet-activating factor induces an increase in intracellular calcium and expression of regulatory genes in human B lymphoblastoma cells. *J Immunol.* **146**: 1914.

Miller B., Sarantis M., Traynelis S.F., Attwell D. (1992) Potentiation of NMDA receptor currents by arachidonic acid. *Nature* **355**:722–5.

Mukherjee P.K., DeCoster M.A., Campbell F.Z., Davis R.J., Bazan N.G. Glutamate receptor signaling interplay modulates stress-sensitive MAP kinases and neuronal cell death (submitted).

Mutoh H., Bito H., Minami M., Nakamura M., Honda Z., Izumi T., Nakata R., Kutachi Y., Terano A. and Shimizu T. (1988) Two different promoters direct expression of two distinct forms of of mRNA of human platelet-activating factor receptor. *FEBS Lett.* **322**: 129.

Mutoh H., Ishii S., Izumi T., Kato S. and Shimizu T. (1994) Platelet-activating factor (PAF) positively auto-regulates the expression of human PAF receptor transcript 1 (leukocyte-type) through NF-kappa B. *Biochem Biophys Res Commun* **205**: 1137.

O'Banion M.K., Sadowski H.B., Winn V., and Young D.A., (1991) A serum- and glucocorticoid-regulated 4-kilobase mRNA encodes a cycooxygenase-related protein. *J. Biol. Chem.* **266**:23261–23267.

O'Dell T.J., Hawkins R.D., Kandel E.R., Arancio O. (1991) Tests of the roles of two diffusible substances in long term potentiation: evidence for nitric oxide as a possible early retrograde messenger, *Proc. Natl Acad. Sci. U S A* **88**: 11285–9.

Pan Z., Kravchenko V V. and Ye R D. (1995) Platelet activating factor stimulates transcription of the heparin-binding epidermal growth factor-like growth factor in monocytes. Correlation with an increased kappa B binding activity. *J Biol Chem.* **270**: 7787.

Panetta T., Marcheselli V L., Braquet P., Spinnewyn B. and Bazan N G. (1987) Effects of a platelet-activating factor antagonist (BN 52021) on free fatty acids, diacylglycerols, polyphosphoinositides and blood flow in the gerbil brain: Inhibition of ischemia-reperfusion induced cerebral injury. *Biochem Biophys Res Commun.* **149**: 580.

Prescott S M., Zimmerman G A. and McIntyre T M. (1990) Platelet-activating factor. *J. Biol. Chem.* **265**: 17381.

Shirasaki H., Adcock I M., Kwon O J., Nishikawa M., Mak J C. and Barnes P J. (1994) Agonist-induced up-regulation of platelet-activating factor receptor messenger RNA in human monocytes. *Eur. J. Pharmacol.* **268**: 263.

Smiley P L., Stremler K E., Prescott S M., Zimmerman G A., and McIntyre, T. M. (1991) Oxidatively-damaged phosphatidylcholines activate human neutrophils through the receptor for phosphatidylcholine. *J. Biol. Chem.* **266**: 11104.

Smith C S. and Shearer W T. (1994) Activation of NF-kappa B and immunoglobulin expression in response to platelet-activating factor in a human B cell line. *Cell Immunol.* **155**: 292.

Snyder F. (1995) Platelet-activating factor and its analogs: metabolic pathways and related intracellular processes. *Biochim. Biophys. Acta.* **1254**: 231.

Sogos V., Bussolino F., Pilia E., Torelli S. and Gremo F. (1990) Acetylcholine-induced production of platelet-activating factor by human fetal brain cells in culture. *J Neurosci Res.* **27**: 706.

Squinto, S.P., Block, A.L., Braquet, P. and Bazan, N.G. (1981a) Platelet-activating factor stimulates a Fos/Jun/AP-1 transcriptional signaling system in human neuroblastoma cells. *J. Neurosci. Res.* **24**:558–66.

Squinto S.P., Block A.L., Braquet P., and Bazan N.G. (1981b) Platelet-activating factor stimulates a Fos/Jun/AP-1 transcriptional signaling system in human neuroblastoma cells, *J Neurosci Res* **24**:558–66.

Squinto S P., Block A L., Braquet P. and Bazan N G. (1989) Platelet-activating factor stimulates a Fos/Jun/AP-1 transcriptional signaling system in human neuroblastoma cells. *J Neurosci Res.* **29:** 558.

Stafforini D M, Prescott S M, Zimmerman G A, and McIntyre T M. (1991) Platelet-activating factor acetylhydrolase activity in human tissues and blood cells. *Lipids* **26**: 1015.

Stafforini D M., Rollins E N., Prescott S M., and McIntyre T M. (1993) The platelet-activating factor acetylhydrolase from human erythrocytes. Purification and properties. *J Biol Chem* **268**: 3857 .

Sugimoto T., Tsuchimochi H., McGregor C G., Mutoh H., Shimizu T. and Kurachi Y. (1992) Molecular cloning and characterization of the platelet-activating factor receptor from human heart. *Biochem Biophys Res Commun.* **189**: 617.

Tao Y., Bazan H E P. and Bazan N G. (1994) Platelet-activating factor (PAF) induces the expression of urokinase-type, but not the tissue-type, plasminogen activator in rabbit corneal epithelium. *Suppl. Invest. Ophthalmol. Vis. Sci.* **35**: 1457.

Tao Y, Bazan H E. and Bazan N G. (1995) Platelet-activating factor induces the metalloproteinases -1 and -9, but not -2 or -3, in the corneal epithelium. *Invest Ophthalmol Vis Sci* **36**: 345.

Tjoelker L W. Eberhardt C. Unger J. Trong H L. Zimmerman G A. McIntyre T M. Stafforini D M. Prescott S M. and Gray P W. (1995) Plasma platelet-activating factor acetylhydrolase is a secreted phospholipase A2 with a catalytic triad. *J Biol Chem* **270**: 25481.

Tripathi Y., Kandala J., Guntaka R., Lim R. and Shukla S. (1991) Platelet-activating factor induces expression of early response genes c-*fos* and TIS-1 in human epidermoid carcinoma-432 cells. Life Sci. **49**: 1761.

Wieraszko A., Li G., Kornecki E., Hogan M.V., and Ehrlich Y.H., (1993) Long-term potentiation in the hippocampus induced by platelet-activating factor, *Neuron* **10**: 553–557.

Williams J.H., Errington L., Lynch M.A., Bliss T.V.P. (1989) Arachidonic acid induces long-term activity-dependent enhancement of synaptic transmission in the hippocampus, *Nature* **341**: 739–42.

Yamagata K., Andeasson K.I., Kaufmann W.E., Barnes C.A., Worley P.F., (1993) Expression of a mitogen-inducible cyclooxygenase in brain neurons: Regulation by synaptic activity and glucocorticoids, *Neuron* **11**:371–86.

Zimmerman G A., Prescott S M. and McIntyre T M. (1994) Oxidatively fragmented phospholipids as inflammatory mediators: the dark side of polyunsaturated lipids. *J. Nutr.* **125**: 1661S.

4

SURFACE PROTEIN PHOSPHORYLATION BY ECTO-PROTEIN KINASES

Role in Neuronal Development and Synaptic Plasticity

Yigal H. Ehrlich, Michael V. Hogan, Zofia Pawlowska, Andrzej Wieraszko, Ethel Katz, Tomasz Sobocki, Anna Babinska, and Elizabeth Kornecki

The Department of Biology and the CSI/IBR Center for Developmental Neuroscience
The College of Staten Island of The City University of New York
2800 Victory Boulevard, Staten Island, New York 10314 and
The Department of Anatomy and Cell Biology
State University of NY Health Science Center at Brooklyn, Brooklyn, New York 11203

INTRODUCTION

Specific patterns of neuronal development in the brain are determined by complex interactions occurring at cell surfaces, but little is known about the enzymatic events that regulate these interactions. Protein phosphorylation, a ubiquitous step in intracellular pathways that produce rapid and transient changes in neuronal activity (Greengard, 1978; Nestler and Greengard, 1983; Ehrlich, 1984), was found to operate also as a key mechanism of molecular adaptation in processes whereby hormonal, pharmacological and behavioral inputs influence neuronal development and induce long-lasting alterations in neuronal function (Ehrlich and Routtenberg, 1975; Ehrlich et al.,1977, 1987; Neary et al., 1981; Kandel and Schwartz, 1982). The discovery of neuronal ecto-protein kinase activity (Ehrlich et al.,1986) has revealed that the powerful regulatory mechanism of protein phosphorylation operates also in the extracellular environment of the nervous system. In this location, the regulation of delicate interactions among cells, and between surface proteins and components of the extracellular matrix, can determine patterns of neuronal connectivity and thus play a significant role in processes underlying synaptic plasticity.

The secretion of ATP from synaptic vesicles of stimulated neurons, described first in the 1950's in peripheral neurons and later in the CNS, is now extensively documented (reviewed by Gordon, 1986). ATP-release from neurons has recently become the subject of intensive investigation with increasing evidence of a role for extracellular ATP in synaptic

Molecular and Cellular Mechanisms of Neuronal Plasticity, edited by Ehrlich. Plenum Press, New York, 1998.

transmission (Edwards et al.,1992; Evans et al.,1992; Silinsky et al.,1992; Shen and North, 1993) and in neuronal events associated with synaptic plasticity and implicated in memory formation (Wieraszko and Seyfried, 1989; Nishimura et al.,1990; Fujii et al.,1995). The mechanisms by which secreted ATP can exert effects on neuronal function include interaction with purinergic receptors (reviewed by Burnstock, 1981,1990), hydrolysis by ecto-ATPase (Nagy et al.,1986, 1995; Nagy, 1997), and the phosphorylation of surface proteins by ecto-protein kinase (reviewed by Ehrlich, 1987; Ehrlich et al.,1990). Studies of cloned neural cell lines of peripheral origin have implicated ecto-protein kinase (ecto-PK) activity in the regulation of events that are initiated at the cell surface, such as Ca^{2+}-influx (Ehrlich et al.,1988) and neurotransmitter uptake (Hendley et al.,1988; Hardwick et al., 1989). Ecto-PK was found to serve as a target for extracellular regulators of neuronal development such as nerve growth factor (NGF) in PC12 cells (Pawlowska et al., 1993), and gangliosides in GOTO neuroblastoma cells (Tsuji et al., 1988). We summarize here research conducted in our laboratories which has provided evidence for the presence of a developmentally-regulated ecto-PK in the central nervous system (CNS), and identified the neuronal surface proteins that serve as endogenous substrates for this activity in brain neurons. In particular, we focus here on the phosphorylation of specific membrane-surface proteins by a brain ecto-protein kinase with catalytic specificity similar to protein kinase C (PKC), and the potential role of this enzymatic system in mechanisms underling synaptic plasticity.

ECTO-PKS IN CELLS THAT SECRETE ATP

Early studies of ecto-PK activities in various cell types, conducted in the 1970s and 1980s, have been reviewed (Ehrlich et al.,1990). Since that review, there have appeared additional reports of ecto-PK and endogenous substrates for this activity in cells shown to release ATP, such as smooth muscle cells (Fedan and Lamport, 1990), platelets (Naik et al.,1991; Babinska et al.,1996) and L6 myoblasts (Chen and Lo, 1991a). Neutrophils (Skubitz et al.,1992;1994) as well as endothelial cells on blood vessel walls (Pirrotton et al., 1992; Hartmen and Schrader, 1992) were also shown to express ecto-PKs on their surface. The phosphorylation-state of surface proteins appears to influence cell communication as well as cell-environment interactions (Pfeifle et al.,1981). The inhibition of the phosphorylation of specific surface proteins by N-CAM antibodies suggested a role for ecto-PK in neuronal development (Ehrlich et al.,1986). This finding also demonstrated that antibodies directed towards substrates of ecto-protein kinase can inhibit their phosphorylation without penetrating cells, and hence provide important tools for the investigation of the precise role of specific surface protein phosphorylation systems in the regulation of biological functions.

The physiological significance of any PK activity is determined by the biological function of its specific protein substrates (Greengard 1978; Ehrlich, 1978; 1984). Ecto-PKs have been shown to phosphorylate two types of substrates: (1) cell surface proteins (Ehrlich et al., 1986; Chen and Lo, 1991a; Vilgrain and Baird, 1991; Volonte et al., 1994), and (2) protein components of the extracellular environment (Kalafatis et al., 1993; Kubler et al.,1992; Vilgrain and Baird, 1991). Chen and Lo (1991b) have shown the involvement of an ecto-PK and its 112K cell surface protein-substrate in the process of myogenesis. Two independent studies (Skubitz and Goueli, 1991;Vilgrain and Baird, 1991) have found that basic fibroblast growth factor serves as a substrate for ecto-PK associated with the outer surface of the target cells. Studies which detail the characterization of a cAMP-dependent ecto-PK capable of phosphorylating atrial natriuretic peptide (Kubler et al.,1992) and the plasma protein fibrinogen (Sonka et al., 1989; Dorner et al.,

1989), have been reported. The major plasma membrane differentiation antigen called PC-1 (Rebbe et al., 1991) was found to have a threonine-specific ecto-protein kinase activity stimulated by acidic fibroblast growth factor (Oda et al., 1991, 1993; Deissler, 1995), similar to the NGF stimulated ecto-phosphorylation of a 53K protein we have described in PC-12 cells (Pawlowska et al, 1993). Phosphorylation and dephosphorylation events involving the extracellularly-acting blood protein, fibrinogen, have been shown to affect thrombin-induced gelation and plasmin degradation of fibrinogen (Forsberg and Martin, 1991; Martin et al., 1991). A releasable surface phosphoprotein substrate of ecto-PK has been implicated in the regulation of cell growth and malignant transformation (Friedberg et al., 1995). More recently, ecto-PK activity has been implicated in the action and processing of β-Amyloid Precursor Protein (βAPP) and its products (Hogan et al., 1995; Walter et al., 1997). All these studies clearly indicate significant roles for ecto-PKs and their endogenous substrates in physiological regulatory processes.

Recent studies demonstrate that ecto-PK activity is not catalyzed by a single enzyme, but is carried-out by several different ecto-protein kinases that can be blocked by inhibitors that interact with the catalytic domain of known, intracellular PKs. Herbert et al (1991) have characterized extracellular binding sites for the PK inhibitor staurosporine on the surface of capillary endothelial cells, and we have found interactions of cyclosporine, a phosphoprotein phosphatase inhibitor, with the surface of platelets (Fernandez et al., 1993). A membrane impermeable protein kinase inhibitor, K-252b, was found to inhibit NGF-induced neuritogenesis in PC12 cells (Nagashima et al., 1991; Pawlowska et al., 1993), to inhibit synaptogenesis in cortical neurons (Muramoto et al., 1988; 1994; Kuroda et al.,1992) and to inhibit the maintenance phase of long-term potentiation (Chen et al., 1994; Fujii et al.,1995a and 1995b). Unique among all known protein kinases is an ecto-PK activity that can phosphorylate neuronal surface proteins in the absence of added Mg^{++} ions (Pawlowska et al., 1993; Volonte et al., 1994; Chen et al., 1994; Merlo et al.,1997). Finally, a casein kinase-like ecto-PK has been purified (Walter et al., 1996), and recently, the use of antibodies directed to the catalytic domain of PKC have identified novel and atypical isozymes of protein kinase C as ecto-enzymes in brain neurons (Hogan et al., 1995; see below). This activity could be inhibited by Alzheimer's β-amyloid peptides (ibid). Furthermore, phosphorylation sites in the ecto-domain of the β-Amyloid Precursor Protein (β-APP) itself were directly demonstrated (Walter et al., 1997) and strongly suggest a neuronal association of a surface-located ecto-PK responsible for the regulation of βAPP activity.

Stimulus-induced exocytosis of vesicular ATP occurs not only in the nervous system, from neurons, but also in the cardiovascular system, from human platelets, where we have described an ecto-protein kinase/ecto-phosphoprotein phosphatase system (Ehrlich, 1987; Ehrlich and Kornecki, 1987; Naik et al., 1991) that plays a role in the cascade of cell surface interactions with extracellular factors that control blood coagulation (Ehrlich, 1987; Ehrlich and Kornecki, 1987; Kalafatis et al.,1993). A more recent report demonstrated that dephosphorylation of a specific ecto-domain in the platelet collagen-receptor (a protein called CD36) controls its ligand specificity by shifting its affinity from collagen binding, in the phosphorylated state, to high-affinity for the extracellular adhesion protein thrombospondin, that CD36 acquires in the de-phosphorylated state (Asch et al.,1993). In an additional extensive study of ecto-PK in circulating cells, the phosphorylation of ecto-domains in the α, β T-cell receptor (TCR) on the surface of T-lymphocytes, was shown to affect TCR-cognate interactions and TCR-multimolecular complex formation (Apasov et al., 1996). Thus, extracellular phosphorylative events can regulate the function of receptors (Ehrlich et al., 1990; Asch et al.,1993), ion channels (Ehrlich et al.,1988), uptake carriers (Hendley et al., 1988; Hardwick et al.,1989), integrins (Ehrlich, 1987; House and Kemp, 1987; Asch et al., 1993), adhesion proteins (Ehrlich et al., 1986, 1990), growth factors (Ehrlich et al., 1990; Nagashima et al., 1991; Pawlowska et al., 1993) and coagulation factors (Kalafatis et al., 1993) in the nervous- and cardiovascular-systems, and thus may play critical roles in homeostasis, hemostasis, thrombosis, neuritogenesis, neuronal adhesion, and synaptogenesis.

A ROLE FOR SECRETED ATP IN PROCESSES UNDERLYING SYNAPTIC PLASTICITY

The accumulation of increasing concentrations of ATP in the synaptic cleft that is induced by a train of repetitive stimuli delivered at high frequency (Silinsky, 1975), and the utilization of ATP by an ecto-PK with a relatively high Km for ATP (Ehrlich, Snider et al., 1988) were proposed (Ehrlich, 1987) as biochemical events that are *unique* to the process of long-term potentiation (LTP) of synaptic strength. Indeed, exocytosis of vesicular ATP induced by high frequency stimulation (HFS) that causes LTP has been demonstrated directly in hippocampal slices (Wieraszko et al., 1989). Furthermore, low-frequency presynaptic stimulation delivered in hippocampal slices *together* with exogenously added ATP has been reported to induce LTP (Wieraszko and Seyfried, 1989). The results of experiments carried-out to determine the effects of several nonhydrolyzable and slowly hydrolyzable analogs of ATP on LTP induced by HFS (Wieraszko and Ehrlich, 1994) have provided evidence that the removal of the gamma phosphate of extracellular ATP is associated with the maintenance phase of LTP, and that this activity has properties consistent more with the phosphotransferase action of a kinase than with the hydrolytic activity of ATPase. We thus hypothesized that the maintenance of stable LTP may involve the activity of ecto-protein kinase. To test this hypothesis, we have initiated an extended investigation designed to determine whether brain neurons in-general, and hippocampal pyramidal neurons, in-particular, have ecto-PK activity, whether selective inhibition of this activity interferes with LTP, and whether the induction- and maintenance- phases of hippocampal LTP are regulated differently by ecto-PK activity. The results of these studies have provided the evidence that ecto-protein kinase is a physiological activity operating on the surface of pyramidal neurons, and that this activity plays a causal role in the maintenance of stable LTP in the hippocampus. These studies are described below.

EVIDENCE FOR THE PRESENCE OF ECTO-PROTEIN KINASE AND PHOSPHOPROTEIN SUBSTRATES ON THE SURFACE OF BRAIN NEURONS

Several criteria set as a requirement for evidence of the presence of ecto-enzymes on the surface of cultured cells by Kreutzberg et al., (1986) have been supplemented with specific requirements for proving the presence of ecto-protein kinases (Ehrlich et al., 1990). These criteria demonstrate directly the presence of phosphorylated sites in ecto-domains of cell-surface proteins, and exclude the contribution by intracellular substrates of protein kinases to the reaction products detected in ecto-PK assays, as follows: (1) Rigorous measurements of the intact state and viability of cells are carried-out, and intact cells are assayed while still attached to the tissue-culture plates in which they are grown. (2) The substrates of ecto-PK phosphorylation by extracellular [γ^{32}P] ATP (ATP does not penetrate through the cell membrane), are carefully compared with those phosphorylated substrates labeled by intracellular ATP following preincubation of cells with inorganic phosphate [^{32}Pi] . (3) Excess concentrations (100–1000 fold) of nonlabeled inorganic phosphate are used during ecto-PK phosphorylation reactions which compete with the uptake of ^{32}P-generated molecules formed as a result of the hydrolysis of ATP. The excess of inorganic phosphate thus eliminates the possibility of radiolabeling of intracellular proteins. (4) The ATP-consuming enzyme, apyrase, is added to the medium to intentionally hydrolyze the radiolabeled ATP, which serves as a control in these experiments. (5) Low concentrations of trypsin are added to surface-phosphorylated cells to demonstrate removal of phosphorylated products from the intact cell surface. Likewise, preincubation of intact cells with trypsin, prior to [γ^{32}P]ATP labeling of the cell surface, is used to inactivate the cell surface ecto-protein kinase and/or its endogenous substrates, and hence to prevent the phosphorylation of surface proteins. (6) Partial permealization of the plasma membrane with low concentrations of nonionic detergents is used to label intracellular proteins. Similarly, addition of lysed cells (and their released contents) to reactions of intact cells carried-out with radiolabeled ATP is used to exclude the possibility that the substrates observed in ecto-protein kinase reactions have emerged from intracellular compartments. (7) The [γ^{32}P]ATP-phosphorylation of exogenously-added substrates, such as casein, myelin basic protein and protamine, by intact cells is another evidence for the presence of ecto-protein kinase activity on the cell surface, and finally, (8) The availability of specific membrane-impermeable inhibitors to various protein kinases, such as specific antibodies directed to the catalytic domain of the kinase, and prototype pseudosubstrate inhibitory-peptides that mimick specific sequences in unique substrates, serve to both demonstrate the surface-location of an ecto-protein kinase and to determine the similarity of its catalytic properties to those of known intracellular protein kinases.

The inhibition of the extracellular phosphorylation of specific proteins by the addition to the medium of antibodies to known surface proteins can also serve as the initial step in identifying the function of substrates of neuronal ecto-protein kinases. The first successful identification of such a substrate of ecto-PK used an antibody to the protein Neuronal Cell Adhesion Molecule, N-CAM (Ehrlich et al.,1986). One of the surface-phosphorylated proteins identified by this antibody migrated in 7–16% gradient polyacrylamide gels with apparent MW of 105K, and displayed solubility properties of an integral transmembrane protein. If this protein is N-CAM, then it should be also phosphorylated in intracellular sites. The results of an experiment carried out to demonstrate this possibility are shown here in Fig.1. In these experiments we used the cloned neural cells NG108–15, reacted for ecto-PK activity as detailed in Ehrlich et al. (1986). Incubation of intact cells for 10 mins with [γ^{32}P] ATP (added to the medium with a 1000 fold excess of cold Pi) served to label phosphorylation sites located exclusively on the outer surface of the cells (see lanes marked A in Fig. 1). Incubation for 60mins with ^{32}Pi served to label intracellular sites (lanes marked P in Fig.1). The protein band of 105K, equally labeled by both procedures, was isolated by electroelution and re-electrophoresed in a 15% gel (see control lanes marked Cntrl., Fig.1). A standard phosphopeptide mapping procedure was carried-out using the proteases trypsin (Tryp.), chymotrypsin (Chym.) and *Stapylococcus-Aureus* V8 protease (V8) in the concentrations indicated at the top of the figure. The autoradiogram depicted in Fig.1 shows marked differences between the phosphopeptide profiles of the A-lanes (proteins labeled by [γ^{32}P] ATP) and the P-lanes (proteins labeled by ^{32}Pi). These differences indicate that the 105K protein is phosphorylated at several different sites along its sequence, both in intra- and extra-cellular domains.

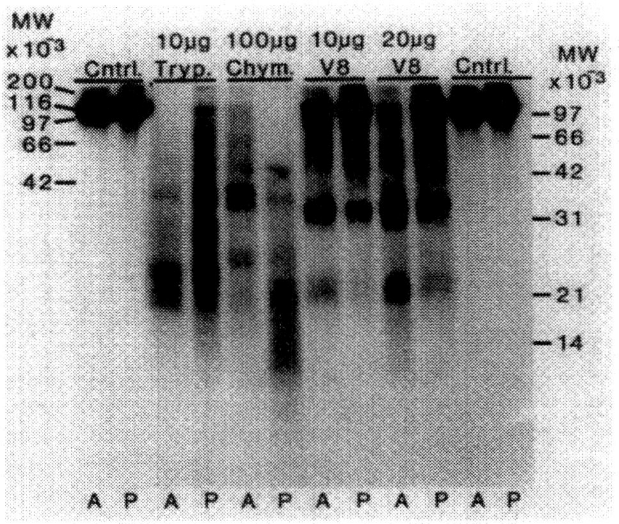

Figure 1. Transmembrane proteins can be phosphorylated both in intra- and extra-cellular domains. Phosphopeptide mapping of a surface-phosphoprotein with apparent MW of 105K in cultured neural cells of the clone NG108–15. Cell-surface proteins were specifically labeled by adding radioactive 0.1µM [γ^{32}P]ATP for 10 min to the medium (lanes marked A), and intracellular proteins were labeled by incubating the cells with radioactive inorganic phosphate for 1 hr (lanes P). Following these phosphorylation reactions, the 105K protein from each reaction was isolated and subjected to phosphopeptide mapping with trypsin (Tryp.), chymotrypsin (Chym.), or V8 protease, or left untreated (Cntrl.). Further details are provided in the text: the methods used are as described previously in Ehrlich et al., (1986)

All of the criteria required as direct evidence for the presence of ecto-protein kinase (ecto-PK) in cultured cells have been fulfilled in experiments conducted with brain neurons that were obtained from embryonic chick telencephalon or mouse hippocampus and maintained in culture in a chemically-defined, serum-free medium. The results obtained by Hogan et al. (1995) with these cells provided evidence that seven surface phosphoproteins with apparent MW of 116K; 105K; 67K; 53K, 17K; 13K, and 12K are phosphorylated on ecto-domains. Of these, the four proteins of MW of 116K, 105K, 67K, and 53K proteins are phosphorylated by **both** *intra*cellular PK and ecto-PK activities. In constrast, the phosphoproteins of 17K, 13K, and 12K are not labeled by inorganic ^{32}Pi (even after 60 min) and thus are phosphorylated *exclusively* by ecto-PK. The neuronal specific ecto-PK substrate of 105K discussed above was previously detected in cloned neural cells of peripheral origin (Ehrlich et al., 1986; Pawlowska et al., 1993). However, the surface phosphoproteins of 13K and 12K appear to be ecto-PK substrates unique to brain neurons labeled exclusively on their ecto-domain.

NOVEL AND ATYPICAL ISOZYMES OF PROTEIN KINASE C AS ECTO-ENZYMES

Fulfillment of three of the criteria for the presence of ecto-PK: (a) the [γ^{32}P]ATP phosphorylation of exogenous protein-substrates by intact cells, (b) the competition of exogenously-added substrates with endogenous cell surface proteins for added ATP, and (c) the selective inhibition of the phosphorylation of ecto-PK substrates by reagents that cannot penetrate through the plasma membranes, such as peptides and antibodies (Ehrlich et al., 1986; 1990; Pawlowska et al., 1993; Ehrlich, 1987; Ehrlich and Kornecki,1987) also have provided important information on the nature of the protein kinases involved. Among other results, we have reported (Hogan et al., 1995) that a specific inhibitor of PKC - the pseudosubstrate peptide PKC 19–36 (added at 10μM) completely inhibited the [γ^{32}P] ATP phosphorylation of the 67K substrate of ecto-PK, and this PKC peptide caused a 74.3 ± 2.4% and 84.2 ± 0.9% inhibition of the phosphorylation of the 13K and 12K proteins, respectively (mean ± SEM, n=7; $p<0.05$). Under the same reaction conditions, the active phorbol ester PMA, an activator of the regulatory domain of PKC, did *not* stimulate the phosphorylation of surface proteins by ecto-PK indicating that ecto-PK is PMA-insensitive. We employed a monoclonal antibody (termed M.Ab. 1.9), developed by Mochly-Rosen and Koshland (1987; 1988) against the catalytic domain of purified rat brain PKC, to shed more light on the similarity to PKC of the neuronal ecto-PK activity. M.Ab. 1.9 inhibits the activity of all PKC isozymes and of the phorbol-*in*sensitive proteolytic product of PKC, PKM, but does *not* inhibit cyclic-AMP or calmodulin-dependent protein kinases (Mochly-Rosen and Koshland, 1987). The addition of 0.1 μg/well of M.Ab. 1.9 to our standard assays with embryonic chick neurons produced a 51.4 ± 7.4% inhibition in the phosphorylation of the 67K protein, and 62.2 ± 4.2% and 44.0 ± 5.6% inhibition of the phosphorylation of the 13K and 12K substrates, respectively (means ± SEM, n=5; $p<0.05$). Polyclonal antibodies that recognize sequences in the catalytic regions of eight different PKC isozymes (α, β, γ, δ, ϵ, θ, η, and ζ) were used to characterize the ecto-PKC activity of brain neurons. Two of these PKC antibodies (anti-δ and anti-ζ) produced significant and selective inhibitory effects. An antibody prepared by Dr. P. J. Parker against the 11 amino-acid sequence at the C-terminus in the catalytic region of the isozyme PKC-δ (used at 1:300) selectively inhibited the endogenous phosphorylation of the 12K and 13K proteins by ecto-PK in intact brain neurons. Quantitation of these results determined that anti-PKC-δ caused a 60.4 ± 2.2% and 55.9 ± 5.1% inhibition (means ± SEM, n= 3; $p<0.05$) in the phosphorylation of the 13K and 12K protein substrates, respectively. This anti-PKC δ antibody, however, did *not* inhibit the endogenous phosphorylation by ecto-PKC of the 67K protein substrate (Hogan et al.,1995). As a control for the specificity of these results, we found that an antibody directed against the V3 region in the regulatory domain of PKC-δ had **no** effect on ecto-PK activity. In contrast to the effects of anti-PKC-δ, another antibody directed to a specific sequence in the C-terminus of PKC ζ (zeta; from Gibco, used at 1:300) caused a 36.8 ± 6.3% inhibition (mean ± SEM, n=3, $p<0.05$) in the surface phosphorylation of the 67K substrate of ecto-PKC in brain neurons, without significant effects on the phosphorylation of the 12K and 13K surface proteins. We conclude that neuronal ecto-protein kinase activity is exerted by several protein kinases, each with a different catalytic specificity. Each ecto-PK phosphorylates different endogenous substrates on the surface of the neuronal plasma membrane, and therefore may be expected to regulate different neuronal functions.

ECTO-PKC ACTIVITY IN BRAIN NEURONS IS DEVELOPMENTALLY REGULATED

The phosphorylation of the 12K and 13K proteins by brain ecto-PKC on the surface of immature neurons (2 to 3 days *in-vitro*; DIV) was found to be much greater than in dif-

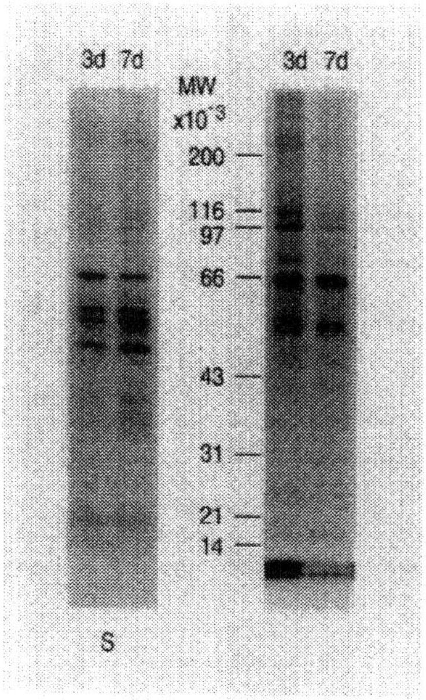

Figure 2. Surface phosphorylation of a 12K/13K protein duplex by ecto-PK is developmentally regulated. Primary neurons were cultured from the telencephalon of chick embryos, maintained up to 7 days *in-vitro* (DIV) and assayed for ecto-PK activity as described in Hogan et al.(1995). Neurons at 3 DIV (marked 3d above the lane) and at 7 DIV (marked 7d) were incubated for 10mins with [γ^{32}P]ATP and the radiolabeled proteins were resolved by SDS-PAGE. The Coomassie-Blue protein staining pattern (S) is shown in the left panel of the figure, and the autoradiogram is shown in the right panel. Note that while the phosphorylation of a 66K protein did not change with development, phosphorylation of the 12K/13K protein duplex (shown below the 14 K MW marker) is far greater at 3 than at 7 DIV.

ferentiated neurons (over 6 DIV). Representative results of such an experiment are shown here in the autoradiogram of Fig. 2. Rapid neurite outgrowth begins in these cells within 1 DIV, and an extensive network of neurites can be seen at 7 DIV. We have determined that the peak in surface phosphorylation of the 12K and 13K proteins by ecto-PK in these cultured neurons occurs between 3–4 DIV, which coincides with the maximal rate of new neurite outgrowth (Hogan et al., 1995). At 7 DIV the surface phosphorylation of the 12K and 13K protein decreases by over 75%. These results suggest that the ecto-PK which phosphorylates these two surface proteins may contribute to molecular mechanisms that regulate neuritogenesis. The preparation of specific antibodies that inhibit selectively the phosphorylation of sites in the ecto-domain of these surface proteins will enable direct experimental testing of the operation of this developmentally regulated process.

ECTO-PKC IN THE CNS: A TARGET FOR ALZHEIMER'S β-AMYLOID PEPTIDES

Physiologically relevant regulators of protein kinase activity are expected to show selectivity towards specific cellular phosphoproteins that are involved in the physiological activity which they regulate (Greengard, 1978; Ehrlich, 1984). Thus, the unique surface-substrates of neuronal ecto-PKC that have been implicated in neuritogenesis are likely candidates for regulation by factors that stimulate or inhibit neurite outgrowth, such as β-amyloid peptides. Using ecto-PKC assays conducted for 10 min with immature brain neurons (3 DIV), we have found (Hogan et al., 1995) that the peptide β-amyloid 25–35 applied at the neurotrophic concentration of 0.1nM (Yankner et al., 1990), significantly stimulated the [γ^{32}P]ATP phosphorylation of the 12K (125 ± 13% of control; mean ± SEM; n=6; $p<0.05$) and the 13K (130 ± 11 %; n=6, $p<0.05$) neuronal surface proteins. The β-amyloid peptide 1–40, when added at 0.1nM, stimulated the [γ^{32}P]ATP phosphorylation of the 13K (134 ± 20 %, n=3; $p<0.05$) surface protein. In contrast, when the β-amyloid peptides 1–40 or 25–35 were added to neurons at 7 DIV, there was a very specific *inhibitory* effect. At 0.1μM, β1–40 inhibited selectively the [γ^{32}P] ATP phosphorylation of the 67K, 13K, and 12K substrates of ecto-PKC in brain neurons. The main effect was over 80% inhibition in surface phosphorylation of the 13K protein. Examination of the structure/function relationships of β-amyloid peptides also revealed specificity. At 0.1 μM, β1–28, a fragment that has no neurotoxic effects, but has been reported to be neurotrophic (Whitson et al., 1989), did not inhibit the extracellular [γ^{32}P] ATP phosphorylation of the 13K and 12K surface proteins but stimulated the phosphorylation of the ecto-PK substrate proteins of 116K and 105K. The β25–35 peptide, on the other hand, has potent neurotoxic effects (Koh et al., 1990; Yankner et al., 1990; Loo et al., 1993; Pike et al., 1993) and we anticipated that it may inhibit the phosphorylation of specific proteins involved in neuritogenesis. The β25–35 peptide indeed inhibited the surface phosphorylation of all three specific substrates of brain ecto-PKC, with the rank order of potency being 13K> 67K> 12K. Interestingly, the inhibition of surface protein phosphorylation in cultured brain neurons by β25–35 which were observed in our laboratory, showed the *same* concentration-dependence as that observed for the neurodegenerative action of this peptide (Koh et al., 1990; Yankner et al., 1990). The extracellular location of ecto-PKC in brain neurons, its selectivity towards developmentally-regulated surface protein substrates, and the specificity of its dual responses (stimulation and inhibition) to different regions within the sequence of β-amyloid protein (1–28 vs 25–35), render this enzyme most suitable as a candidate biochemical mediator for the neurotrophic and neurotoxic actions of Alzheimer's amyloid peptides.

IDENTIFICATION OF NEURONAL SURFACE PHOSPHOPROTEINS ASSOCIATED WITH NEURITOGENESIS AND SYNAPTOGENESIS.

As described above, we have determined that embryonic chick-brain neurons express proteins of MW 12K and 13K whose surface phosphorylation is maximal during the onset of neuritogenesis. We have also detected surface phosphorylation of this low MW protein duplex early in the development of pyramidal neurons cultured from embryonic mouse hippocampus (2–5 DIV), and this activity decreased significantly after 8 DIV, when neuritogenesis was completed. As in chick brain neurons, the surface phosphorylation of this duplex in pyramidal mouse neurons was also inhibited by an antibody directed to the C-terminus of PKC-delta. Upon maturation and during synaptogenesis, the surface phosphorylation of another protein duplex (MW of 48K and 50K) appeared in pyramidal neurons, and reached maximum at 18–21 DIV. As anticipated from the results listed above, we found that a hybrid, pyramidal-neurons cell line (HN9) derived from embryonic cells (provided by Dr. H. Wainer, Albert Einstein School of Medicine) exhibits a high degree of [γ^{32}P] ATP surface phosphorylation of the 12K/13K protein duplex associated with neuritogenesis. The ecto-PK phosphorylation of the 12K/13K duplex in HN9 cells was found to have the same properties as the ecto-PKC of embryonic chick neurons and primary pyramidal cells from mouse hippocampus assayed at 3–5 DIV. A more mature cell-line called HN33 displayed surface phosphorylation of the 48K-50K protein duplex associated with synaptogenesis, but HN33 cells displayed low levels of the 12K/13K duplex phosphorylation (Pawlowska et al, 1994). The functions of these two surface protein phosphorylation systems can now be studied in homogenous populations of cloned neural cells.

ECTO-PROTEIN KINASE ON THE SURFACE OF HIPPOCAMPAL PYRAMIDAL NEURONS

The involvement of ATP secreted during tetanic stimulation in the induction of long term potentiation (LTP) of synaptic strength in the hippocampus has been reported previously. The main role of ATP appears to be in the maintenance phase of LTP (Wieraszko and Ehrlich, 1994), which is associated with sprouting and *de-novo* synapse formation. Ecto-protein kinase activity that may mediate this effect was studied in pyramidal neurons cultured from the hippocampus of 16-day mice embryos, maintained for 3 weeks in serum-free Neurobasal medium with supplement B27 (Gibco), until achieving maturation and synaptogenesis (Chen et al.,1994; 1996). All the criteria detailed above as evidence for the presence of ecto-PK in intact neurons were fulfilled in experiments carried out with pyramidal neurons cultured from the hippocampus of embryonic mice. The surface phosphoprotein substrates of ecto-PK in the hippocampus were identified as components migrating with apparent molecular mass of: 105K; a 66K/68K duplex; a 48K/50K duplex; 43K; 32K; 17K and a 12K/13K duplex. Phosphorylation by ecto-PK of the 12K/13K duplex on the surface of pyramidal cells was detected only in cells younger than 8 DIV, and that of the 48K/50K began to appear at 8 DIV and reached maximum at 18–21 DIV (Chen et al., in-preparation). The surface phosphorylation of the 48K/50K duplex in 18–21 DIV pyramidal neurons was extensively characterized in our laboratory. We found that the phosphorylation of this duplex increased with reaction time and reached a maximum after 2 hrs. However, even after 2 hrs labeling with inorganic ^{32}Pi this duplex was not labeled, indicating that in pyramidal neurons at 18–21 DIV these proteins are phosphorylated *only* on ecto-domains, exclusively by ecto-PK.

The hydrophilic membrane impermeable protein kinase inhibitor K-252b was shown to inhibit ecto-PK *without* inhibiting intracellular protein phosphorylation in intact PC12 cells (Pawlowska et al; 1993). In cultured pyramidal cells from mouse hippocampus assayed at 18 DIV, K-252b inhibited surface protein phosphorylation in a dose-dependent, selective manner. A K-252b concentration of 0.2μM caused about 50% inhibition of the surface phosphorylation of the 48K and 50K proteins but had *no* effect on phosphorylation of a 17K protein by ecto-PK. In hippocampal slices incubated with ATP added to the medium at a concentration of 500nM (0.5μM), K-252b at 0.2μM caused about 50% inhibition of ATP-induced LTP (Chen et al.,1994). We then ncubated hippocampal slices under the same conditions with 500nM of radiolabeled [γ-^{32}P] ATP added to the extracellular medium, and found that under these reaction conditions the same 48K/50K duplex was the major phosphorylated component in these hippocampal slices (Chen et al., 1996). These findings provided the basis for the experimental use of hippocampal slices in studies on the utilization of secreted ATP by ecto-PK in mechanisms underlying synaptic plasticity.

ROLE FOR THE HYDROLYSIS OF EXTRACELLULAR ATP IN THE MAINTENANCE OF LTP

The involvement of a purinergic system in the mechanisms of ATP- and electrically induced long-term potentiation (LTP) has been investigated in mouse hippocampal slices (Wieraszko and Ehrlich, 1994). Application of extracellular ATP (100–500 nM) or its slowly hydrolyzable analogue adenosine 5'-0-(3-thiotriphosphate)(ATP-γ-S; 2.5 μM), together with *low* -frequency stimulation, amplified permanently the magnitude of the population spike. This effect was antagonized by adenylimidodiphosphate (AMPPNP), a nonhydrolyzable analogue of ATP. AMPPNP, other ATP analogues [2-methylthioadenosine triphosphate (2-MeSATP) and α, β-methyleneadenosine 5'-triphosphate (α, β-ATP)], or a purinergic receptor antagonist (Cibacron Blue 3G) tested in the concentration range of 3–40 μM did not exert agonistic activity similar to that of ATP or ATP-γ-S, suggesting that ATP hydrolysis is required to exert this effect. All the tested nonhydrolyzable analogues reduced or prevented the establishment of stable, nondecremental LTP induced by high frequency electrical stimulation, without blocking the short-lasting increase in the magnitude of the population spike seen immediately after electrical stimulation (short-term potentiation). These results indicate that ATP released by high-frequency stimulation contributes to the maintenance of stable LTP. The underlying mechanism operating in this process may involve a new type of ATP receptor or hydrolysis by ecto-ATPase. However, the finding that ATP-γ-S is less potent than ATP and that other ATP analogues known to act as agonists of purinergic receptors did not induce LTP but rather inhibited its maintenance, is more consistent with the involvement of an ecto-protein kinase, using extracellular ATP as a co-substrate, in mechanisms underlying synaptic plasticity.

ECTO-PROTEIN KINASE ACTIVITY IS REQUIRED FOR THE MAINTENANCE OF STABLE LTP

Phosphorylation of a 48K/50K surface protein duplex by extracellular ATP was identified in cultured pyramidal neurons and in hippocampal slices. An inhibitory monoclonal antibody termed M.Ab.1.9, directed against the catalytic domain of PKC, selectively blocked this protein phosphorylation activity. To study the involvement of this surface phosphorylation system in LTP, M.Ab.1.9 was added to hippocampal slices 2hrs prior to the induction of LTP by high frequency stimulation (HFS). The experiment has been conducted as follows: slices were transferred to the interface in a recording chamber maintained at 33°C. A bipolar, stimulating electrode was placed on Schaffer collateral-commissural fibers (Sch. coll.) and recording electrodes were guided into pyramidal or radiatum layer of CA1 to record population spike or excitatory-post-synaptic potentials (EPSP). The pulse duration was fixed at 150 μsec. The magnitude of the population spike was measured as an average distance between the highest negativity and preceding and following positivities. The initial slope of the negative wave was taken as a measure of field EPSP. The strength of the stimulation in all tested slices was in the range of 0.01–0.05 mA and was adjusted at the beginning of each experiment to obtain a population spike of approximately 1 mV. The potential was monitored at this level for the next 15–20 minutes, and experiments were performed only on slices demonstrating stable responses to low frequency (0.03 Hz) stimulation. The electrical LTP was then induced by stimulation of the Sch. coll. with 100 Hz three times for 1 sec, with 10 sec intervals. M.Ab.1.9 was added directly to the recording chamber with a Hamilton syringe in a volume of 10 μl two hours prior to induction of LTP. The final concentration of M.Ab. 1.9 in the chamber was 20μg/ml. The increase in the size of the potential and in the slope of EPSP was expressed as percentage of change in the control value observed before the treatment. The magnitude of amplification that was evaluated at 1 and 30 minutes after the application of HFS provided the results that are depicted here in Figure 3 and Figure 4, and summarized in the next paragraph.

The influence of M.Ab.1.9 on the slope of EPSP and on the population spike was found to be as follows: compared to the slope of EPSP in control slices which was permanently amplified

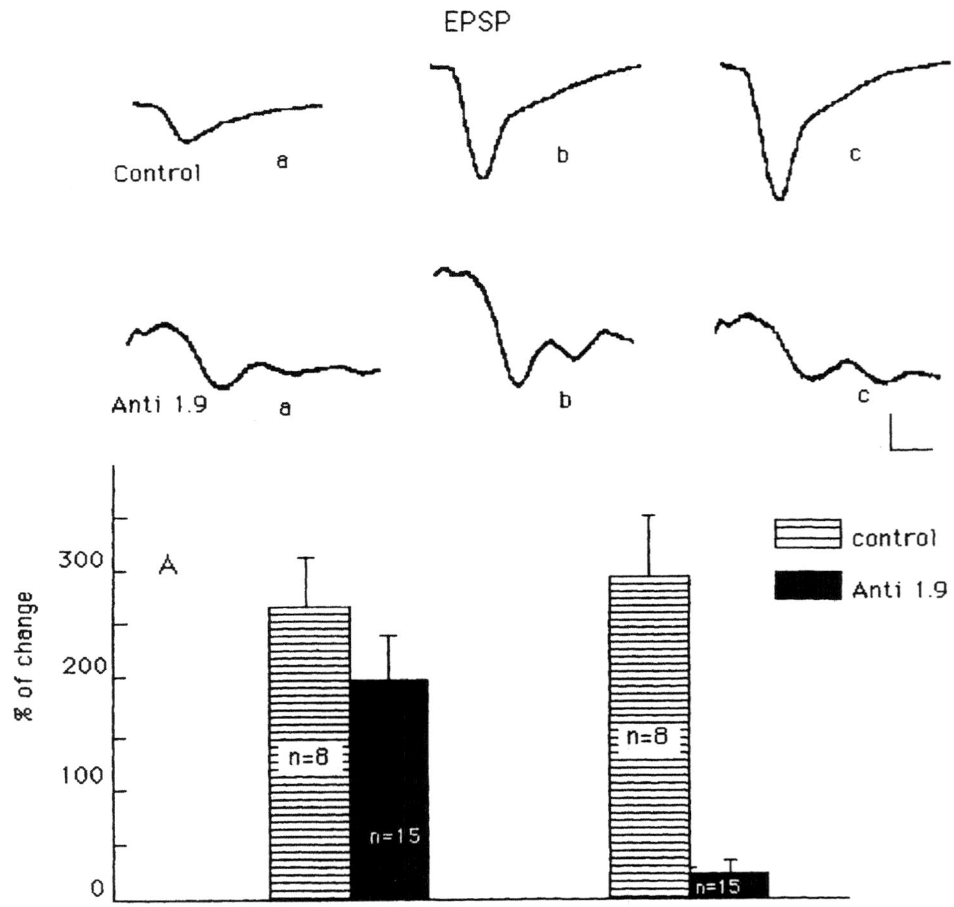

Figure 3. Inhibition of ecto-PK by an anti-PKC antibody blocks the maintenance of LTP: Effect of M.Ab.1.9 on EPSP. The excitatory post-synaptic potentials (EPSP) were recorded from hippocampal slices following high frequency electrical stimulation (HFS) as detailed in Wieraszko and Ehrlich (1994) under control conditions, and following preincubation with an anti-PKC monoclonal antibody, M. Ab.1.9 (Anti 1.9), as described in the text. The potentials marked by letters a, b, and c at the upper part of the figure are representative recordings taken 5 mins before, and 10 mins and 30 mins after application of HFS, respectively. The values for the means ± SEM for the indicated number (n's) of the percent of change from prestimulation values as recorded 1min (left side) and 30 mins (right side) after the application of HFS, are depicted in the histogram at bottom of figure. Data adapted from Chen et al. (1996).

by the HFS, in M.Ab.1.9-treated slices only a brief period of amplification was observed immediately after HFS . This was followed by a sharp decrease of EPSP and return to the control, prestimulus value within 30 to 40 minutes after HFS application. Parallel changes were recorded in the magnitude of the population spike: the increase in population spike measured immediately after HFS applied in the presence of M.Ab.1.9 was identical to that of control slices, but the potential recorded 30 min later was 40% lower than that of control slices. In slices incubated with a control antibody that binds to PKC but that cannot inhibit its activity (M.Ab.1.12), LTP was the

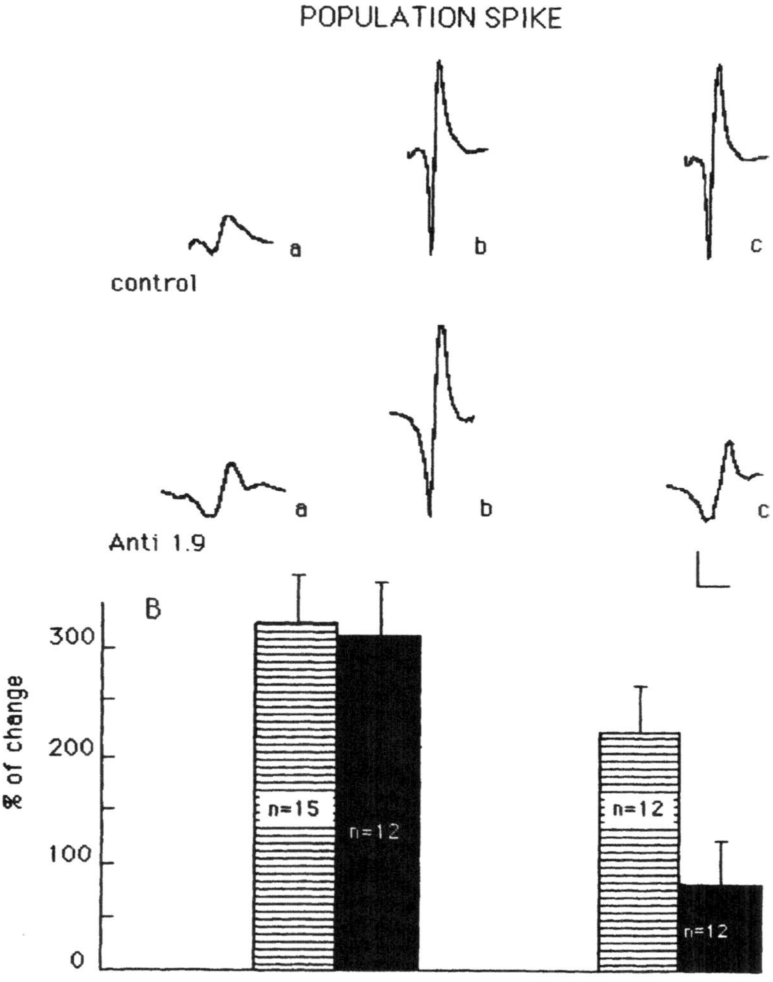

Figure 4. Inhibition of ecto-PK by an anti-PKC antibody blocks the maintenance of LTP: Effect of M.Ab.1.9 on the population spike. Details of the experimental design and conditions are the same as for Fig. 3, except that the depicted data are as described above, but for the amplification of the population spike inuced by HFS in control slices, and following preincubation with M.Ab.1.9 (Anti 1.9). Data adapted from Chen et al., (1996).

same as in slices tested without any antibody (Control slices). Furthermore, both the slope of the EPSP and the magnitude of the population spike measured just before the application of HFS were not altered during very long preincubations with M.Ab. 1.9, and remained the same as those recorded before adding the antibody, which did not differ from those seen in the control slices. Thus, M.Ab.1.9 does *not* influence basal synaptic transmission. Finally, the size and shape of the potentials recorded 30 min after HFS was applied in the presence of the antibody returned to the very same values observed before HFS, again confirming that M.Ab.1.9 does not interfere with basal activity. These selective inhibitory effects of the PKC antibody M.Ab.1.9 have provided the first direct evidence of a causal role for ecto-protein kinase in the maintenance of stable LTP. We anticipate that our ongoing studies, focusing on the purification and cloning of the ecto-PK inhibited by M.Ab.1.9, and on the selective 48K and 50K surface phosphoproteins substrates of ecto-PK of brain neurons, will reveal novel biochemical mechanisms operating in the regulation of processes underlying synaptic plasticity.

INTRAJUNCTIONAL LOCALIZATION BY IMMUNO-ELECTRON-MICROSCOPY OF ECTO-PKC WITHIN THE SYNAPTIC CLEFT

The activity of the ecto-PK involved in the maintenance of LTP is inhibited by antibody M.Ab.1.9; this antibody recognizes an epitope in the catalytic domain of PKC and inhibits PKC activity. Therefore, we would expect that when this monoclonal antibody is applied to permealized cells and fixed synaptosomes, it should recognize epitopes of *intra*cellular PKC that are localized in all intracellular membranes and on synaptic vesicles. When applied to intact cells or to unfixed synaptosomes, M.Ab. 1.9 binding should be restricted to ecto-PKC, and observed as binding to externally-oriented epitope(s) localized on the outer surface of the synaptic plasma membrane. Studies designed to examine these events were carried out by Dr. Robert Lasher from the University of Colorado Medical Center. As reported by Ehrlich et al., (1997), immunoperoxidase staining with M.Ab.1.9 localized a labeled epitope on the external surface of the junctional plasma membrane of unlysed synaptosomes, and this staining was restricted to locations within the synaptic cleft. When M.Ab.1.9 was added to permealized synaptosomes, after fixation, this antibody recognized intracellular epitopes associated with membranes, synaptic vesicles and mitochondria, in addition to the staining of junctional plasma membranes. Thus, immunoperoxidase localization of ecto-PKC revealed that the externally-oriented epitope recognized by M.Ab.1.9 is present primarily on the surface of pre- and post-synaptic membranes within the synaptic cleft .

Antibody M.Ab.1.9 recognizes an epitope in the catalytic region of *all* PKC isozymes. Another antibody termed M6 (M.Ab.M6) recognizes specifically an epitope in the catalytic domain of PKC type III (Leach et al., 1988), a PKC alpha-type isozyme. Dr. Lasher has determined that when M.Ab.M6 is applied to fixed synaptosomes, it binds to intracellular epitopes localized on vesicles, mitochondria, and the plasma membrane, similar to M.Ab.1.9. However, *no* specific labeling of extracellular sites with M.Ab.M6 could be seen in unfixed synaptosomes. These control experiments demonstrated the unique specificity of the catalytic region of the ecto-protein kinase recognized by M.Ab.1.9 which is located on the surface of the synaptic plasma membrane within the synaptic cleft, precisely the region where secreted ATP accumulation is induced by high-frequency stimulation (HFS), and the location where M.Ab.1.9 acts to prevent the stabilization of LTP induced by HFS (Chen et al., 1996).

IDENTIFICATION OF AN ATP-BINDING PROTEIN RECOGNIZED BY M.AB.1.9 ON THE NEURONAL CELL SURFACE

A significant goal of our continued research in this area is the purification and cloning of the gene of the neuronal ecto-protein kinase that is required for the maintenance of stable LTP. The critical first step in achieving this goal is the identification of the protein that serves as ecto-PKC. This ecto-PK should be a kinase that operates on the surface of intact hippocampal pyramidal neurons and catalyzes reactions carried-out with low concentrations (50–100 nM) of extracellular ATP. Preincubation with M.Ab.1.9 that binds to the catalytic domain of this kinase, and inhibits its activity as well as blocks the stabilization of LTP, should inhibit the binding of ATP to this ecto-PK. We have successfully accomplished the identification of an ecto-PK with these properties by using the procedures for photoaffinity labeling of ATP binding proteins with [α^{32}P]8-azido ATP as described by Nagy and Shuster (1995). We have used, for photoaffinity labeling, chick telencephalon neurons and mature hippocampal pyramidal neurons. The neurons were labeled as intact cells, while still attached to plates under conditions identical to those in which we assay ecto-PK activity in these cells (Pawlowska et al., 1993; Hogan et al., 1995; Chen et al., 1996). Using these procedures we have determined that a protein migrating with apparent molecular mass of 43Kda is labeled when this membrane-impermeable photoaffinity probe is added to the extracellular medium at a concentration of 100 nanomolar (Ehrlich et al., 1997); identical to the concentration of ATP used in our routine ecto-PK assays. In one of the experiments conducted to prove the extracellular orientation of this ATP-binding protein, we incubated intact, attached neurons pre-labeled with [α^{32}P]8-azido ATP, with very low concentrations of trypsin (as described in Ehrlich et al.,1986). As expected from a surface protein, this mild proteolysis degraded the 43K binding-protein without harming cell viability. As expected from the ATP binding site of a protein kinase, an excess of nonradioactive ATP fully competed with the labeling of this 43K surface protein with 100nM [α^{32}P]8-azido ATP.

Our investigation of the relationships between the high-affinity binding of ATP to the 43K surface protein, and the activity of the pyramidal ecto-PK which is involved in the maintenance of LTP was initiated by using several selective monoclonal-antibodies to PKC. Preincubation with 1μg/ml of M.Ab.1.9 (the concentration used to inhibit the phosphorylation of the 48K and 50K substrates of ecto-PK in pyramidal neurons, Chen et al., 1996), completely inhibited the binding of 100nM [α^{32}P]8-azido ATP to the 43K protein on the neuronal cell- surface. The correspondence in the ability of several specific PKC antibodies to inhibit the phosphorylation of the 48K and 50K proteins on the surface of pyramidal neurons by ecto-PKC, *and* to block the binding of [α^{32}P]8-azido ATP to the surface 43K protein of these neurons, *and* to stain intrajunctional surface epitopes within the synaptic cleft, *and* to prevent the stabilization of LTP in the hippocampus, collectively identify the ecto-protein kinase molecule (ecto-PKC) and the specific surface protein substrates that play a role in synaptic plasticity.

CARDIOVASCULAR IMPLICATIONS

Stimulus-induced exocytosis of vesicular ATP occurs not only in the nervous system, from neurons, but also in the cardiovascular system, from platelets. We have described in human platelets an ecto-protein kinase/ectophosphoprotein phosphatase system (Ehrlich, 1987; Ehrlich and Kornecki, 1987; Naik et al., 1991) that plays a role in the cascade of cell surface interactions with extracellular factors that control blood coagulation (Ehrlich, 1987; Ehrlich and Kornecki, 1987; Kalafatis et al.,1993). The ecto-PK activity of platelets could be inhibited by the PKC antibody M.Ab.1.9. The effect was dose-dependent with 50% inhibition exerted by about 5μg of M.Ab.1.9 per 5 x 10^8 platelets. These results indicated that, like in neurons, one of the ecto-PKs acting on the surface of platelets has the catalytic specificity of PKC. Most interestingly, incubation of washed, intact human platelets with antibody that inhibits the platelets ecto-PKC, antibody M.Ab-1.9, resulted in direct activation of the platelets, inducing aggregation and granule-secretion (Babinska et al.,1996). Two control antibodies (M.Ab.1.12 and 1.3) directed to the regulatory domain of PKC did *not* activate platelets. Increasing the concentration of M.Ab.1.9 had no effect on the extent of aggregation, but in a dose-dependent manner shortened the latency of antibody-induced activation, similar to the manner induced by the platelet-stimulating antibody M.Ab.F11(Kornecki et al; 1990). Using procedures of flow cytometry and immunofluorescence, we have shown that the ecto-PKC whose inhibition by M.Ab.1.9 causes platelet aggregation is localized on the external surface of the platelet plasma membrane (Babinska et al., 1996). Normal hemostasis maintains platelets in a discoid shape and prevents cells from aggregating spontaneously in the circulation. Constitutive phosphorylation of surface-proteins by a platelet ecto-PKC with a low Km for ATP (ATP concentration in circulating plasma is $<1\mu M$) may play a role in hemostasis by maintaining a steady-state of phosphorylation of these platelets surface proteins. M.Ab.1.9 may thus cause platelet aggregation because it disrupts this steady state. This hypothesis was tested by using a membrane-impermeable, potent inhibitor of phosphatase activity, microcystin-LR (Honkanene et al., 1990). Microcystin-LR is a cyclic heptapeptide which inhibits type 1 and type 2A protein phosphatases. The addition of microcystin to the extracellular medium prevented the induction of platelet aggregation caused by M.Ab.1.9. This result provided evidence that the inhibition of surface ecto-protein kinase *together* with continuing ecto-protein phosphatase activity causes an agonist-independent aggregation of human platelets by M.Ab.1.9. Thus, the platelet activation process which is trigerred by M.Ab.1.9 begins with changes in the phosphorylation state of surface proteins. In the same study we have determined that the next step in this process involves membrane rearrangements that bring about the exposure of fibrinogen-binding sites on the platelet surface (Babinska et al., 1996). This represents a novel homeostatic mechanism in which the maintenance of a proper steady-state of phosphorylation of proteins on the cell surface controls the state of cell activity. In the circulation, this mechanism protects platelets from spontaneous aggregation. We propose that in the nervous system, rearrangements of plasma-membrane proteins that are controlled by surface phosphorylation events may serve as signals, or regulators, of neuritogenesis and synaptogenesis.

CONCLUSIONS, HYPOTHESIS AND FUTURE DIRECTIONS

The study of neuronal development and neuronal plasticity has been advanced greatly in recent years by the demonstration of *activity-dependent neuronal events* in which release from presynaptic nerve terminals triggers and influences the connectivity among cells and between components of the extracellular matrix and cell surfaces. These processes depend on neuronal stimulation and neurotransmitter release from nerve terminals. Neuronal exocytotic release delivers not only neurotransmitters, but also the nucleotide ATP which is secreted to the synaptic cleft. Nucleotides represent an ubiquitous class of important signalling molecules in all tissues, including the nervous system. However, the need for the secretion of the high energy nucleotide, ATP, to the extracellular space was not clear. The discovery of neuronal ecto-protein kinase suggested that ATP is secreted by neurons to fulfill a regulatory role exerted by extracellular protein-phosphorylation systems. Accordingly, in molecular mechanisms operating in activity-dependent neuronal events, the ATP co-secreted together with the neurotransmitter molecules into the synaptic cleft is utilized by ecto-protein kinases for the regulation of cellular interactions involved in neuronal development and underlying synaptic plasticity.

Our studies of the extracellular phosphorylation of unique proteins on the surface of brain neurons have implicated a specific, developmentally-regulated, ecto-PK activity involved in neuritogenesis and synaptogenesis. This ecto-PK activity is inhibited selectively by extracellular application of monoclonal antibodies directed to the catalytic domain of protein kinase C (PKC) and by PKC pseudosubstrate peptides. We termed this enzyme ecto-PKC. The onset of neuritogenesis correlated temporally with phosphorylation by ecto-PKC of 12K and 13K surface proteins, whereas the onset of synaptogenesis corresponded with phosphorylation of 48K and 50K surface proteins by ecto-PKC. In electrophysiological studies of the induction of long-term potentiation (LTP) in hippocampal slices, a role for extracellular ATP has been established. The critical involvement of ecto-PK in molecular mechanisms operating in LTP was revealed by use of the PKC inhibitory antibody M.Ab.1.9. Application of M.Ab.1.9 inhibited the phosphorylation of the 48K and 50K proteins on the surface of pyramidal neurons and blocked the stabilization of LTP induced by high-frequency stimulation in hippocampal slices. Photoaffinity labeling by [α^{32}P]8-azido-ATP identified ecto-PKC as a 43K cell-surface protein whose binding of ATP was blocked by M.Ab.1.9. The externally-oriented epitope of the ecto-PKC recognized by M.Ab.1.9 was localized by immuno-electron microscopy of synaptosomes in intrajunctional- locations; demonstrating that ecto-PKC operates within the synaptic cleft - where ATP is secreted upon neuronal stimulation. Studies of the ecto-PKC of human platelets revealed that changes in the phosphorylation state of surface proteins can trigger the membrane rearrangements associated with cellular activation.

Based on all these findings we have formulated a working **hypothesis** stating that alterations in the phosphorylation state of ecto-domains in surface proteins of developing neurons induce conformational changes leading to rearrangements of membrane components which trigger the onset of neuritogenesis and/or synaptogenesis. We propose that a similar process operates in the mature brain in LTP: the phosphorylation of specific surface proteins by ecto-PKC, which occurs uniquely after the enhanced release of presynaptic ATP by high-frequency stimulation, signals for the initiation of cellular events involved specifically in the formation of long-lasting alterations in neuronal structure and function, including the dendritic-sprouting and *de-novo* synaptogenesis involved in memory formation.

We anticipate that future studies based on the synthesis of pseudosubstrate-peptides that inhibit specifically the phosphorylation of ecto-domains in the 12K, 13K, 48K and 50K surface protein- substrates of ecto-PKC, the preparation of new antibodies specific to different ecto-PKs, and studies of the expression of genes to be cloned for all these proteins, will reveal novel mechanisms with critical roles in the regulation of neuronal development and synaptic plasticity, and will provide a new generation of therapeutic agents for the treatment of developmental disabilities, dementia, and certain neurodegenerative disorders.

ACKNOWLEDGMENTS

The research in the authors laboratory described in this chapter has been supported by grants from the National Institute of Health, the U.S. Air Force Office for Scientific Research, the American Heart Association, and by research awards from the PSC-CUNY.

REFERENCES

Asch, A.S. et al. (1993). Analysis of CD36 binding domains: Ligand specificity controlled by dephosphorylation of an ectodomain. *Science*. 262. 1436–1440.

Apasov, S.G., Smith, P.T., Jelonek, M.T., Margulies, D.H., and Sitkovsky, M.V. (1996) Phosphorylation of extracellular domains of T-lymphocyte surface proteins. *J. Biol. Chem*. 271: 25677–25683.

Babinska, A., Ehrlich, Y.H. and Kornecki, E. (1996) Activation of human platelets by protein kinase C antibody: role of ecto-protein kinase in platelet homeostasis. *Amer. J. Physiol*. 27: H2134–2144.

Burnstock, G. J. (1981). Neurotransmitters and trophic factors in the autonomic nervous system. J.*Physiol*. 313, 1–35.

Burnstock, G. (1990). Purinergic mechanisms. *Ann. NY Acad. Sci*., 603, 1–18.

Chen, W., Hogan, N.V., Wieraszko, A., Pawlowska, Z., Soifer, D. and Ehrlich, Y.H. (1994) NMDA-regulated ecto-protein kinase in hippocampal neurons: role in LTP. *Society for Neuroscience Abstracts*, 20: 263, no. 119.6.

Chen, W., Wieraszko, A., Hogan, M.V., Yang, W., Kornecki, E. and Ehrlich, Y.H. (1996) Surface protein phosphorylation by ecto-protein kinase is required for the maintenance of hippocampal long-term potentiation. *Proc.Natl.Acad.Sci. US*, 93:8688–93.

Chen, X.Y. and Lo, T.C. (1991a) Phosphorylation of a cell surface 112 Kd protein by an ecto- protein kinase in rat L6 myoblasts. *Biochem J*. 279:467–474.

Chen, X.Y. and Lo, T.C. (1991b) Involvement of a cell surface protein and an ecto-protein kinase in myogenesis. *Biochem. J*. 279:475–482.

Deissler et al., (1995) Affinity purification and cDNA cloning of rat neural differentiation and tumor cell surface antigen gp130RB13–6 reveals relationship to human and murine PC-1. *J.Biol.Chem*. 270:9849–9855.

Dorner, T., Gagelmann, M., Geller, S., Herbst, F. and Forssmann, W.G. (1989) Phosphorylation and dephosphorylation of the natriuretic peptide urodilatin (CDD-ANP-95–126) and the effect onbiological activity. *Biochem. Biophys. Res. Commun*. 163:830–835.

Edwards, F.A., Gibb, A.J. and Colquhoun, D. (1992). ATP receptor - mediated synaptic currents in the central nervous system. *Nature*. 359, 144–147.

Ehrlich, Y.H., and Routtenberg, A. (1975). Phosphorylation differences in brain membranes of three related mice strains. *Adv. Cycl. Nucl. Res*. 5, 813–815.

Ehrlich, Y.H., Rabjohns, R. and Routtenberg, A. (1977). Experiential-input alters the phosphorylation of specific proteins in brain membranes. *Pharm., Biochem. and Behavior* ,6, 169–174.

Ehrlich, Y.H. (1984). Protein phosphorylation. Role in the function, regulation, and adaptation of neural receptors. In *Handbook of Neurochemistry*, 2nd edition, (NY : Plenum Press) Vol.6, pp. 541–574.

Ehrlich, Y.H., Davis, T.B., Bock, E., Kornecki, E. and Lenox, R.H. (1986). Ecto protein kinase activity on the external surface of intact neural cells. *Nature* 320, 67–69.

Ehrlich, Y.H., Lenox, R.H., Kornecki, E. and Berry, W. eds. (1987). Molecular mechanisms of neuronal responsiveness. *Adv. Exper. Med. Biol*., vol. 221, (New York: Plenum Press).

Ehrlich, Y.H. (1987). Molecular mechanisms of neuronal responsiveness. Adv. *Exp. Med. Biol*. 221, 187–199.

Ehrlich, Y.H. and Kornecki, E. (1987). In: Mechanisms of Signal Transduction by Hormones and Growth Factors. Cabot. M.C. and McKeehan. W.L., eds. (A. Liss, N.Y.), pp. 193–205.

Ehrlich, Y.H., Snider, R.M., Kornecki, E., Garfield, M.K. and Lenox, R.H.(1988). Modulation of neuronal signal transduction system by extracellular ATP. *J. Neurochem*. 50,295–301.

Ehrlich, Y.H., Hogan, M.V., Pawlowska, Z., Naik, U. and Kornecki, E. (1990). Ectoprotein kinase in the regulation of cellular responsiveness to extracellular ATP. *Ann. NY Acad. Sci*. 603, 401–416.

Ehrlich, Y.H. (1997) Lasher, R., Hogan, M.V., Babinska, A., Wieraszko, A. And Kornecki, E. Molecular identification and synaptic localization of the ecto-protein kinase required for the stabilization of LTP. *Society for Neuroscience Abstracts*. 23: 1393, no. 549.1.

Evans, R., Derkach, V. and Surprenant, A. (1992). ATP mediates fast synaptic transmission in mammalian neurons. *Nature* 357, 503–505.

Fedan, J.S. and Lamport, S.J. (1990) P2-purinoceptor antagonists. *Ann. New York Acad. Sci*. 603:182–97.

Fernandez, J., Naik, U., Ehrlich, Y.H. and Kornecki, E. (1993) Comparative investigation of the effects of the immunosuppressants cyclosporine A, cyclosporine G and FK-506 on platelet activation. *Cell. Mol. Biol. Res*.39:257–264.

Forsberg, P.O. and Martin, S.C. (1991) Phosphorylation/dephosphorylation and the regulation of fibrinogen and complement factor C3. *Upsala J. Med. Sci*. 96:75–93.

Friedberg, H., Belzer, I., Ogad-Plesz, O., and Kuebler, D. (1995) Activation of cell growth inhibitor by ectoprotein kinase-mediated phosphorylation in transformed mouse fibroblasts. *J. Biol. Chem*. 270: 20560–20567.

Fujii, S., Kato,H., Furuse, H, Osada,H.and Kuroda, Y. (1995a) The mechanism of ATP-induced long-term potentiation involves extracellular phosphorylation of membrane proteins in guinea-pig hippocampal CA1 neurons. *Neuroscience Lett.* 187:130–132.

Fujii,S., Osada,H., Hamaguchi, Kuroda,Y.Kato, H. (1995b) Extracellular phosphorylation of membrane protein in hippocampal slices. *Neuroscience Lett.* 187:133–136.

Gordon, J.L. (1986) Extracellular ATP: effects, sources and fate. *Biochem. J.* 233:309–319.

Greengard, P. (1978). Cyclic Nucleotides, Phosphorylated Proteins and Neuronal Function. (New York : *Raven Press*).

Hardwick, J.C., Ehrlich, Y.H. and Hendley, E.D. (1989). Extracellular ATP stimulates norepinephrine uptake in PC12 cells. *J. Neurochem.* 53, 1512–1518.

Hartmann, M., Shrader, J. (1992) Exo-protein kinase release from intact cultured aortic endothelial cells. *Biochim.Biophys.Acta.* 1136 (2) 189–195.

Hendley, E.D., Whittemore, S.R., Chaffee, J.E. and Ehrlich, Y.H. (1988). Regulation of norepinephrine uptake by adenine nucleotides and divalent cations: Role for extracellular protein phosphorylation. *J. Neurochem.* 50, 263–273.

Herbert, J.M., Daviet, I. and Maffrand, J.P. (1991) Characterization of extracellular binding for[^{3}H]-staurosporine on capillary endothelial cells. *Cell Bio.Int.Rep.* 15:883–890.

Hogan, M.V.,Pawlowska, Z., Yang, H.A., Kornecki, E. and Ehrlich, Y.H. (1995) Surface phosphorylation by ecto-protein kinase C in brain neurons: a target for Alzheimer's β- amyloid peptides. *J. Neurochem.* 65: 2022–2030.

Honkanene,R.E., Zwiller, Boynton, A.L. (1990) Characterization of microcystin-LR, a potent inhibitor of type 1 and type 2A protein phosphatases. *J. Biol. Chem.* 265:19401–404.

House, C. and Kemp, B.E. (1987). Protein kinase C contains a pseudosubstrate prototope in its regulatory domain. *Science* 238, 1726–1728.

Kandel, E.R. and Schwartz, J.H. (1982). Molecular biology of learning: modulation of transmitter release. *Science* 218, 433–442.

Kalafatis, M., Rand, M.D., Jenny, R.J., Ehrlich, Y.H. Mann, K.G. (1993). Phosphorylation of Factor Va and Factor VIIIa by activated platelets. *Blood* 81, 704–719.

Kornecki, E., Walkowiak, B., Naik, U. and Ehrlich, Y.H. (1990) Activation of human platelets by a stimulatory monoclonal antibody. *J.Biol.Chem.*265:10042–10048.

Koh, J., Yang, L.L. and Cotman, C.W. (1990). β-Amyloid protein increases the vulnerability of cultured cortical neurons to excitotoxic damage. *Brain Res.* 533, 315–320.

Kreutzberg, G.W., Reddington, M., and Zimmermann, H. Cellular Biology of Ectoenzymes: Proceedings of the International Erwin-Riesch-Symposium on ectoenzymes. Springer- Verlag, Berlin, (1986).

Kubler, D., Reinhardt, D., Reed, J., Pyerin, W. and Kinzel, V. (1992) Atrial natriuretic peptide is phosphorylated by intact cells through cAMP-dependent ecto-protein kinase. *Eur. J.Biochem.* 206:10638–10645.

Kuroda,Y., Ichikawa, M. et al., (1992) Block of synapse formation by a protein kinase inhibitor. *Neurosci. Lett.* 135: 255–258.

Loo, D.T. et al. (1993). Apoptosis is induced by b-amyloid in cultured central nervous system neurons. *Proc. Natl. Acad. Sci. USA* 90, 7951–7955.

Martin, S.C., Forsberg, P.O. and Eriksson, S.D. (1991) The effects of in vitro phosphorylation and dephosphorylation on the thrombin-induced gelation and plasmic degradation of fibrinogen. *Thromb.Res.*61:243–252.

Merlo, D., Anelli, R., Calissano, P., Ciotti, M.T. And Volonte, C. (1997) Characterization of an ecto-phosphorylated protein of cultured cerebellar granule neurons. *J. Neurosci. Res.* 47: 500–508.

Mochly-Rosen, D. and Koshland, D.E. (1987). Domain structure and phosphorylation of protein kinase C. *J. Biol. Chem.* 262, 2291–2297.

Mochly-Rosen, D. and Koshland, D.E. (1988). A general procedure for screening inhibitory antibodies: Application for identifying anti-protein kinase C antibodies. *Anal. Biochem.* 170, 31–37.

Muramoto, K., Kobayashi, K., Nakanishi, S. and Kuroda, Y. (1988). Functional synapse formation between cultured neurons of rat cerebral cortex. *Proc. Jpn. Acad. Sci.* [Ser. B] 64, 319–322.

Muramoto, K., Kawahara, M., Kobayashi, K., Taniguchi, H. and Kuroda, Y. (1992) Phosphorylation of extracellular domains of microtubule-associated protein 1B may be involved in synapse formation between cortical neurons. *Neurosci.Lett.*, 135, 255.

Nagashima, K., Nakanishi, S. and Matsuda, Y. (1991). Inhibition of nerve growth factor - induced neurite outgrowth of PC12 cells by a protein kinase inhibitor which does not permeate the cell membrane. *FEBS Lett.* 293, 119–123.

Nagy, A.K., Shuster, T.A. and Delgado-Escueta, A.V. (1986). Ecto-ATPase of mammalian synaptosomes. Identification and enzymic characterization. *J. Neurochem.* 47, 976–86.

Nagy, A.K. and Shuster, T.A. (1995) ATP-binding proteins on the external surface of synaptic membranes: identification by photoaffinity labeling. *J. Neurochem.* 65: 1849–1858.

Nagy, A.K. (1997) Ecto-ATPases of the Nervous System. In: Ecto-ATPases. Recent Progress on Structure and Function. Plesner, L., Kirley, T.L. and Knowles, A.F., Eds. Plenum Press, NY.

Naik, U.P., Kornecki, E. and Ehrlich, Y.H. (1991). Phosphorylation and dephosphorylation of human platelet surface proteins by an ecto-protein kinase/phosphatase system. *Biochem. Biophys. Acta* 1092, 256–264.

Neary, J., Crow, T. and Alkon, D.L. (1981). Changes in a specific phosphorylation following associative learning in Hermissenda. *Nature* 293, 658–660.

Nestler, E.J., and Greengard, P. (1983). Protein phosphorylation in the brain. *Nature* 305, 583–588.

Nishimura, S. et al.(1990). Excitatory and inhibitory effects of adenosine on neurotransmission in the hippocampal slices of guinea pig. *Brain Res.* 525, 165–169

Oda, Y., M. Kuo.S.S. Huang, and J.S. Huang. (1993) Acidic fibroblast growth factor receptor purfied from bovine stimulated phosphoprotein from bovine liver plasma membranes has aFGF-stimulated kinase, autoadenlylation, and alkaline nucleotide phosphodiesterase activities. *J. Biol. Chem.* 268(36): 27318–27326.

Oda, Y., M. Ko, S.S. Huang, and J.S. Huang. (1991) The plasma cell membrane glycoprotein, PC-1, is a threonine-specific protein kinase stimulated by acidic fibroblast growth factor. *J. Biol. Chem.* 266(25): 16791–16795.

Pawlowska, Z., Hogan, M.V., Kornecki, E. and Ehrlich, Y.H. (1993). Ecto-protein kinase and surface protein phosphorylation in PC12 cells: Interactions with nerve growth factor. *J. Neurochem.* 60, 678–686.

Pawlowska, Z., Chen, W., Hogan, M.V., Yang, H-A., and Ehrlich, Y.H. (1994) Neuronal surface phosphoproteins associated with neuritogenesis and synaptogenesis. *Society for Neuroscience Abstracts.* 20: 1298, no. 534.5.

Pfeifle, J., Hagmann, W. and Anderer, A. (1981) Cell adhesion-dependent differences in endogenous protein phosphorylation on the surface of various cell lines. *Biochim.Biophys.Acta* 670:274.

Pike, C.J., Burdick, D. and Walencewicz, A.J. (1993). Neurodegeneration induced by b-amyloid peptides in-vitro: The role of peptide assembly state. *J. Neurosci.* 13 (4), 1676–1687.

Pirotton, S., Boutherin-Falson, O., Robaye, B., Boeynaems, J.M., (1992) Ecto-phosphorylation on aortic endothelial cells. Exquisite sensitivity to staurosporine. *J. Biochem.*, 285, 585–591.

Rebbe, N.F., B.D. Tong, E.M. Finley, and S. Hickman. (1991) Identification of nucleotide pyrophosphatase/alkaline phosphodiesterase I activity associated with the mouse plasma cell differentiation antigen PC-1. *Proc. Natl. Acad. Sci. USA* 88:5192–5196.

Shen, K.Z. and North, R.A. (1993). Excitation of rat locus coeruleus neurons by adenosine 5'- triphosphate: ionic mechanism and receptor characterization. *J. Neurosci.* 13, 894–99.

Silinsky, E.M. (1975) On the association between transmitter secretion and the release of adenine nucleotides from mammalian motor nerve terminals. *J. Physiol.* 247: 145–62.

Silinsky, E.M., Gerzanichne, V. and Vanner, S.M. (1992). ATP mediates excitatory synaptic transmission in mammalian neurons. *Brit. J. Pharm.* 106, 762–763.

Skubitz, K.M., Ehresmann, D.D. (1992) The angiogenesis inhibitor beta-cyclodextrin tetradecasulfate inhibits ecto-protein kinase activity. *Cell Mol. Biol.* 38 (6), 543–60.

Skubitz, K.M., Goueli, S.A. (1994) CD31 (PECAM-1), CDw32 (Fc gamma RII), and anti-HLA class I monoclonal antibodies recognize phosphotyrosine-containing proteins on the surface of human neutrophils. *J. Immunol.* 152 (12), 5902–5911.

Sonka, J., Kubler, D. and Kinzel, V. (1989) Phosphorylation by cell surface protein kinase of bovine and human fibrinogen and fibrin. *Biochim.Biophys.Acta*, 997:268–277.

Tsuji, S., Yamashita, T. and Nagai, Y. (1988). A novel, carbohydrate signal-mediated cell surface protein phosphorylation: ganglioside GQ1b stimulates ecto-protein kinase activity on the cell surface of a human neuroblastoma cell line, GOTO. *J. Biochem.* (Japan) 104, 498–503.

Vilgrain, I. and A. Baird. (1991) Phosphorylation of basic fibroblast growth factor by a protein kinase associated with the outer surface of a target cell. *Molec. Endocrinol.* 5:1003–12.

Volonte, C., Merlo, D., Ciotti, M.T., and Calissano, P. (1994) Identification of an ectokinase activity in cerebellar granule primary neuronal cultures. *J. of Neurochem.* 63: 2028–2037.

Walter, J., Schnolzer, M., Pyerin, W., Kinzel, V. and Kubler, W. (1996) Induced release of cell surface protein kinase yields CK1- and CK2-like enzymes in tandem. *J. Biol. Chem.* 271:111–119.

Walter, J., Capell, A., Hung, A.Y., Langen, H., Schnolzer, M., Thinakaran, G., Sisodia, S.S., Selkoe, D.J., and Haass, C. (1997) Ectodomain phosphorylation of β-amyloid precursor protein at two distinct cellular locations. *J. Biol. Chem.* 272: 1896–1903.

Whitson, J.S., Selkoe, D.J. and Cotman, C.W. (1989). Amyloid β-protein enhances the survival of hippocampal neurons in vitro. *Science* 243, 1488–1490.

Wieraszko, A., Goldsmith, G., and Seyfried, T.N. (1989) Stimulation-dependent release of adenosine triphosphate from hippocampal slices. *Brain Res.* 485: 244–250.

Wieraszko, A. and Seyfried, T. (1989). ATP - induced synaptic potentiation in hippocampal slices. *Brain Res.* 491, 356–359.

Wieraszko, A. and Ehrlich, Y.H. (1994). On the role of extracellular ATP in the induction of Long-Term Potentiation (LTP) in the hippocampus. *J. Neurochem.* 63:1731–1738.

Yasuda, I. et al., (1990). A synthetic peptide substrate for selective assay of protein kinase C. *Biochem.Biophys. Res. Comm.* 166, 1220–1227.

Yankner, B.A., Duffy, L.K. and Kirschner, D.A. (1990). Neurotrophic and neurotoxic effects of amyloid β -protein: Reversal by tachykinin neuropeptides. *Science* 250, 279–282.

5

EXTRACELLULAR ATP-INDUCED APOPTOSIS IN PC12 CELLS

Albert Y. Sun* and Yong-Mei Chen

Department of Pharmacology
University of Missouri
Columbia, Missouri 65212

ABSTRACT

Studies in our laboratory indicate that extracellular ATP $(ATP)_o$ may induce cell death by reactive oxygen insults. We have also shown that the Ca^{2+}-induced oxidative stress as elicited by ATP may lead to an activation of a specific AP-1 activity. Since early impairment of mitochondria constitutes a critical event of the apoptotic cell death, we have examined whether $(ATP)_o$ will affect mitochondrial damage and cell injury by using mitochondrial specific probes, dihydrorhodamine and 3-(4,5-dimethylthiazo-2-yl)-2,5-diphenyl tetrazolium bromide (MTT). We have found that $(ATP)_o$ induced cell death in a concentration dependent manner by MTT assay. The $(ATP)_o$ induced cell death correlated well with the reactive oxygen species (ROS) generation in mitochondria, since $(ATP)_o$ enhanced both cell death and ROS production and antioxidant blocked both of these processes. We found $(ATP)_o$ treatment led to apoptotic cell death by examining DNA laddering and the TUNEL assay. Interestingly, vitamin C and vitamin E combined treatment appeared to attenuate the $(ATP)_o$-induced apoptosis. Results indicated that $(ATP)_o$ may cause oxidative damage of mitochondria leading to apoptotic cell death. Antioxidants may be useful in preventing apoptosis by preventing ROS formation in mitochondria.

1. INTRODUCTION

ATP has been shown to be copackaged with various neurotransmitters and appears to be coreleased with them in millimolar concentrations (Burnstock 1986; Von Kugelgen and Starke 1994). Once released, ATP may interact with P_2 purinoceptors or be degraded to adenosine by ectoenzymes such as ecto-ATPase and ecto-nucleotidases (Zimmerman et al.

* Address correspondence to Dr. A.Y. Sun, Dept. of Pharmacology, Univ. of Missouri, Columbia, MO 65212 Telephone: (573) 882–1565 and Fax (573) 884–4558 E-mail: Pharmays@muccmail.missouri.edu

Molecular and Cellular Mechanisms of Neuronal Plasticity, edited by Ehrlich. Plenum Press, New York, 1998.

1979; Lin 1989), or may serve as a substrate for ecto-protein kinases (Hendley et al. 1988; Wixom and Sun 1988; Hogan et al. 1995; Chen et al. 1996).

ATP is considered to be a neurotransmitter in both peripheral and central neurons (Jahr and Jessell 1983; Inoue et al. 1992; Valera et al. 1994; Zimmermann 1994). By activating specific P_2 purinoceptors, ATP may open a variety of cation-permeable ion channels (Bean 1992) leading to the influx of Na^+ and Ca^{2+} and efflux of K^+ (Krishtal et al. 1988; Bean 1992). By depolarization of the cell, Ca^{2+} may also enter through the voltage-sensitive Ca^{2+} channels (Illes and Norenberg 1993). It has been suggested that an increase in $[Ca^{2+}]i$ may contribute to tissue injury (Nicotera et al. 1992). Although the mechanisms by which Ca^{2+}-induced cell damage are mostly unknown, activation of Ca^{2+}-dependent protease (calpains) has been shown to play an important role, especially in the presence of catalytic amounts of Fe^{2+} (Cheng et al. 1994). It is possible that limited proteolysis by calpains may activate xanthine oxidase, which in turn may enhance free radical formation via the purinergic metabolic pathway, leading to cellular damage and cell death (McCord 1987).

Recently we have demonstrated that extracellularly added ATP enhanced the increased membrane lipid peroxidation in PC12 cells in the presence of $FeCl_2$/DTP (Cheng et al. 1996). P_2 purinoceptor agonists, αβ-methylene ATP and 2-methylthio-ATP also induced PC12 cell lipid peroxidation, but to a lesser extent than ATP. We have attributed the increased lipid peroxidation in PC12 cells primarily to the increase in intracellular Ca^{2+}. More recently we have also shown that the nuclear signalling cascade may be involved in cell death by extracellular ATP (Chen and Sun 1997). $(ATP)_o$ leads to the activation of transcription factor AP-1 binding to the promotor region of targeted genes. Since apoptosis is known to be required for normal development of the nervous system and is associated with many neurodegenerative diseases (Buttke and Sandstrom 1994; Petit et al. 1997; Draczynska-Lusiak et al. 1997), excessive release of ATP may trigger the apoptotic cell death, leading to various neurodegenerative processes.

More recently, attention has been paid to the involvement of early impairment of mitochondria in the apoptotic cascade (Du et al. 1997). The permeability transition (PT) may lead to the release of protease activators that can promote apoptotic nuclear degradation. Since the production of reactive oxygen species in mitochondria was implicated in the transduction pathway (Francelanord et al. 1997), we also examined whether $(ATP)_o$ may lead to neuronal cell death through mitochondrial damage. We have used the MTT method to assess the mitochondrial damage and cell death. We have also used dihydrorhodamine (DHR), which is specifically taken up into mitochondria, as a probe to detect the production of reactive oxygen species (ROS). DHR is nonfluorescent, but can be converted into fluorescent rhodamine 123 in the presence of ROS, especially superoxides (Emmendorffer et al. 1990; Rothe et al. 1988), and is therefore a very useful tool in studying the mitochondrial involvement in oxidative stress events and in the apoptotic cascade process.

2. MATERIALS AND METHODS

2.1. Cell Culture and Treatment

Rat pheochromocytoma cells (PC12 cells) were grown routinely at 37°C under an atmosphere of 5% CO_2/95% air in Dulbecco's modified Eagle's medium (DMEM) containing 10% horse serum, 5% fetal calf serum and 50 units/ml penicillin/ streptomycin as described previously (Chen and Sun 1997). Various agents were added to the culture medium after it was replaced with serum-free DMEM.

2.2. MTT Calorimetric Assay

This method is based on the reduction of the yellow dye, tetrazolium salt MTT (3-(4,5-dimethylthiazo-2-yl)-2,5-diphenyltetrazolium bromide) into blue formazan product, mainly by the mitochondrial enzyme dehydrogenase (Draczynskz-Lusiak et al. 1997). The amount of formazan produced is proportional to the number of living cells. Cells were maintained in 96-well plates. After the appropriate treatment, MTT dye was added (0.5 mg/ml) and the plates were incubated at 37°C for 3 hrs. Then the solution was removed and the blue formazan produced was solublized in dimethyl sulfoxide. The optical density of the blue formazan formed was determined at 540 nm using an Anthos Labtec plate reader and expressed as percent of control.

2.3. Fluorescence Measurement of Reactive Oxygen Intermediates (ROIs) with Dihydroxyrhodamine 123 (DHR)

The method for the measurement of ROIs was modified from that described by Emmendorffer et al. (1990). In brief, PC12 cells were collected, washed and resuspended in KRB solution. For each experiment, 5×10^5 cells were preincubated with 2 µg/ml of DHR for 10 min at 37°C. Immediately after addition of ATP (with or without 5 min preincubation of antioxidants as indicated), the fluorescence intensity was monitored continuously with a Perkin-Elmer MPF-66 fluorescence spectrophotometer at the excitation wavelength of 503 nm and emission wavelength of 532 nm. The data at each time point was expressed as a percentage of the fluorescence intensity at zero time.

2.4. DNA Fragmentation Analysis

2.4.1. DNA Extraction and Electrophoresis. A modified method of Lyons et al. (1992) was used to assess the fragmentation of PC12 cell DNA. After treatment, PC12 cells were washed with cold PBS and pelleted at 4°C. The cells were resuspended in 500 µl of buffer containing 20 mM EDTA, 5% Triton X-100, 5 mM Tris-HCl, pH 8.0, and stored at 4°C for 20 min. Intact nuclei were pelleted and eliminated by centrifugation at top speed in the microfuge for 10 min at 4°C. The low-molecular weight oligonucleotide fragments of soluble DNA in the supernatant were extracted with phenol:chloroform:isoamyl alcohol (25:24:1). The nucleic acids were then precipitated by isopropanol, resuspended in TE, and digested with RNase (20 µg/ml). DNA was loaded on a 1.5% agarose gel, which was stained with 0.8 µg/ml ethidium bromide after migration The gel was visualized and photographed under transmitted UV light with a Polaroid camera.

2.4.2. In Situ Detection of DNA Fragmentation. The 3' OH ends resulting from DNA degradation were detected with the TUNEL (terminal deoxynucleotidyl transferase-mediated dUTP-X 3' nick end-labeling) technique as described by Gavrieli et al. (1992). The experiment was conducted using the In Situ Cell Death Detection Kit (Boehringer Mannheim), according to the manufacturer's instructions. Briefly, PC12 cells were plated at a density of 70×10^3 cells/cm^2 on coverslips (16x16 mm). After treatment, cells were washed with PBS, fixed with 4% paraformaldehyde, pH 7.4, and permeabilized with 0.1% Triton X-100. Cells were then covered with a labeling reaction mixture containing terminal deoxynucleotidyl transferase (TdT) and fluorescein-deoxyuridine triphosphate (dUTP), and incubated at 37°C for 60 min. Subsequently, cells were stained with antifluo-

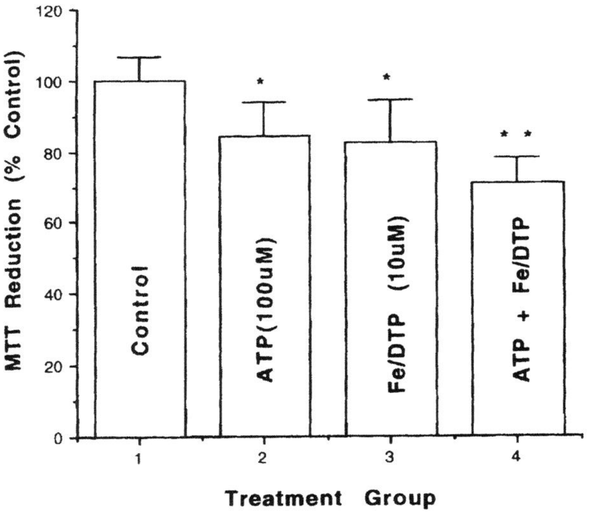

Figure 1. The combined action of $(ATP)_o$ and Fe^{2+}/DTP on cell viability. The MTT assay is based on the ability of intact mitochondrial dehydrogenase to reduce MTT to blue formazan product as measured by the optical density at 540 nm. The amount of formazan produced is proportional to the number of living cells. ATP (100 μM) and/or $FeCl_2$/DTP (10 μM, at equimolar concentrations) were added to the culture medium and incubated for 45 hrs. The ability of the cells to reduce MTT was assayed as described in Methods. Data represent mean ± SD from quadruple experiments. $*p<0.05$, $**p<0.01$.

rescein antibody conjugated with horseradish peroxidase, followed by washing, and incubating with diaminobenzidine substrate (DAB) to produce a dark brown precipitate. The morphology of the cells was examined under light microscope and photographed.

2.5. Materials

All other reagents were obtained from Sigma Chemical Company (St. Louis, MO).

3. RESULTS

3.1. Effect of $(ATP)_O$ on Mitochondrial Injury and Cell Death

We have used the MTT assay method to assess mitochondrial injury and cell death. Basically, the intact mitochondria will reduce the yellow dye tetrazolium salt MTT into blue formazan product. The amount of formazan produced is proportional to the number of living cells. As shown in Fig. 1., $(ATP)_o$ alone produced some mitochondrial damage and cell death. Fe/DTP at 10 μM concentration also induced the same degree of injury. However, when both of these were added into the incubation medium, the mitochondrial damage and cell death were greatly increased. The enhancement of these injuries was proportional to the concentration of $(ATP)_o$ as shown in Fig. 2. At 1 mM $(ATP)_o$, which is possible through a sudden release of ATP upon excessive stimulation (Cheng et al. 1994), 50% of the cells suffered injury or cell death after $(ATP)_o$ treatment for 48 hrs.

3.2. Effect of $(ATP)_O$ on Mitochondrial ROS Generation

To investigate the possible free radical generation in mitochondria following $(ATP)_o$ treatment, we have used another dye, dihydrorhodamine (DHR), which is specifically

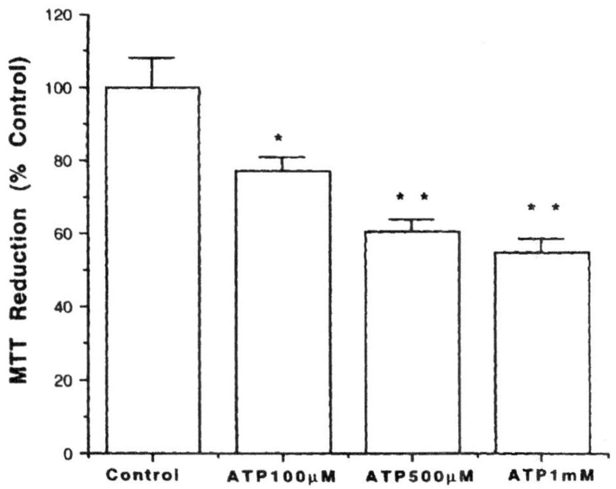

Figure 2. $(ATP)_o$ concentration dependence in Fe^{2+}-induced cell death. PC12 cells were incubated with Fe^{2+}/DTP (10 μM) and various concentrations of $(ATP)_o$ for 45 hrs. The ability of mitochondrial dehydrogenase to reduce MTT to formazan was measured as described. Data represent mean ± SD from quadruple experiments. $^*p<0.05$; $^{**}p<0.01$.

taken up by mitochondria. DHR is a non-fluorescent probe, but can be converted into fluorescent rhodamine123 in the presence of ROS, especially superoxides (Cheng et al. 1996). We have demonstrated the generation of ROS in mitochondria upon $(ATP)_o$ treatment. Fig. 3 shows that ROS generation increased almost 300%. Pretreatment of cells with vitamin C or vitamin E alone for 5 min did not attenuate the mitochondrial ROS production. However, combination of vitamin C and vitamin E treatment 5 min before the addition of $(ATP)_o$ and Fe/DTP did suppress the mitochondrial ROS generation.

3.3. Protection of PC12 Cells from $(ATP)_O$-Induced Cell Injury by Antioxidants

We have again used antioxidants to examine whether they protect the cells against $(ATP)_o$-induced cell injury by MTT assay. As shown in Fig. 4, the combination of vitamin C and vitamin E did ameliorate the oxidative damage and cell death induced by the combined action of ATP and Fe/DTP.

3.4. $(ATP)_O$ Induced Apoptosis in PC12 Cells

Cellular apoptosis is characterized morphologically by cell shrinkage, chromatin condensation and DNA breakage through a cascade of molecular and biochemical events. We have used two methods to assess the apoptosis induced by $(ATP)_o$: 1) Conventional DNA laddering methods were used to examine DNA breakage. After cells were treated with $(ATP)_o$ (500 μM) and/or Fe/DTP (10 μM) for 24 hrs DNA laddering was clearly shown in Fig. 5, especially in the presence of both $(ATP)_o$ and Fe/DTP (Lane 6). We have also demonstrated that the combination of vitamin C and vitamin E will alleviate the

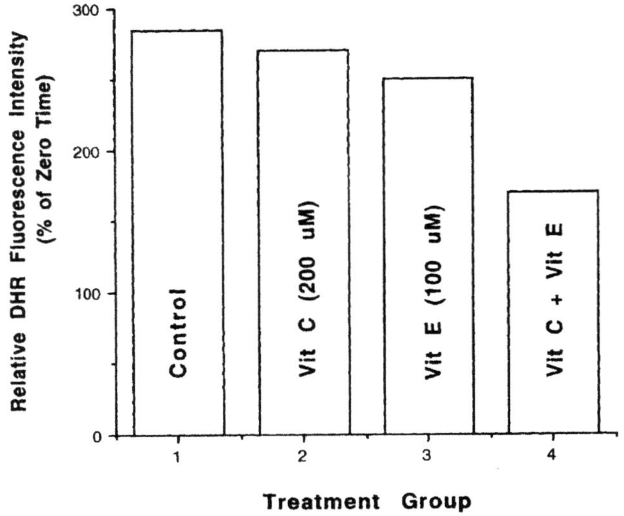

Figure 3. Effect of vitamin C/vitamin E on ATP-induced mitochondrial ROS formation in PC12 cells. PC12 cells were preincubated for 3 hrs with vitamin C (200 μM) or vitamin E (100 μM) or both before addition of Fe^{2+}/DTP (0.1 mM) and ATP (1 mM). Data represent mean increases in rhodamine 123 fluorescent intensity in 10 min from 3 determinations.

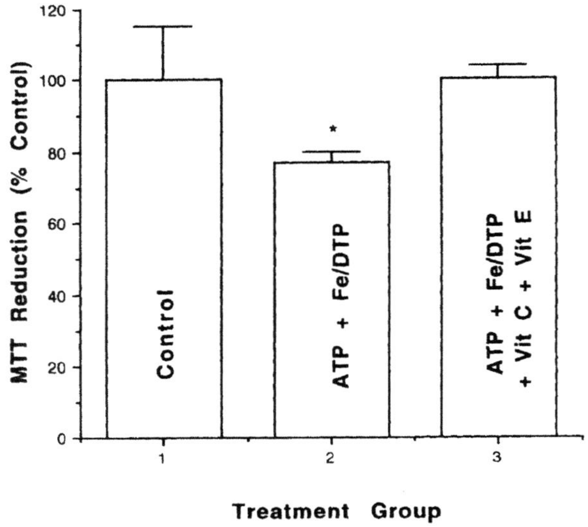

Figure 4. Effect of vitamin C and vitamin E on cell viability after ATP/Fe/DTP insults. The cell viability was assessed by measuring the ability of mitochondrial dehydrogenase to reduce MTT to formazan blue as described in Fig. 2. Combined vitamin C and vitamin E were added 3 hrs before the addition of ATP/Fe/DTP (Group III). Data represent mean ± SD from quadruple experiments. $*p<0.05$.

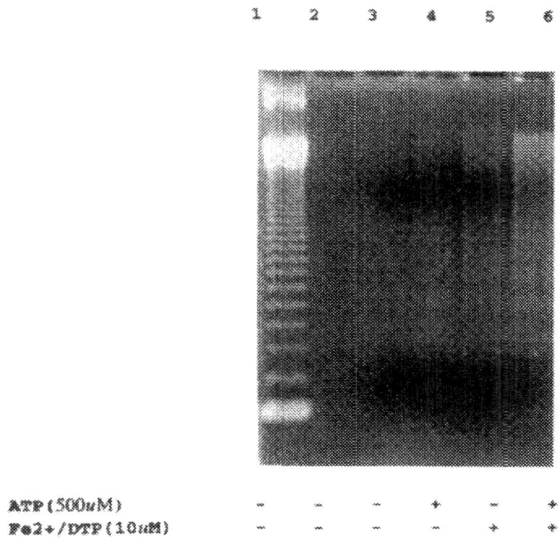

Figure 5. $(ATP)_o$ and Fe^{2+}/DTP-induced apoptosis in PC12 cells. PC12 cells were treated with ATP (100 μM) and/or $FeCl_2$/DTP (10 μM) for 24 hrs and determination of DNA laddering was performed as described in Methods. ATP (Lane 4); $FeCl_2$/DTP (Lane 5); ATP+$FeCl_2$/DTP (Lane 6); Lane 3: control cells without any treatment; and Lane 1: DNA standards of different sizes.

$(ATP)_o$/Fe/DTP-induced apoptosis as shown in Fig. 6. 2) In situ detection of DNA fragmentation was used to assess the apoptosis. After PC12 cells were treated with $(ATP)_o$ and Fe/DTP in the presence or absence of antioxidants (combined vitamin C and vitamin E), we used the TUNEL method to end-label the fragments of DNA by labeling the exposed 3'OH ends of DNA with terminal deoxynucleotidyl transferase, followed by the application of horseradish peroxidase. After this treatment, healthy cells appeared in green, while apoptotic cells or dying cells appeared in brown. We have counted the viable cells and dying cells in a defined viewing area and assessed the apoptosis as a percent of viable cells remaining after the treatment. As shown in Fig. 7, $(ATP)_o$ and Fe/DTP treated cells showed the greatest number of apoptotic cells.

4. DISCUSSION

Previous studies in our laboratory have indicated that the presence of trace amounts of ferrous ion greatly enhanced oxidative stress (Cheng et al. 1994; Chanvitayapongs et al. 1997). It is possible that most of the oxidative stress is initiated by superoxide, H_2O_2 or fatty acyl hydroperoxides. In the presence of iron, hydroxyl or alkoxyl radicals may be generated via the iron-catalyzed Haber-Weiss-like cycle (Burkitt and Gilbert 1990) and induce cell death and neurodegenerative processes. In this study, we have demonstrated that $(ATP)_o$ induced mitochondrial damage and cell death in a dose-dependent manner in the presence of Fe^{2+}/DTP (Figs. 1 and 2). It appears that the ROS generated in mitochondria play an important

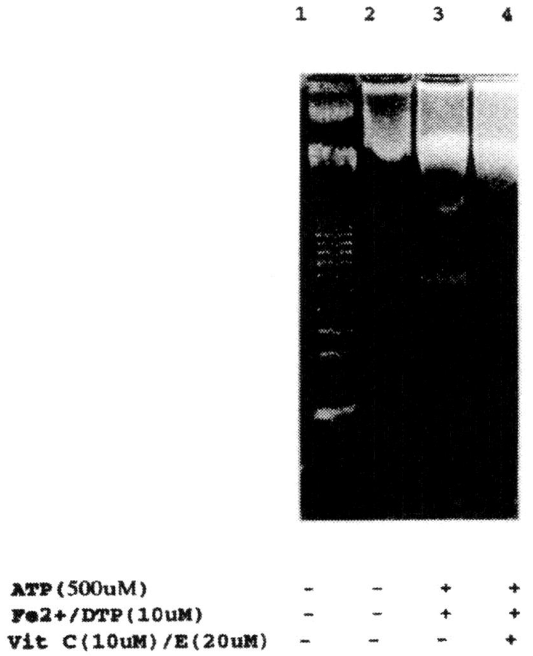

Figure 6. The effect of vitamin C/vitamin E on ATP-induced apoptosis. Same conditions as in Fig. 5, except vitamin C/vitamin E were added 3 hrs before the addition of ATP/Fe/DTP.

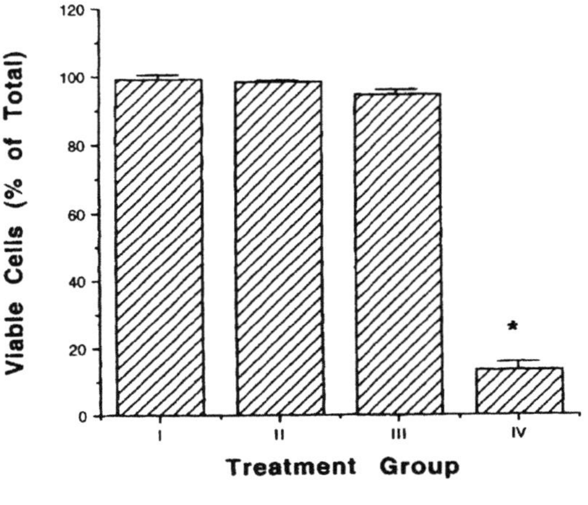

Figure 7. In situ characterization of apoptosis as induced by $(ATP)_o$ and Fe/DTP. The conditions for the induction of apoptosis were the same as described in Fig. 5. Group I: control; Group II: $(ATP)_o$-treated; Group III: Fe/DTP-treated; Group IV: $(ATP)_o$+Fe/DTP. Data represent mean ± SD. *$p<0.0001$.

role in mitochondrial damage and cell death, since the combined vitamin C and vitamin E treatment was shown to suppress the ROS production in mitochondria (Fig. 3) and to protect against cell death (Fig. 4). In this regard, vitamin C was ineffective and vitamin E alone had very little protective effect. Only the combination of vitamin C and vitamin E showed some effect (Fig. 3). It probably depends on their lipid solubility and their ability to pass through membranes to the site to exert protection. The protective effect may be attributed to the ability of vitamin C and vitamin E to scavenge the ROS generated in mitochondria. Only through the concerted action of vitamin C and vitamin E the antioxidants may exert a protective effect on the cell as a whole and to mitochondria in particular.

Most evidence indicates that (ATP)o-induced intracellular Ca^{2+} $(Ca^{2+})_i$, accumulation may be critical in lipid peroxidation and cell death (Cheng et al. 1994). ATP may mobilize Ca^{2+} from both intracellular and extracellular pools (Zimmermann 1994; Cheng et al. 1994). Excessive $(Ca^{2+})_i$ accumulation has been shown to trigger various pathological pathways that lead to cell degeneration (Nicotera et al. 1992). The increase in $(Ca^{2+})_i$ not only leads to the activation of Ca^{2+} dependent proteases, NO-synthase and phospholipase A2 with the consequence of generation of some detrimental ROS (Wang and Yuen 1994), but also causes the accumulation of Ca^{2+} in mitochondria, which may uncouple oxidative phosphorylation and lead to superoxide leakage from the electron transport chain. Recently, the mitochondrial cyclosporin-sensitive pore opening has been shown to be modulated by the redox state of the pyridine nucleotide and the superoxide formation (Chernyak 1997). This permeability transition caused by reactive oxygen species may lead to the release of protease activators that can promote apoptotic nuclear degradation (Francelanord et al. 1997; Hirsch et al. 1997). In the meantime, the mitochondria may be injured by the production of ROS or by the activation of protease, leading to the cell death detected by the MTT method (Fig. 1 and 2). It is of interest to note that the combined vitamin C and vitamin E not only attenuate the production of mitochondrial ROS (Fig. 4) but also slow down the apoptotic process (Fig. 6 and 7). It is possible that mitochondria play a very important role (Von Kugelgen et al. 1994). Excessive release of ATP could be caused by overstimulation of the sympathetic nerve (Von Kugelgen et al. 1994) or cholinergic neurons (Richardson and Brown 1989). Electric stimulation (Von Kugelgen and Stark 1994; and exposure to PAF (Kornecki and Ehrlich 1991) may also lead to abnormal release of ATP, especially when brain trauma and injury have damaged the blood-brain barrier, resulting in the exposure of neurons to relatively high levels of PAF. All these circumstances may cause apoptotic cell death.

Several studies have suggested that oxidative stress may be involved in the pathogenesis of several neurodegenerative diseases, including Parkinson's disease and Alzheimer's disease (Lyras et al. 1997; Sayre et al. 1997; Olanow 1992). More recently significant increases in TUNEL-positive cells were observed in the frontal and hippocampal cortices of Alzheimer's patients when compared with control (Li et al. 1997). It is possible that excessive release of ATP may induce apoptotic cell death leading to the manifestation of various degenerative diseases.

The fact that antioxidants are effective in preventing ROS formation in mitochondria and protect the cells from apoptotic cell death indicates that these antioxidants may be beneficial in preventing neurodegenerative diseases.

ACKNOWLEDGEMENTS

We appreciate the help of Dr. Pat Wixom in the preparation of this manuscript. This work has been supported in part by NIH Grant #AA02054.

REFERENCES

Bean B.P. (1992) Pharmacology and electrophysiology of ATP-activated ion channels. *Trends Pharmacol. Sci.* **13**, 87–90.

Burkitt M.J. and Gilbert B.C. (1990) Model studies of the iron-catalysed Haber-Weiss cycle and the ascorbate-driven Fenton reaction. *Free Radic. Res. Commun.* **10**, 265–280.

Burnstock G. (1986) The changing face of autonomic transmission. *Acta Physiol. Scand.* **126**, 76–91.

Buttke T.M. and Sandstrom P.A. (1994) Oxidative stress as a mediator of apoptosis. *Immunol. Today* **15**, 7–10.

Chanvitayapongs S., Draczynska-Lusiak B. and Sun A.Y. (1997) Amelioration of oxidative stress by antioxidants and resveratrol in PC12 cells. *NeuroReport* **8**, 1499–1502.

Chen W., Wieraszko A., Hogan M.V., Yan H.A., Kornecki E. and Ehrlich Y.H. (1996) Surface protein phosphorylation by ecto-protein kinase is required for the maintenance of hippocampal long-term potentiation. *Proc. Natl. Acad. Sci. USA* **93**, 8688–8693.

Chen Y. and Sun A.Y. (1997) Activation of transcription factor AP-1 by extracellular ATP in PC12 cells. *Neurochem. Res.* (In press).

Cheng Y., Wixom P., James-Kracke M.R. and Sun A.Y. (1994) Effects of extracellular ATP on Fe^{2+}-induced cytotoxicity in PC12 cells. *J. Neurochem.* **63**, 895–902.

Cheng Y., Chen M., James-Kracke M.R., Wixom P. and Sun A.Y. (1996) Enhanced lipid peroxidation by extracellular ATP in PC12 cells. *Neurochem. Res.* **21**, 27–33.

Chernyak B.V. (1997) Redox regulation of the mitochondrial permeability transition pore. *Biosci. Reports* **17**, 293–302.

Draczynska-Lusiak B., Chen Y.M. and Sun A.Y. (1997) Oxidized lipoproteins activate NF-κB binding activity and apoptosis in PC12 cells. *NeuroReport* (In press).

Du Y.S., Dodel R.C., Bales K.R., Jemmerson R., Hamiltonbyrd E. and Paul S.M. (1997) Involvement of a caspase 3-like cysteine protease in 1-methyl-4-phenylpyridinium-mediated apoptosis of cultured cerebellar granule neurons. *J. Neurochem.* **69**, 1382–1388.

Emmendorffer A., Hecht M., Lohmann-Matthes M.-L. and Roesler J. (1990) A fast and easy method to determine the production of reactive oxygen intermediates by human and murine phagocytes using dihydrorhodamine 123. *J. Immunol. Methods* **131**, 269–275.

Francelanord V., Brugg B., Michel P.P., Agid Y. and Ruberg M. (1997) Mitochondrial free radical signal in ceramide-dependent apoptosis - a putative mechanism for neuronal death in Parkinson's disease. *J. Neurochem.* **69**, 1612–1621.

Gavrieli Y., Sherman Y. and Ben-Sasson S.A. (1992) Identification of programmed cell death in situ via specific labeling of nuclear fragmentation. *J. Cell Biol.* **119**, 493–501.

Hendley E.D., Whittemore S.R., Chaffee J.E. and Ehrlich Y.H. (1988) Regulation of norepinephrine uptake by adenine nucleotides and divalent cations: role for extracellular protein phosphorylation. *J. Neurochem.* **50**, 263–273.

Hirsch T., Marzo T. and Kroemer G. (1997) Role of the mitochondrial permeability transition pore in apoptosis. *Biosci. Reports* **17**, 67–76.

Hogan M.V., Pawlowska Z., Yang H.-A., Kornecki, E. and Ehrlich Y.H. (1995) Surface phosphorylation by ecto-protein kinase C in brain neurons. A target for Alzheimer's β-amyloid peptide. *J. Neurochem.* **65**, 2022–2030.

Illes P. and Norenberg W. (1993) Neuronal ATP receptors and their mechanism of action. *Trends Pharmacol. Sci.* **14**, 50–54.

Inoue K., Nakasawa K., Fujimori K., Watano T. and Takanaka A. (1992) Extracellular adenosine 5'-triphosphate-evoked glutamate release in cultured hippocampal neurons. *Neurosci. Lett.* **134**, 215–218.

Jahr C.E. and Jessell T.M. (1983) ATP excites a population of rat dorsal horn neurones. *Nature* **304**, 730–733.

Kornecki E. and Ehrlich Y. (1991) Calcium ion mobilization in neuronal cells induced by PAF. *Lipids* **26**, 1243–1246.

Krishtal O.A., Marchenko S.M. and Obukhov A.G. (1988) Cationic channels activated by extracellular ATP in rat sensory neurons. *Neuroscience* **27**, 995–1000.

Li W.P., Chan W.Y., Lai H.W.L. and Yew D.T. (1997) Terminal dUTP nick end labeling (TUNEL) positive cells in the different regions of the brain in normal and Alzheimer's patients. *J. Mol. Neurosci.* **8**, 75–82.

Lin S.-H. (1989) Localization of the ecto-ATPase (Ecto-nucleotidase) in the rat hepatocyte plasma membrane. *J. Biol. Chem.* **264**, 14402–14407.

Lyons A.B., Samuel K., Sanderson A. and Meddy A.H. (1992) Simultaneous analysis of immunophenotype and apoptosis of murine thymocytes by single laser flow cytometry. *Cytometry* **13**:809–821.

Lyras L., Cairns N.J., Jenner A., Jenner P. and Halliwell B. (1997) An assessment of oxidative damage to proteins, lipids and DNA in brain from patients with Alzheimer's disease. *J. Neurochem.* **68**, 2061–2069.

McCord J.M. (1987) Oxygen-derived radicals: a link between reperfusion injury and inflammation. *Fed. Proc.* **46**, 2402–2406.

Nicotera P., Bellomo G. and Orrenius S. (1992) Calcium-mediated mechanisms in chemically induced cell death. *Annu. Rev. Pharmacol. Toxicol.* **32**, 449–470.

Olanow C.W. (1992) An introduction of the free radical hypothesis in Parkinson's disease. *Ann. Neurol.* **32** (Suppl.), S2-S9.

Petit P.X., Zamzami N., Vayssiere J.L., Mignotte B., Kroemer G. and Castedo M. (1997) Implication of mitochondria in apoptosis. *Mol. Cell. Biochem.* **174**, 185–188.

Richardson P.J. and Brown S.J. (1989) ATP release from affinity purified rat cholinergic nerve terminals. *J. Neurochem.* **48**, 622–630.

Rothe G., Oser A. and Valet G. (1988) Dihydrorhodamine 123: a new flow cytometric indicator for respiratory burst activity in neutrophil granulocytes. *Naturwissenschaftern* **75**, 354.

Sayre L.M., Zalasko D.A., Harris P.L.R., Perry G., Salomon R.G. and Smith M.A. (1997) 4-hydroxyronenal-derived advanced lipid peroxidation end product are increased in Alzheimer's disease. *J. Neurochem.* **68**, 2092–2097.

Valera S., Hussy N., Evans R.J., Aclamp N., North R.A., Surprenant A. and Buell G. (1994) A new class of ligand-gated ion channel defined by P2x receptor for extracellular ATP. *Nature* **371**, 516–527.

Von Kugelgen I. and Starke K. (1994) Co-release of noradrenaline and ATP by brief pulse trains in guinea pig vas deferens. *Naunyn-Schmiedeberg Arch. Pharmacol.* **350**, 123–129.

Von Kugelgen I., Allgaier C., Schubert A. and Starke K. (1994) Co-release of noradrenaline and ATP from cultured sympathetic neurons. *Neuroscience* **61**, 199–202.

Wang K.W. and Yuen P.W. (1994) Calpain inhibition: An overview of its therapeutic potential. *TIPS.* **15**, 412–419.

Wixom P.M. and Sun A.Y. (1988) Effect of ethanol on Ca2+ dependent phosphorylation of a 45 kDa synaptosomal protein, in *Biomedical and Social Aspects of Alcohol and Alcoholism* (Kuriyama K., Takada A. and Ishii H., eds), pp311–314, Excerpta Medica, Amsterdam.

Zimmerman H., Dawdall M.J. and Lane D.A. (1979) Purine salvage at the cholinergic nerve endings of the *Torpedo* electric organ: the central role of adenosine. *Neuroscience* **4**, 979–993.

Zimmermann H. (1994) Signalling via ATP in the nervous system. *TINS.* **17**, 420–426.

6

THE ROLE OF THE NEURAL GROWTH ASSOCIATED PROTEIN B-50/GAP-43 IN MORPHOGENESIS

L. H. J. Aarts,[1] P. Schotman,[1] J. Verhaagen,[3] L. H. Schrama,[1*] and W. H. Gispen[2]

[1]Rudolf Magnus Institute for Neurosciences
Laboratory of Physiological Chemistry
[2]Department of Medical Pharmacology Stratenum
Universiteitsweg 100
3584 CG Utrecht, The Netherlands.
[3]Netherlands Institute for Brain Research
Meibergdreef 33, 1105 AZ
Amsterdam, The Netherlands

1. THE AXONAL GROWTH CONE, NEURITE GROWTH AND PATHFINDING

Axonal guidance during development and regeneration can largely be attributed to a specialized structure at the distal end of a neurite called the growth cone (Ramón y Cajal, 1966). Structurally, these highly dynamic swellings can be divided into three distinct regions: a base, a central region and a peripheral region. The base is formed by the transition from growth cone to neurite, where bundles of microtubules diverge and extend into the central region. This region consists of a microtubular skeleton and is rich in membranous organelles such as mitochondria, dense and clear cored vesicles and endoplasmic reticulum like structures. The flattened leading edge of the growth cone, the peripheral region, is filled with a dense meshwork of actin filaments and is virtually devoid of microtubules and organelles. Growth cone and neurite architecture and motility are supported by a complex network of cell adhesion molecules, microtubules, actin, intermediate filaments and specific interacting regulatory proteins.

* Phone: 31-30-2532891; Fax: 31-30-2539032; E-mail: L.H. Schrama@med.uu.nl

Molecular and Cellular Mechanisms of Neuronal Plasticity, edited by Ehrlich. Plenum Press, New York, 1998.

1.1. The Actin Cytoskeleton

The actin cytoskeleton plays a crucial role in axonal guidance processes. Treatment of neurons with low concentrations of cytochalasins (inhibitors of actin polymerization) disrupts actin based structures within growth cones and allows invasion of microtubules and organelles in the peripheral domain. Although this does not substantially affect neurite extension it completely abolishes directed pathfinding (Bentley and Toroian-Raymond, 1986).

Actin filaments (three G-actin molecules form a F-actin subunit (Pollard, 1990)) are assembled at nucleation sites along the leading edge of the growth cone followed by retrograde translocation and disassembly in the growth cone body. The net movement of the leading edge of the growth cone thus depends on the balance of actin assembly and the retrograde movement of actin filaments. This process of actin treadmilling is a fundamental property of directed growth processes and cell migration (Cramer et al., 1994). Actin polymerization and depolymerization is highly regulated by actin binding proteins. Profilins are responsible for binding and sequestration of actin monomers, gelsolin caps, severs or stabilizes actin while proteins such as spectrin and focal adhesion proteins crosslink actin to membrane associated proteins. The rearward flow of microfilaments has recently been shown to depend at least in part on myosins (Lin et al., 1996).

Short bundles of uniformly polarized (barbed ends distally) actin filaments are often found protruding from the surface of the growth cone, to produce microspikes and filopodia (O'Connor and Bentley, 1993). Sheet-like extensions called lamellipodia, composed of a highly compact meshwork of less polarized actin filaments, are often found extending between two filopodia. Several studies have indicated that filopodia play a key role in signal detection by acting as autonomous sensors conveying information to the growth cone proper (Davenport et al., 1993). They are the most distal structures extended by growth cones and are thus ideally placed to interact with collapse-inducing and growth-promoting substances. Upon receptor-binding these ligands trigger second messenger systems in the appropriate region of the growth cone, leading to changes in cytoskeletal organization and function (Goodman, 1996; Davenport et al., 1993).

The direction of growth cone motility can be altered by a single filopodial contact (O'Connor et al., 1990). Filopodia that extend in the direction of a positive stimulus are stabilized. Actin filaments thicken and move down into the body of the growth cone. Lamellipodia subsequently form between stable filopodia and remain stationary, followed by microtubule extension and protrusion of the central cytoplasmic domain of the growth cone. Repulsive or inhibitory signals on the other hand lead to loss of F-actin at the leading edge of the growth cone followed by a lateral movement, a stop or retraction of the neurite (Fan and Raper, 1995). Growth cone movement thus appears to be regulated by extension of actin based structures followed by consolidation that involves microtubules.

1.2. Microtubules and MAPs

Microtubules play an important role in neurite growth, providing structural support for the neurite and serving as a basis for axonal transport. Substances that depolymerize microtubules inhibit neurite elongation of cultured neurons and when substrate attachment is low may even lead to neurite retraction (Gordon-Weeks, 1991a). Antisense oligonucleotides directed against tubulin mRNA block neurite extension in culture (Teichmann-Weinberg et al., 1988).

Microtubules are formed by heterodimeric tubulin subunits, consisting of an α- and ß-chain and are oriented with their "plus" ends (the preferred site of subunit addition) distal from the cell soma. The addition of new tubulin subunits thus occurs in or near the axonal growth cone (Gordon-Weeks, 1991b). The highly dynamic assembly and disassembly ("dynamic instability") of microtubules in the neurite is controlled by posttranslational modification of tubulin and by a group of proteins referred to as microtubule associated proteins (MAPs). These proteins represent a family that includes low- to intermediate molecular mass proteins such as MAP-2c and tau and high molecular mass proteins like MAP-1a, MAP-1b (MAP-5), MAP-1c (dynein), MAP-2, MAP-3 and MAP-4. Neuronal MAPs have been found compartmentalized with MAP1 being widely distributed while MAP-2 is predominantly a dendritic and tau an axonal protein. Most MAPs studied so far stabilize microtubules while preserving their dynamic character (reviewed in Maccioni and Cambiazo, 1995).

There is increasing evidence that MAPs play an important role in linking microtubules to other components of the neuronal cytoskeleton such as neurofilaments or actin (Maccioni and Cambiazo, 1995). Although microtubules are typically excluded from the peripheral domain of the growth cone, occasionally distal microtubules can be found aligned with stable filopodia. It has been proposed that these stable filopodia can arrest unstable microtubules and stabilize them by preventing shrinkage (Gordon-Weeks, 1991a). Target oriented growth of axons may thus depend on direction-selective retention of microtubules by stable bundles of actin microfilament in filopodia. These distal microtubules can subsequently serve as transport tracks for the advance of membrane organelles into more peripheral growth cone regions thereby initiating the conversion of the dynamic F-actin rich cytoplasm to a more stable organelle-rich cytoplasm (Bridgeman, 1991). Motor proteins such as the kinesin family and other ATP dependent motor proteins, cytosolic dyneins and MAP related proteins interact with specific domains on the surface of microtubules to serve as a driving force for the transport of these organelles.

1.3. Cell Adhesion Molecules

The ability of neurons to extend axons and innervate their targets in an appropriate manner depends to a large extent on expression of specific cell adhesion molecules (CAMs). CAMs are transmembrane macromolecules controlling cell-cell and cell-extracellular matrix interactions during development of the nervous system. They regulate processes as neuronal adhesion and migration, neurite outgrowth, fasciculation and synaptogenesis. The extracellular region of CAMs binds to macromolecules on neighboring cells or in the extracellular matrix while the intracellular region interacts with the cortical cytoskeleton (Fields and Itoh, 1996; Vaughn and Bjorkman, 1996). Functionally, CAMs are classified into a calcium dependent (cadherins) and a calcium independent group (integrins and the Ig superfamily of CAMs).

Molecules of the Ig superfamily of CAMs are involved in a variety of adhesive interactions including homophilic and heterophilic binding with other Ig family members, recognition by integrins and binding to extracellular matrix molecules. Their extracellular region is built up out of one or more domains with sequence similarities to antibodies (Ig-like domains) in tandem with fibronectin-III like domains (Vaughn and Bjorkman, 1996). Oligosaccharides consisting of polysialic acids regulate adhesive interactions and the proportion of these molecules varies with molecular species and with posttranslational modification. Cadherins are transmembrane proteins that bind homophilically to identical cadherins on neighboring cells in the absence of calcium, forming adherens junctions. The cytoplasmic domain of cadherins interacts with actin filaments via catenin proteins

Integrins make up a large family of heterodimeric (α and β subunit) proteins that mediate cell-cell and cell-extracellular matrix adhesive connections. Extracellular matrix molecules such as fibronectin, laminin, collagen and vitronectin, cluster integrins leading to the formation of focal adhesions, large intracellular complexes linking integrins to bundles of actin filaments. The type of integrin that is found in a focal adhesion is generally dictated by the extracellular matrix to which cells adhere. Actin binding proteins that colocalize with integrins in focal adhesions include talin, α-actinin, tensin and vinculin (Gilmore and Burridge, 1996b). Evidence for a role of focal adhesion proteins in growth cone function comes from a study by Varnum-Finney and Reinhardt (1994), who generated PC12 cells expressing one fifth of the normal vinculin level. These vinculin deficient cells extended unstable filopodia and lamellipodia and as a consequence had a reduced rate of neurite outgrowth (Varnum-Finney and Reichardt, 1994). Focal inactivation of talin or vinculin function by chromophore assisted laser inactivation (CALI), resulted in cessation of filopodial extension and retraction or in filopodial buckling and bending respectively, arguing for the importance of these proteins in filopodial integrity and motility (Sydor et al., 1996). Analogous to nonneuronal cells, focal adhesions may thus play important roles in regulating growth cone morphology and mobility and/or serve as structural frameworks for the association of signaling proteins to induce signal transduction pathways leading to changes in cell behavior (reviewed in Clark and Brugge, 1995).

Paradoxically, CAMs increase cell adhesion and axon fasciculation but they also increase cell motility and neurite outgrowth. These antagonistic effects are regulated by differences in expression levels, isoforms, posttranslational modifications and intracellular signaling (Fields and Itoh, 1996). A number of CAM-receptors are prominently present on filopodia especially at filopodial tips such as integrin and L1 whereas others such as N-cadherin and N-CAM are uniformly distributed throughout the neurite and the growth cone. It is believed that N-cadherin and N-CAM are involved in stable attachment of the neurite to the substratum while integrins and L1 are implicated in the exploratory behavior of growth cones (Letourneau, 1992).

2. SIGNAL TRANSDUCTION AND GROWTH CONE GUIDANCE

Steering decisions appear to be regulated through a complex interplay of guidance cues, cell adhesion molecules and specific signal transduction proteins leading to changes in the growth cone cytoskeleton. Almost every known second messenger system has been implicated in aspects of neurite outgrowth and growth cone function including receptor and nonreceptor tyrosine kinases and phosphatases, serine-threonine kinases and phosphatases, calcium, calmodulin, G-proteins, phospholipids, cAMP, nitric oxide and rho-like GTPases.

2.1. Tyrosine Kinases and Phosphatases

Receptor and non-receptor protein tyrosine kinases (PTKs) have been shown to transduce a variety of extracellular signals to accomplish neurite initiation, elongation and pathfinding. Soluble factors like neurotrophins (including nerve growth factor (NGF), brain derived growth factor (BDNF), neurotrophin-3 (NT-3), NT-4/5 and NT-6) and fibroblast growth factors (aFGF and bFGF) can promote survival and differentiation of a variety of neurons by binding and activation of transmembrane receptor PTKs (Chao, 1992; Schlessinger and Ullrich, 1992). In addition, activation of distinct receptor PTKs induced qualitively different morphogenetic activity of growth cones. Whereas NGF caused extensive growth cone spreading, IGF-1 induced substantial growth cone branching (Aigner and Caroni, 1995). Various cell surface ligands also promote neurite outgrowth through activation of receptor PTKs NCAM, L1 and N-cadherin induced outgrowth was shown to require activation of the FGF-receptor (Williams et al., 1994). Members of the Eph family of receptor PTKs transduce signals of membrane anchored ligands and may function in axon pathfinding and fasciculation in the developing nervous system (Cheng et al., 1995; Drescher et al., 1995).

In general, ligand binding to receptor PTKs induces receptor di(or oligo)merization and subsequent autophosphorylation of intrinsic tyrosines. These phosphorylated residues then serve as docking sites for a variety of adaptor proteins (for example p85, Shc, Crk and Grb2) or enzymatic proteins (for example PLCγ and PI3-kinase) including non-receptor tyrosine kinases (for example Src). Complex networks of interactions between these intracellular signalling molecules and their target proteins ultimately result in cellular responses such as growth or differentiation (reviewed by Pawson, 1995).

Non-receptor PTKs encoded by proto-oncogenes such as Src, Yes, Abl and Fyn have also been localized to growth cones, some of which are expressed at high levels during periods of neurite outgrowth and regeneration (Maness et al., 1992). Src and Fyn have been implicated specifically in neurite outgrowth induced by L1 and NCAM respectively (Ignelzi, et al., 1994; Beggs et al., 1994). Recently, another non-receptor PTK called focal adhesion kinase (FAK) was shown to be expressed in neuronal growth cones (Stevens et al., 1996). Activation of this kinase occurs among others upon integrin crosslinking and might regulate the formation of growth cone focal adhesions when encountering specific substrates (Richardson and Parsons, 1995). These studies suggest that individual non receptor PTKs function downstream from specific guidance cues.

There is increasing evidence that transmembrane and non transmembrane tyrosine phosphatases also play important roles in the control of growth cone guidance. Receptor protein tyrosine phosphatases (PTPs) were found to be expressed on *Drosophila* axons and their growth cones (Zinn, 1993; Stoker et al., 1995). Drosophila embryos expressing mutant forms of the transmembrane RTPs DPTP69D, DPTP99A or DLAR displayed impaired pathfinding and defasciculation of motor axons in the periphery arguing for their importance in growth cone function (Krueger et al., 1996; Desai et al., 1996).

The tyrosine phosphorylation state of protein tyrosine residues is an important factor in the regulation of growth cone filopodia dynamics. *Aplysia* growth cones growing on a polylysine substratum exhibited high levels of phosphotyrosine immunoreactivity on filopodial tips, whereas growth cones growing on *Aplysia* haemolymph displayed a strong decrease in phosphotyrosine immunoreactivity accompanied with the appearance of larger and more stable filopodia. Reduction of tyrosine phosphorylation at these tips by administration of genistein (an inhibitor of tyrosine kinase activity) resulted in an increase in filopodial length and stability (Wu et al., 1996). Similarly, neurite growth in developing chick neurons was potentiated by genistein (Bixby and Jhabvala, 1992). The mechanism underlying these changes in filopodial dynamics may include differences in actin filament treadmilling and (dis)assembly of multimolecular adhesion complexes.

2.2. Calcium/Calmodulin

Rises in growth cone calcium levels, either from extracellular sources or intracellular calcium pools, have been implicated to be involved in many aspects of growth cone behavior including changes in filopodial length, lamellipodial extent, growth cone collapse and turning towards a gradient (Goodman, 1996). Controversial results on the influence of intracellular calcium on neurite outgrowth have led to the formulation of the calcium set-point theory stating that an optimal level of $[Ca^{2+}]_i$ is required for outgrowth and concentrations above or below this concentration inhibit outgrowth (Lankford and Letourneau, 1991). The effect of increasing levels of $[Ca^{2+}]_i$ on neurite outgrowth has been studied extensively in PC12 cells. Depolarization induced calcium influx through voltage dependent calcium channels causes extensive neurite outgrowth through a signal transduction mechanism involving activation of the MAPK cascade (Rosen et al., 1994). The mechanism by which changes in calcium concentrations may lead to outgrowth inhibition is however still elusive.

Protein kinase C (PKC) isozymes, a subfamily of which is activated upon calcium influx, play a major function in growth cone physiology. Introduction of a PKC pseudosubstrate in cortical neurons to specifically block PKC activity, caused collapse of growth cones (Théodore et al., 1995). Among targets of PKC are proteins that regulate tubulin polymerization such as MAP2 and tau, as well as the calmodulin and actin binding proteins B-50 and MARCKS.

Calmodulin is a major, ubiquitous, Ca^{2+}-binding regulatory protein mediating calcium actions in a diversity of biological events including neuronal proliferation and differentiation of nerve cells. Alterations in levels of calcium and calmodulin cause changes in assembly and organization of microfilaments, intermediate filaments and microtubules via proteins like caldesmon, a protein controlling actin-myosin interactions (Kira et al., 1995) and MARCKS, an F-actin crosslinking protein (Hartwig et al., 1992). In addition, Ca^{2+}-calmodulin can activate kinases and phosphatases such as CaMKII and protein phosphatase 2b (calcineurin). Overexpression of CaMKII in PC12 cells caused changes in adhesion properties and inhibited NGF induced differentiation (Massé et al., 1997). Focal inactivation of calcineurin, which has been reported to dephosphorylate amongst others B 50, tau and dynamin, was shown to cause localized retraction of filopodia and lamellipodia and to influence the direction of subsequent growth (Chang et al., 1995).

Undoubtedly, calmodulin plays a crucial role in transducing many of the calcium signals during growth cone guidance processes. Calmodulin antagonists, applied to explants of goldfish retina caused loss of growth cone filopodia and lamellipodia and subsequent growth cone arrest. Moreover, disruption of calcium-calmodulin function in transgenic *Drosophila* growth cones caused stalls in axon extension as well as defects in fasciculation and midline crossing errors (VanBerkum and Goodman, 1995). However, the large amount of calmodulin regulated processes, including cyclic nucleotide metabolism, energy metabolism, secretion and cytoskeletal dynamics makes a clear mechanistic description on the role of calmodulin in growth cone morphology and pathfinding at present impossible.

2.3. G-Proteins

Heterotrimeric G-proteins are highly expressed in growth cones and may play an important role in growth cone physiology (Strittmatter et al., 1990). The type of activated G-protein and the downstream intracellular signalling repertoire determine the eventual growth cone response. This may include activation (through G_s) or inhibition (through G_i) of adenylate cyclases and subsequent changes in cAMP levels, a potent stimulator of neurite outgrowth. Several known activators of G-protein coupled receptors, such as dopamine (Lankford et al., 1988), thrombin (Suidan et al., 1992) and LPA (Jalink et al., 1994) caused growth cone collapse and subsequent neurite retraction. Growth cone collapse induced by membrane associated components of embryonic brain and by central nervous system myosin could be prevented by pretreatment with pertussis toxin, a substance that specifically blocks the function of G_o and G_i proteins through ADP-ribosylation of the α-subunit (Igarashi et al., 1993). N-CAM and L1 induced neurite outgrowth in PC12 cells could also be inhibited by pertussis toxin (Doherty et al., 1991). However, it has been shown recently that pertussis toxin can generate many of these results by a mechanism independent of G-protein inactivation thereby questioning whether G-proteins are really involved in transduction of these various signals (Kindt and Lander, 1995).

2.4. Rho like GTPases

In fibroblasts, members of the family of rho-like GTPases, comprising Rho, Rac, Cdc42, TC10, RhoG and RhoE, appear to be key players in the control of the actin cytoskeleton (Nobes and Hall, 1995; Foster et al., 1996). Activation of cdc42 triggers formation of filopodia, whereas rac and rho control formation of lamellipodia and stress fibers respectively, while each of them can stimulate the assembly of distinct multimolecular focal adhesions (Nobes and Hall, 1995). Since there is a striking resemblance between the actin cytoskeleton of spreading fibroblasts and advancing neuronal growth cones, these GTPases may also function in linking extracellular guidance signals to the growth cone actin cytoskeleton (Mackay et al., 1996). Several recent reports indeed imply a crucial role for rho-like GTPases in growth cone function. Expression of constitutively active or dominant negative Drac1 or Dcdc42 in *Drosophila* resulted in severe defects in the development and pathfinding of axons and dendrites of peripheral neurons (Luo et al., 1994). Transgenic mice, expressing constitutively active rac in the cerebellum, displayed a clear reduction in axon terminals and a severely altered dendritic spine morphology (Luo et al., 1996). The rho-GTPase is specifically implicated in mediating transduction of growth cone collapse inducing signals. Differentiated N1E-115 neuroblastoma cells and PC12 cells responded to addition of serum (or its active component lysophosphatidic acid (LPA)) with a rapid collapse of their growth cones and subsequent neurite retraction. These effects are mediated via receptor coupled heterotrimeric G-proteins and subsequent activation of the Rho-GTPase (Jalink et al., 1994). Effector molecules through which rho-like GTPases may exert their function in cell and growth cone morphology and physiology include proteins like PKN, PLD, PI 5-kinase, PI 3-kinase, PLA_2 and MLCK.

3. B-50 AS A MARKER FOR NEURITE OUTGROWTH

Initiation of neurite outgrowth is correlated with strikingly high levels of expression of a relatively small subset of growth associated proteins (GAPs). They include essential building components for neurite formation such as actin, α-tubulin, microtubule associated proteins and proteins thought to play a role in neurite outgrowth competence, such as the nervous tissue specific protein B-50. Initially, B-50 was isolated from the synaptic plasma membrane as a phosphoprotein involved in synaptic function. In subsequent years, proteins were isolated to identify components that are involved in axonal regeneration (GAP43; GAP48) growth cone function (pp46), synaptic plasticity (F1) and calmodulin binding (P-57 or neuromodulin). Molecular cloning experiments revealed that these proteins represented an identical protein, which we from now on refer to as B-50 (reviewed in Oestreicher et al., 1997; Benowitz et al., 1996).

Expression of the B-50 protein is highly enhanced during neuronal development, axonal regeneration and reactive synaptogenesis following nerve injury. The earliest expression of B-50 is detected in post-mitotic neurons which are committed to fiber outgrowth. From this stage of differentiation on, when cells obtain features which are essential for neuronal function, such as neurite outgrowth and morphological polarization, spontaneously occurring action potentials and neurotransmitter release, B-50 (mRNA and protein) is highly expressed, almost ubiquitously in the nervous system. Downregulation of B-50 expression to essentially undetectable levels coincides with the regression of terminal arborization and maturation of functional synapses and with the maturation of axon-glia interactions. In mature neurons, expression is relatively low except for neurons that are believed to be involved in synaptic remodeling such as in human associative brain areas and rat hippocampal and olfactory areas. In situations of disruption of the nervous tissue architecture, expression of B-50 is increased, not only in neurons, but also in glial cells *in situ*, though expression is at a lower level. The high correlation between extension of neuritic processes and increased B-50 expression levels has made B-50 expression as an often used tool to mark neurite outgrowth (Oestreicher et al., 1997).

4. MOLECULAR STRUCTURE OF B-50

The rat B-50 protein is a 226 amino acid, highly acidic protein with a rod like structure that is synthesized on free ribosomes in the cytosol. Upon attachment to Golgi-derived vesicles the protein is transported to the axonal growth cone via fast axonal transport where it accumulates at the internal surface of the plasma membrane. In undifferentiated PC12 cells, B-50 immunoreactivity was found associated with intracellular membrane-rich and vesicular structures, including organelles of the lysosome family and the Golgi apparatus. Differentiation with NGF or db-cAMP induced an upregulation of B-50 expression accompanied with a translocation of B-50 immunoreactivity to the plasma membrane where it is believed to exert its function (Van Hooff et al., 1989). A large number of mainly biochemical and mutational studies have enabled the assignment of several functional domains within the B-50 protein.

4.1. Membrane Association

The propensity of B-50 to be efficiently coupled to membranes resides within the N-terminal 10 to 20 amino acids of the protein. This sequence is able to target a number of otherwise cytosolic proteins to the Golgi apparatus and the plasma membrane, including growth cone membranes (Zuber et al., 1989; Liu et al., 1991; Liu et al., 1994). Site directed mutagenesis of two cysteine residues at positions 3 and 4 within the N-terminus resulted in a cytosolic distribution (Fig. 1.; Zuber et al., 1989; Liu et al., 1991; Liu et al., 1994; Aarts et al., 1995). In addition, a short stretch of positively charged amino acids (Arg6, Arg7 and Lys9) may be required for optimal membrane linkage since mutation of these residues significantly reduced membrane binding (Liu et al., 1994).

Anchorage of B-50 to membranes may be achieved through palmitylation of the N-terminal cysteines, and subsequent hydrophobic interaction (Liu et al., 1991). B-50 can be partially released from synaptic plasma membranes by 2-mercaptoethanol in vitro, giving further evidence for a thioester linkage of B-50 to membranes (De Graan et al., 1993). The concentrations of 2-mercaptoethanol used were however not sufficient to directly cleave the thioester bonds linking the fatty acid residue(s) to B-50. It is therefore likely that the 2-mercaptoethanol-induced release of B-50 from the membrane involves reduction of di-

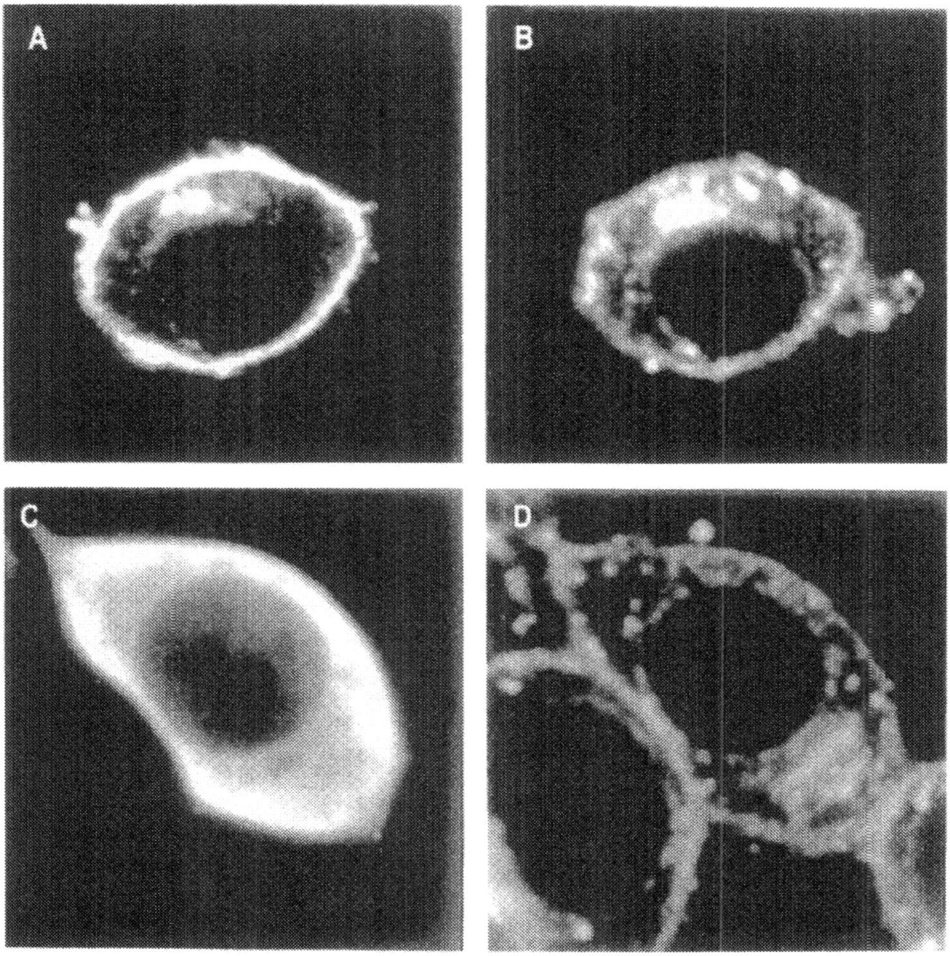

Figure 1. Targeting of B-50 to the Golgi apparatus and the plasma membrane dependent of Cys_3 and Cys_4. PC12 cells were transfected with wild type (a,b) or Ser_3Gly_4 (c,d) B-50 cDNA and stained for B-50 (a,c) or the golgi marker WGA-FITC, two days after transfection (b,d). Bar = 5 μm (reproduced from Aarts et al., 1995).

sulfide bonds with a putative membrane-located B-50 anchoring protein (De Graan et al., 1993), suggesting an alternative, unknown mechanism for B-50 membrane association. This may explain the fact that B-50 immunoreactivity is often distributed in a punctate pattern along neuronal plasma membranes as well as in growth cones (Zuber et al., 1989; Meiri and Gordon-Weeks, 1990).

In vitro, B-50 can be ADP-ribosylated mainly on the two N-terminal cysteines but also at minor sites (probably Arg_{211} or His_{225}; Coggins et al., 1993). This ADP-ribosylation leads to a redistribution of B-50 from the plasma membrane to the cytosol. How this process proceeds is not known, but the first step might be ongoing depalmitoylation of the N-terminal cysteine residues and release of B-50 from the membrane, followed by ADP-ribosylation of B-50 (Milligan et al., 1995).

4.2. G-Protein Interaction

B-50 has been shown to regulate the activity of the heterotrimeric G-proteins G_0 and G_i in vitro by stimulating the binding of GTPγS and by increasing their GTPase activity (Strittmatter et al., 1991; Sudo et al., 1992). In addition, microinjection of B-50 into *Xenopus* oocytes increased the sensitivity of G-protein-coupled receptor transduction (Strittmatter et al., 1993). The increase in G_0 activation by B-50 could only be observed when B-50 was not membrane-bound, i.e. palmitoylation blocks G_0 activation by B-50 (Sudo et al., 1992). This would suggest that the cytosolic form of B-50 is the active form with respect to G-protein activation. The minimal B-50 N-terminal fragment that is still active in stimulating G_0 is the 1–10 peptide. This sequence comprises a stretch of amino acids with some similarity to the cytoplasmic tail of G-protein coupled receptors and has therefore been proposed to exhibit a similar G-protein activating capacity (Sudo et al., 1992; Strittmatter et al., 1994). The activation of these G-proteins could not be blocked however by pertussis toxin nor affected by G-protein βγ subunits or phospholipids marking B-50 as a different guanine nucleotide releasing protein.

4.3. PKC-Phosphorylation and Calmodulin Binding

Calmodulin can bind to a family of proteins with an IQ (with core consensus IQXXXRGXXXR) domain of which B-50 is the prototype (Neel and Young, 1994). In contrast to most of these members, B-50 binds calmodulin under low ambient calcium concentrations (Andreasen et al., 1983). In rat B-50, the calmodulin binding domain spans amino acids 43 to 51 adjacent to a major PKC phosphorylation site (Ser_{41}; Apel et al., 1990). Phosphorylation of this site inhibits binding of calmodulin and vice versa calmodulin binding inhibits phosphorylation by PKC (Apel et al., 1990; Chapman et al., 1991). These findings have led to the hypothesis that B-50 could concentrate calmodulin in specific regions of unstimulated nerve terminals (Estep et al., 1990). Elevation of the internal calcium concentration or PKC-mediated phosphorylation would lead to disruption of this complex and subsequent activation of nearby calmodulin dependent target proteins. Indeed, *in vitro*, unphosphorylated B-50 inhibited the activity of calmodulin dependent nitric oxide synthase under low "physiological" calcium concentrations. This inhibition could be relieved by raising the calcium concentration or upon phosphorylation by PKC (Slemmon and Martzen, 1994). Recent crosslinking experiments in primary hippocampal neurons provided evidence that a calcium and PKC-dependent interaction between B-50 and calmodulin indeed occurs *in situ* (Gamby et al., 1996).

In a yeast interaction trap, binding of calmodulin to B-50 was increased upon deletion of C-terminal two-thirds of the B-50 protein (Chao et al., 1996). A number of additional phosphorylation sites have been identified within this C-terminus including two major casein kinase II phosphorylation sites at positions 191 and 192. Phosphorylation of these sites could be prevented *in vitro* upon interaction of calmodulin with the IQ domain in the N-terminal part of B-50 (Apel et al., 1990). The significance of these latter findings for B-50 function remains however obscure.

4.4. Cytoskeletal Interaction

In subcellular fractionation studies of isolated growth cones from fetal rat brain, B-50 has been detected in the actin-enriched membrane skeleton, together with proteins like talin, α-actinin and fodrin (Meiri and Gordon-Weeks, 1990). Direct interaction of B-50 and actin was demonstrated in two separate studies (Hens et al., 1993; Strittmatter et al., 1992). B-50, either in the PKC phosphorylated or in the dephosphorylated form, co-sedimented with filamentous actin (F-actin) in the absence and presence of calcium (Hens et al., 1993). However, though B-50 bound to F-actin, it did not affect actin polymerization and bundling, in contrast to synapsin I, a synaptic vesicle specific protein (Bähler and Greengard, 1987), nor cross-linking actin, in contrast to MARCKS (Hartwig et al., 1992). The interaction site of B-50 with F-actin is currently unknown.

5. INTERFERENCE WITH B-50 EXPRESSION INHIBITS GROWTH CONE SPREADING, ADHERENCE AND PATHFINDING

One approach to establish a function for B-50 in neurite outgrowth and growth cone activity has been to analyze the effects of B-50 depletion in cell cultures as well as in intact mice. Interference with B-50 expression by antisense oligonucleotides against B-50 mRNA completely impaired NGF induced neurite outgrowth of PC12 cells (Jap Tjoen San et al., 1992). Moreover, introduction of anti B-50 antibodies into neuroblastoma cells via lipid transfer was shown to arrest cAMP-induced neurite outgrowth, arguing for a crucial role for B-50 in neuritogenesis (Shea et al., 1991). However, suppression of B-50 transport to primary neuronal growth cones, by using antisense oligonucleotides against the fast axonal transport protein kinesin, did not interfere with neurite elongation (Ferreira et al., 1992) and a B-50 deficient clonal PC12 cell line (PC12B) was still capable of extending neurites upon NGF induced differentiation (Baetge and Hammang, 1991).

In primary dorsal root ganglion neurons, the effects of B-50 depletion by antisense oligonucleotides on neurite growth were shown to be substrate dependent. Neurite outgrowth and growth cone formation were completely blocked on a polyornithine substratum, whereas B-50 depleted DRG neurons grown on laminin/polyornithine still extended neurites that, however, appeared to be longer, thinner and less branched with drastically reduced growth cones (Aigner and Caroni, 1993). Similarly, clonal PC12 cell lines with B-50 overexpression extended highly branched neurites with complex growth cones whereas neurites of PC12 cells with low endogenous B-50 ended in blunt growth cones (Fig. 2; Jap Tjoen San et al., 1995).

The morphological and functional properties of B-50 depleted DRG growth cones differed significantly from B-50 expressing growth cones in that they displayed poor substrate adherence, highly dynamic but unstable lamellar extensions and a striking absence of local F-actin accumulations. Furthermore, NGF-induced spreading as well as IGF-1 (insulin like growth factor-1)-induced branching were impaired in B-50 depleted growth cones. B-50 expressing neurons were more resistant to growth cone collapse and subsequent neurite retraction induced by central nervous system myelin-derived liposomes compared to B-50 depleted neurons (Aigner and Caroni, 1995). These observations together imply that B-50 depletion causes changes in the cytoskeleton leading to diminished neurite and growth cone adhesiveness and stability. This is in agreement with observations that B-50 deficient PC12B cells exhibited greatly reduced adherence which could be restored upon reintroduction of B-50 cDNA (Meiri et al., 1996). Moreover, the blocking effect of the anti B-50 antibodies on neuritogenesis in neuroblastoma cells was substrate-dependent since it could be overruled by using strongly adhesive substrates, giving further evidence for a role of B-50 in membrane-substrate adhesiveness (Shea and Benowitz, 1995).

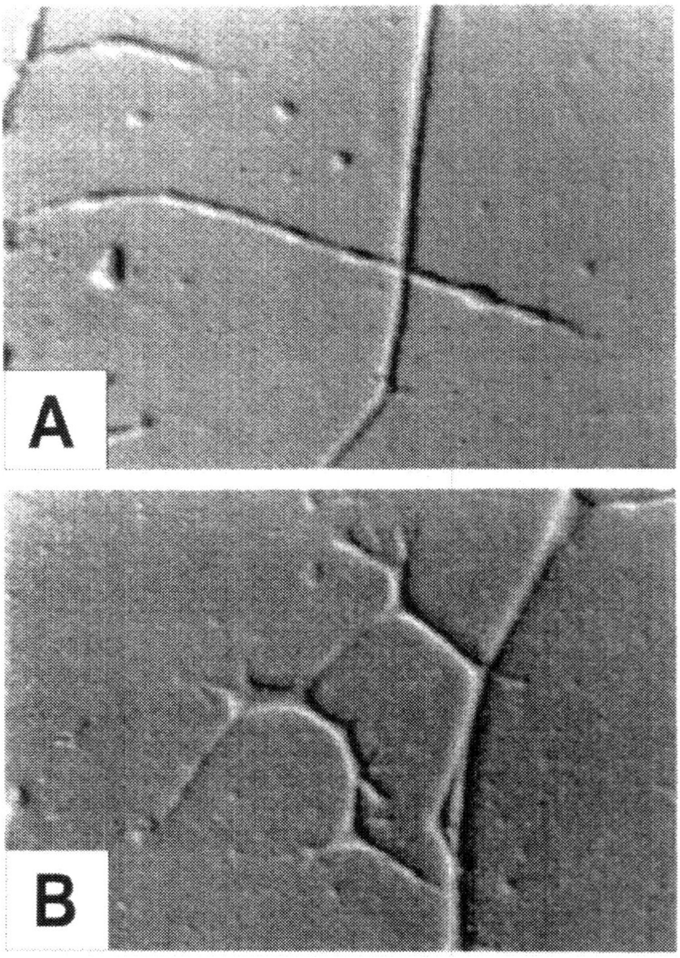

Figure 2. Complexity of neurites and growth cones in PC12 cells under control of B-50. Clonal PC12 cells expressing low levels of B-50 (PC-B2) extend straight neurites with varicose and blunt neurite endings (a) whereas clonal PC12 cells expressing high levels of B-50 (+3B) extend branced neurites with well spread, elaborated growth cones (b) upon differentiation with NGF for seven days. Bar = 15 μm. (Reproduced from Jap Tjoen San et al., 1996).

The consequence of B-50 depletion in developing mice was studied by Strittmatter and coworkers (1995), who generated mice lacking exon two of the B-50 gene. Although these mice developed a grossly normal nervous system they were found to die soon after birth, presumably due to starvation. Closer examination of critical developmental phases in the homozygous knockout mice by means of fluorescent dye tracing revealed that B-50 depleted retinal axons became trapped in the optic chiasm for 6 days, before crossing over. A more severe phenotype was observed in mice homozygous for a targeted deletion of exon 1 of the B-50 gene (Zhu and Julien, personal communication). The majority (>97%) of the exon 1 knock-out mice died either in the very first week after birth or around the weaning period. These mice displayed severely retarded postnatal growth and visual system defects. In contrast to the exon 2 knock-out mice, in the exon 1 knock-out mice the majority of retinal axons was not trapped in the optic chiasm during embryonic development but rather extended in a large proportion towards the ipsilateral and hypothalamic regions. Both studies indicate however that although B-50 is not essential for axonal outgrowth and nervous system development, the protein might play a key role in making pathfinding decisions in response to guidance signals during development of the nervous system.

6. B-50 OVEREXPRESSION CAUSES ENHANCED PROCESS OUTGROWTH COMPETENCE

Further evidence supporting a fundamental role for B-50 in morphogenetic processes has been obtained in experiments where nonneuronal or neuronal cells in culture, or subpopulations of cells in intact mice were forced to express high levels of wild type or mutant forms of the B-50 protein.

Stable or transient transfection of B-50 in a variety of nonneuronal cells (NIH 3T3, Cos-7, L6 and Rat1cells) caused spontaneous formation of numerous thin filopodia and a decreased initial cell spreading (Zuber et al., 1989; Widmer and Caroni, 1993; Aarts et al., 1996). B-50 immunoreactivity was mainly associated with the cell cortex and colocalized to a great extent with actin filaments. Moreover, cells expressing high levels of B-50 displayed prominent radial arrays of peripheral actin filaments (Widmer and Caroni, 1993). These effects were all shown to be transient and lasted for several hours after replating on coated coverslips. A more dramatic morphological change was observed when B-50 was introduced into Vero or rabbit skin cells, using a herpes simplex virus construct. Virtually all B-50 expressing cells exhibited membrane ruffling, formation of filopodia and extensive cell spreading, persisting for several days in culture. In addition, a large number of cells extended long neurite-like processes with growth cone-like endings whereas control, β-gal expressing cells displayed no morphological changes (Fig. 3; Verhaagen et al., 1994).

Stably transfected PC12 and Neuro 2a neuroblastoma cells exhibited an accelerated neurite outgrowth in response to differentiating agents (Yankner et al., 1990; Morton and Buss, 1992) and an enhanced maintenance of neurites in response to withdrawal of such stimuli (Kumagai et al., 1992). In the absence of a differentiating stimulus, transient B-50 transfection resulted in the formation of numerous filopodia and blebs on the surface of PC-B2 cells (a virtually B-50 deficient PC12 clone) and did not lead to spontaneous neurite outgrowth (Fig 4 a,b). As in nonneuronal cells, B-50 immunoreactivity colocalized to a great extent with cortical F-actin (Fig. 4 c,d; Nielander et al., 1993).

In order to assign functional domains for the morphogenetic activity of B-50, transfection experiments have been performed with several mutant B-50 constructs. Two B-50 phosphorylation site mutants (chicken Ser_{42} 6 Ala preventing and Ser_{42} 6 Asp mimicking phosphorylation by PKC) had prominent effects on cell morphology. Non-neuronal COS-7 and L6 cells transfected with the Ser_{42} 6 Ala mutant spread poorly and contained numerous short filopodia, whereas cells that were transfected with Ser_{42} 6 Asp spread extensively and displayed large irregular membranous extensions with few filopodia (Widmer and Caroni, 1993). In another study, stable expression of the Ser_{41} 6 Ala mutant in B-50 deficient PC12B cells resulted in the appearance of irregular plasma membranes with blebs, protrusions and thin neurites. Moreover, whereas stable transfection of wild type B-50 increased the initial attachment of PC12B cells on plastic, Ser_{41} 6 Ala transfected cells dis-

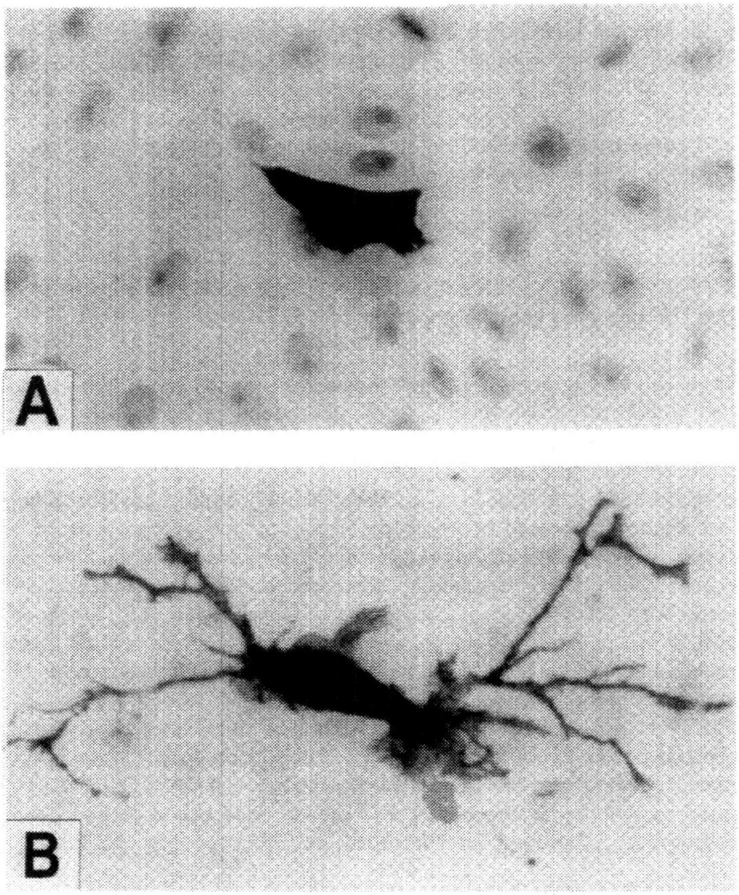

Figure 3. B-50 induces outgrowth of neurite-like processes in nonneuronal Vero cells. Vero cells were infected with adenoviral constructs to obtain high expression levels of lacZ (a) or B-50 (b) protein. Cells expressing high levels of B-50 extended long, neurite like processes with growth cone-like endings whereas control cells (expressing lacZ) exhibited no morphological changes 48 (a) or 72 (b) hours after infection. Bar = 50 μm. (Reproduced from Verhaagen et al., 1994).

played strongly reduced adherence compared to non- and mock transfected controls. Initial attachment on laminin was not impaired but long term adhesion was drastically reduced (Meiri et al., 1996).

Mutation of the N-terminal cysteine residues (Cys_3Cys_4), important for correct membrane targeting of the B-50 protein (see section 4), rendered B-50 morphogenetically inactive in spreading COS-7, L-6 and Rat1 cells, indicating that interactions with the cell cortex are required for these effects (Widmer and Caroni, 1993; Aarts et al., 1996). Transfection of COS-7 cells with constructs containing the first 10 amino acids of B-50 fused to chloramphenicol acetyl transferase induced similar morphological changes as those described for full length B-50. These shape changes depended not only on cysteine 3 and 4 but also on the presence of the basic residues Arg_6 and Lys_9, indicated to be crucial for G-

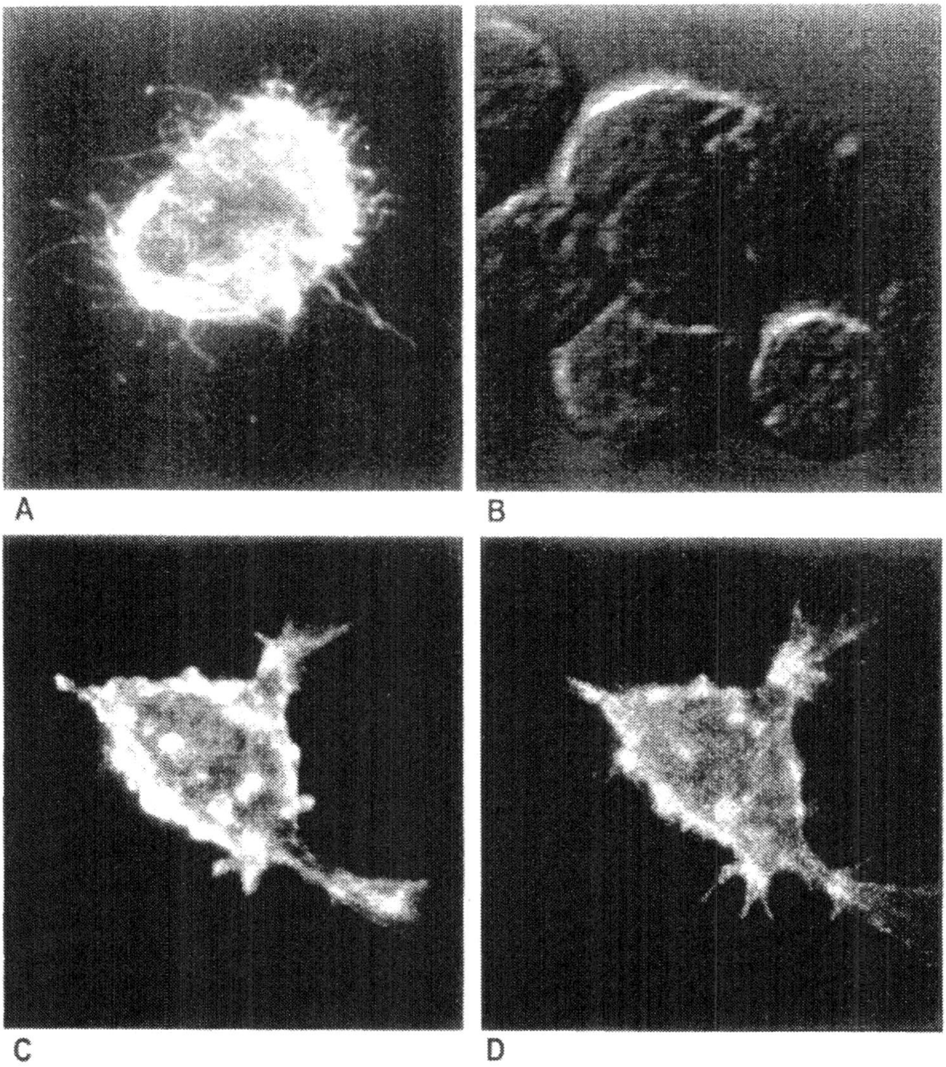

Figure 4. Spontaneous formation of filopodia and blebs by B-50, colocalizing with F-actin. PC-B2 cells, a clonal PC12 cell line with low endogenous B-50 expression, were transfected with B-50 cDNA and stained for B-50 (a and c) or F-actin (d) two days after. Cells were analyzed by confocal laser scanning microscopy using nomarski optics as well (b). The cells in a and b as well as cells in c and d represent identical cells. Cells expressing B-50 are round extending long, thin filopodia to all sides (a), or become flattened with filopodin attached to the substratum and blebs and with B-50 immunoreactivity colocalizing with F-actin (c and d). Scale bar = 10 μm. (Reproduced from Nielander et al., 1993).

protein activation (Strittmatter et al., 1994). Moreover, B-50 amino terminal (1–10) peptides were shown to modulate growth cone morphology and neurite outgrowth in primary neurons and neuroblastoma cells via a pertussis toxin sensitive mechanism. Again, peptides with substitutions of Cys_3, Cys_4, Arg_6 and Lys_9 were shown to be inactive (Strittmatter et al., 1994).

To study the consequence of constitutive overexpression of B-50 in the nervous system of intact mice, two classes of transgenic mice have been created driving chick or rat B-50 expression under the control of regulatory sequences of the Thy-1.2 (Aigner et al., 1995) or olfactory marker protein (OMP; Holtmaat et al., 1995) promoter, respectively. Chick B-50 expression in the Thy-1.2 transgene lines was detected in a variety of neuronal populations in the CNS and the PNS. These mice displayed spontaneous nerve sprouting in the terminal fields of hippocampal mossy fibers and at the neuromuscular junction. Moreover, injury-induced sprouting and terminal arborization were greatly potentiated in B-50 overexpressing mice. Overexpression of non-phosphorylatable B-50 (Ser_{42} 6 Ala) resulted in significantly less spontaneous and induced sprouting, whereas Ser_{42} 6 Asp (mimicking permanently phosphorylated B-50) B-50 transgenes had a comparable sprout promoting activity as compared to transgenic mice overexpressing the wildtype protein. Remarkably, a gross immunostaining analysis of various typical nervous tissue proteins did not reveal striking differences in the brain and spinal cord between transgenic and wildtype mice (Aigner et al., 1995). B-50 expression in the various OMP transgene lines was restricted to olfactory neurons, which normally express OMP and are devoid of endogenous B-50. Striking differences were found in the morphology of primary olfactory axons, which formed enlarged bulbous olfactory axon endings in the glomeruli. Immuno EM analysis revealed that numerous olfactory axon profiles contained extended axolemma organized in remarkable concentric sheaths (called labyrinths), which in serial sections could be labeled both by antibodies to B-50 and to OMP (Holtmaat et al., 1995). Similar labyrinths could be observed in olfactory axon endings upon overexpression of B-50 but not lacZ, accomplished via infection with adenovirus constructs (Holtmaat et al., 1997). Axonal labyrinth formation occurred predominantly in the vicinity of the glomerular boundary formed by juxtaglomerular neurons astrocytes and oligodendrocytes. Axonal growth into and beyond the juxtaglomerular layer may be prevented by growth inhibitory molecules. Taking the results of these studies together, they show that enhanced B-50 expression promotes a vigorous neurite outgrowth response but that *in vivo* the presence of a predetermined neuroanatomical boundary forms an obstacle for neurite elongation.

7. CONCLUDING REMARKS

Although B-50 expression per se is not a prerequisite for axonal outgrowth, the presence of B-50 seems to be essential for proper growth cone function during development. Several lines of evidence indicate that B-50 may play a critical role in membrane-substrate adhesiveness. First, B-50 depletion in numerous neuronal cell cultures caused a dramatic reduction in growth cone spreading and adhesion (Aigner and Caroni, 1993; Shea, 1994; Jap Tjoen San et al., 1995). Second, when attached isolated growth cones were dislodged with gentle streams of culture medium highly adhesive, B-50 immunoreactive membrane patches were left behind (Meiri and Gordon-Weeks, 1990). Similar B-50 immunoreactive spots have been observed in B-50 transfected PC12 cells probably left behind upon neurite retraction (Nielander et al., 1993). Third, the blocking effects of B-50 depletion on neurite outgrowth were shown to be substrate dependent, with most dramatic effects on less adhesive substrata (Aigner and Caroni, 1993; Shea and Benowitz, 1995). This suggests that any requirement for B-50 in adhesiveness may be reduced by treatments that increase adherence (Shea, 1995). Fourth, the reduced substrate adhesiveness of B-50 deficient PC12B cells could be restored upon reintroduction of B-50 (Meiri et al., 1996). Related results were obtained in nonneuronal cells that displayed enhanced spreading upon B-50 transfection (Widmer and Caroni, 1993; Verhaagen et al., 1994).

It is tempting to speculate that B-50 influences growth cone spreading and adherence through interaction with the actin cytoskeleton. Introduction of B-50 in Rat1 fibroblasts resulted in formation of focal adhesions, multimolecular complexes linking the actin cytoskeleton to the extracellular matrix (Aarts et al., 1996). Whether the highly adhesive B-50 immunoreactive spots observed in neuronal cells (Meiri and Gordon-Weeks, 1990; Nielander et al., 1993) represent similar adhesion sites remains to be established. Interestingly, stably transfected Ser_{41} 6 Ala PC12B cells displayed a drastically reduced amount of $\alpha 1\beta 1$ integrin on the cell surface which may account for their strongly decreased adherence capacity on laminin (Meiri et al., 1996). Whereas B-50 depleted growth cones exhibited striking absence of local F-actin accumulations (Aigner and Caroni, 1993), introduction of B-50 in nonneuronal cells resulted in the formation of numerous filopodia containing high levels of F-actin (Widmer and Caroni, 1993; Aarts et al., 1996). Association of B-50 with the cortical cytoskeleton could be disrupted with cytochalasins suggesting that this interaction is dynamic: actin filaments may retain B-50 at the cell cortex and high levels of B-50 may promote or stabilize submembranous actin filaments (Widmer and Caroni, 1993; Aarts et al., 1996). The absence of B-50 in spreading growth cones may compromise the formation of a well spread, adhesive growth cone and extension of filopodia and hence cause a reduced capacity to respond to local guidance cues, as was observed in the B-50 knock out mice (Strittmatter et al., 1995; Zhu and Julien, personal communication). Alternatively, increased stability of F-actin structures such as filopodia and lamellipodia at the nerve terminal would allow enhanced neurite outgrowth and sprouting capacity, as was observed in various B-50 transfected neuronal cell lines and in transgenic mice (Yankner et al., 1990; Morton and Buss, 1992; Aigner et al., 1995; Holtmaat et al., 1995).

The molecular mechanism through which B-50 influences the actin cytoskeleton is still obscure. Although B-50 was demonstrated to bind to F-actin in a cosedimentation assay, no effects of B-50 on actin polymerization or F-actin organization could be observed *in vitro* (Hens et al., 1993). B-50 may however affect actin dynamics by modulating second messenger systems in the growth cone. A specific role for the N-terminal part of the B-50 protein has been suggested with regard to the regulation of the activity of heterotrimeric G-proteins (Strittmatter et al., 1992). Control of the availability of calmodulin near the growth cone membrane by B-50, depending on internal calcium concentrations and phosphorylation state, may indirectly affect the actin cytoskeleton through calmodulin dependent kinases (f.e. CaMKII) or phosphatases (f.e. calcineurin) (Gamby et al., 1996). Alternatively, calmodulin binding to B-50 may serve to regulate other functions of the B-50 protein. B-50 has been reported to regulate the turnover of phosphatidyl 4,5-biphosphate (PIP_2) (Van Dongen et al., 1985; Van Hooff et al., 1988), a lipid that has been implicated in controlling actin polymerization and formation of focal adhesions (Gilmore and Burridge, 1996a). Interestingly, we have recently observed that transient B-50 transfection caused spontaneous formation of filopodia and focal adhesions in spreading fibroblasts (Aarts et al., 1996). Currently, we are examining the possible involvement of rho-like proteins in mediating B-50 induced morphological changes, with the objective to obtain more clarity in the molecular mechanisms underlying growth cone function.

ACKNOWLEDGMENTS

This research was supported by NWO-grants 903-42-006, 903-47-050 and 030-94-142.

REFERENCES

Aarts L.H.J., Van der Linden J.A.M., Hage W.J., Van Rozen A.J., Oestreicher A.B., Gispen W.H., and Schotman P. (1995) N-terminal cysteines essential for Golgi sorting of B-50 (GAP-43) in PC12 cells. *Neuroreport* **6**, 969–972.

Aarts L.H.J., Oestreicher A.B., Schrama L.H., Gispen W.H., and Schotman P. (1996) B-50 (GAP-43) induces spontaneous filopodia and focal adhesion-like complexes in Rat1 fibroblasts. *Soc Neurosci Abstr.* **22**, 1709.

Aigner L., Arber S., Kapfhammer J.P., Laux T., Schneider C., Botteri F., Brenner H.R., and Caroni P. (1995) Overexpression of the neural growth-associated protein GAP-43 Induces nerve sprouting in the adult nervous system of transgenic mice. *Cell* **83**, 269–278.

Aigner L. and Caroni P. (1993) Depletion of 43-kD Growth-associated protein in primary sensory neurons leads to diminished formation and spreading of growth cones. *J Cell Biol* **123**, 417–429.

Aigner L. and Caroni P. (1995) Absence of persistent spreading, branching, and adhesion in GAP-43-depleted growth cones. *J Cell Biol* **128**, 647–660.

Andreasen T.J., Luetje C.W., Heideman, W., and Storm D.R. (1983) Purification of a novel calmodulin binding protein from bovine cerebral cortex membranes. *Biochemistry* **22**, 4615–4618.

Apel E.D., Byford M.F., Au D., Walsh K.A., and Storm D.R. (1990) Identification of the protein kinase C phosphorylation site in neuromodulin. *Biochemistry* **29**, 2330–2335.

Baetge E.E. and Hammang J.P. (1991) Neurite outgrowth in PC12 cells deficient in GAP-43. *Neuron* **6**, 21–30.

Bahler M. and Greengard P. (1987) Synapsin I bundles F-actin in a phosphorylation dependent manner. *Nature* **326**, 704–707.

Beggs H.E., Soriano P., and Maness P.F. (1994) NCAM-dependent neurite outgrowth is inhibited in neurons from Fyn-minus mice. *J Cell Biol* **127**, 825–833.

Bentley D. and O'Connor T.P. (1994) Cytoskeletal events in growth cone steering. *Curr Opin Neurobiol* **4**, 43–48.

Bentley D. and Toroian-Raymond A. (1986) Disoriented pathfinding by pioneer neurone growth cones deprived of filopodia by cytochalasin treatment. *Nature* **323**, 712–715.

Bixby J.L. and Jhabvala P. (1992) Inhibition of tyrosine phosphorylation potentiates substrate-induced neurite growth. *J Neurobiol* **23**, 468–480.

Bridgeman, P.C. (1991). Functional anatomy of the growth cone in relation to its role in locomotion and neurite assembly. In P.C. Letourneau, S.B. Kater & F.R. Macagno (Eds.). *The Nerve Growth Cone* (pp. 39–53). New York: Raven Press

Burns M.E. and Augustine G.J. (1995) Synaptic structure and function: Dynamic organization yields architectural precision. *Cell* **83**, 187–194.

Chang H.Y., Takei K., Sydor A.M., Born T., Rusnak F., and Jay D.G. (1995) Asymmetric retraction of growth cone filopodia following focal inactivation of calcineurin. *Nature* **376**, 686–690.

Chao M.V. (1992) Neurotrophin receptors: a window into neuronal differentiation. *Neuron* **9**, 583–593.

Chao S., Benowitz L.I., Krainc D., and Irwin N. (1996) Use of a two-hybrid system to investigate molecular interactions of GAP-43. *Mol Brain Res* **40**, 195–202.

Chapman E.R., Au D., Alexander K.A., Nicolson T.A., and Storm D.R. (1991) Characterization of the calmodulin binding domain of neuromodulin. *J Biol Chem* **266**, 207–213.

Cheng H.J., Nakamoto M., Bergemann A.D., and Flanagan J.G. (1995) Complementary gradients in expression and binding of ELF-1 and Mek4 in development of the topographic retinotectal projection map. *Cell* **82**, 371–381.

Clark E.A. and Brugge J.S. (1995) integrins and signal transduction pathways: the road taken. *Science* **268**, 233–238.

Coggins P.J., McLean K., Nagy A., and Zwiers H. (1993) ADP-ribosylation of the neuronal phosphoprotein B-50/GAP-43. *J Neurochem* **60**, 368–371.

Cramer L.P., Mitchison T.J., and Theriot J.A. (1994) Actin-dependent motile forces and cell motility. *Curr Opin Cell Biol* **6**, 82–86.

Davenport R.W., Dou P., Rehder V., and Kater S.B. (1993) A sensory role for neuronal growth cone filopodia. *Nature* **361**, 721–724.

De Graan P.N.E.. Moritz A.. Dewit M.. and Gispen W.H. (1993) Purification of B-50 by 2-mercaptoethanol extraction from rat brain synaptosomal plasma membranes. *Neurochem Res* **18,** 875–881.

Desai C.J., Gindhard J.G.. Goldstein L.S.B., and Zinn K. (1996) Receptor tyrosine phosphatases are required for motor axon guidance in the Drosophila embryo. *Cell* **84,** 599–609.

Drescher U.. Kremoser C.. Handwerker C., Loschinger J., Noda M., and Bonhoeffer F. (1995) In vitro guidance of retinal ganglion cell axons by RAGS, a 25 kDa tectal protein related to ligands for Eph receptor tyrosine kinases. *Cell* **82,** 359–370.

Estep R.P., Alexander K.A., and Storm D.R. (1990) Regulation of free calmodulin levels in neurons by neuromodulin: relationship to neuronal growth and regeneration. *Curr Topics Cell Reg* **31,** 161–180.

Fan J. and Raper J.A. (1995) Localized collapsing cues can steer growth cones without inducing their full collapse. *Neuron* **14,** 263–274.

Fang K.S., Sabe H.. Saito H., and Hanafusa H. (1994) Comparative study of three protein-tyrosine phosphatases. Chicken protein-tyrosine phosphatase lambda dephosphorylates c-Src tyrosine 527. *J Biol Chem* **269,** 20194–20200.

Ferreira A., Niclas J., Vale R.D.. Banker G., and Kosik K.S. (1992) Suppression of kinesin expression in cultured hippocampal neurons using antisense oligonucleotides. *J Cell Biol* **117,** 595–606.

Fields R.D. and Itoh K. (1996) Neural cell adhesion molecules in activity-dependent development and synaptic plasticity. *Trends Neurosci* **19,** 473–480.

Foster R., Hu K.Q., Lu Y.. Nolan K.M., Thissen J., and Settleman J. (1996) Identification of a novel human Rho protein with unusual properties: GTPase deficiency and in vivo farnesylation. *Mol Cell Biol* **16,** 2689–2699.

Gamby C., Waage M.C., Allen R.G., and Baizer L. (1996) Analysis of the role of calmodulin binding and sequestration in neuromodulin (GA1–43) function. *J Biol Chem* **271,** 26698–26705.

Gilmore A.P. and Burridge K. (1996a) Regulation of vinculin binding to talin and actin by phosphatidyl-inositol-4–5-bisphosphate. *Nature* **381,** 531–535.

Gilmore A.P. and Burridge K. (1996b) Molecular mechanisms for focal adhesion assembly through regulation of protein-protein interactions. *Structure* **4,** 647–651.

Goodman C.S. (1996) Mechanisms and molecules that control growth cone guidance. *Annu Rev Neurosci* **19,** 341–377.

Gordon-Weeks P.R. (1991a) Control of microtubule assembly in growth cones. *J Cell Sci* **S15,** 45–49.

Gordon-Weeks P.R. (1991b) Microtubule organization in growth cones. *Biochem Soc Trans* **19,** 1080–1085.

Hartwig J.H., Thelen M., Rosen A., Janmey P.A., Nairn A.C., and Aderem A. (1992) MARCKS is an actin filament crosslinking protein regulated by protein kinase C and calcium-calmodulin. *Nature* **356,** 618–622.

Hens J.J.H., Benfenati F., Nielander H.B., Valtorta F., Gispen W.H., and De Graan P.N.E. (1993) B-50/GAP-43 binds to actin filaments without affecting actin polymerization and filament organization. *J Neurochem* **61,** 1530–1533.

Holtmaat A.J.G.D.. Dijkhuizen P.A., Oestreicher A.B., Romijn H.J., Van der Lugt N.M.T., Berns A., Margolis F.L., Gispen W.H., and Verhaagen J. (1995) Directed expression of the growth-associated protein B-50/GAP-43 to olfactory neurons in transgenic mice results in changes in axon morphology and extraglomerular fiber growth. *J Neurosci.* **15,** 7953–7965.

Holtmaat, A.J.G.D., Hermens. W.T.J.M.C., Sonnemans, M.A.F., Giger, R.J., Van Leeuwen, F.W., Kaplitt, M.G., Oestreicher, A.B., Gispen, W.H., & Verhaagen, J. (1997). Adenoviral vector-mediated expression of B-50/GAP-43 induces alterations in the membrane organization of olfactory axon terminals in vivo. *J Neurosci* **17,** 6575-6586

Igarashi M.. Strittmatter S.M., Vartanian T., and Fishman M.C. (1993) Mediation by G-Proteins of Signals That Cause Collapse of Growth Cones. *Science* **259,** 77–79.

Ignelzi M.A., Jr., Miller D.R., Soriano P., and Maness P.F. (1994) Impaired neurite outgrowth of src-minus cerebellar neurons on the cell adhesion molecule L1. *Neuron* **12,** 873–884.

Jalink K., van Corven E.J., Hengeveld T., Morii N., Narumiya S., and Moolenaar W.H. (1994) Inhibition of lysophosphatidate- and thrombin-induced neurite retraction and neuronal cellrounding by ADP ribosylation of the small GTP-binding protein Rho. *J Cell Biol* **126,** 801–810.

Jap Tjoen San E.R.A., Van Rozen A.J., Nielander H.B., Oestreicher A.B., Gispen W.H., and Schotman P. (1995) Expression levels of B-50/GAP-43 in PC12 cells are decisive for the complexity of their neurites and growth cones. *J Mol Neurosci* **6,** 185–200.

Jap Tjoen San E.R.A., Schmidt-Michels M.H., Oestreicher A.B., Gispen W.H., and Schotman P. (1992) Inhibition of nerve growth factor-induced B-50/GAP-43 expression by antisense oligomers interferes with neurite outgrowth of PC12 cells. *Biochem Biophys Res Commun* **187,** 839–846.

Kindt R.M. and Lander A.D. (1995) Pertussis toxin specifically inhibits growth cone guidance by a mechanism independent of direct G protein inactivation. *Neuron* **15,** 79–88.

Kira M., Tanaka J., and Sobue K. (1995) Caldesmon and low Mr isoform of tropomyosin are localized in neuronal growth cones. *J Neurosci Res* **40**, 294–305.

Krueger N.X., Van Vactor D., Wan H.I., Gelbart W.M., Goodman C.S., and Saito H. (1996) The transmembrane tyrosine phophatase DLAR controls motor axon guidance in Drosophila. *Cell* **84**, 611–622.

Kumagai C., Tohda M., Isobe M., and Nomura Y. (1992) Involvement of growth-associated protein-43 with irreversible neurite outgrowth by dibutyryl cyclic AMP and phorbol ester in NG 108–15 cells. *J Neurochem* **59**, 41–47.

LaBate M.E. and Skene J.H.P. (1989) Selective conservation of GAP-43 structure in vertebrate evolution. *Neuron* **3**, 299–310.

Lankford K.L., DeMello F.G., and Klein W.L. (1988) D1-type dopamine receptors inhibit growth cone motility in cultured retina neurons: evidence that neurotransmitters act as morphogenic growth regulators in the developing central nervous system. *Proc Natl Acad Sci U S A* **85**, 4567–4571.

Lankford K.L. and Letourneau P.C. (1991) Roles of actin filaments and three second-messenger systems in short-term regulation of chick dorsal root ganglion neurite outgrowth. *Cell Motil Cytoskel* **20**, 7–29.

Letourneau, P.C. (1992). Integrins and N-cadherin are adhesive molecules involved in growth cone migration. In P.C. Letourneau, S.B. Kater & F.R. Macagno (Eds.). *The nerve growth cone* (pp. 181–193). New York: Raven Press

Lin C.H., Espreafico E.M., Mooseker M.S., and Forscher P. (1996) Myosin drives retrograde F-actin flow in neuronal growth cones. *Neuron* **16**, 769–782.

Lin C.H. and Forscher P. (1993) Cytoskeletal remodeling during growth cone-target interactions. *J Cell Biol* **121**, 1369–1383.

Liu Y., Chapman E.R., and Storm D.R. (1991) Targeting of neuromodulin (GAP-43) fusion proteins to growth cones in cultured rat embryonic neurons. *Neuron* **6**, 411–420.

Liu Y., Fisher D.A., and Storm D.R. (1994) Intracellular sorting of neuromodulin (GAP-43) mutants modified in the membrane targeting domain. *J Neurosci* **14**, 5807–5817.

Luo L., Liao Y.J., Jan L.Y., and Jan Y.N. (1994) Distinct morphogenetic functions of similar small GTPases: *Drosophila* Drac1 is involved in axonal outgrowth and myoblast function. *Genes Development* **8**, 1787–1802.

Luo L.Q., Hensch T.K., Ackerman L., Barbel S., Jan L.Y., and Jan Y.N. (1996) Differential effects of the Rac GTPase on Purkinje cell axons and dendritic trunks and spines. *Nature* **379**, 837–840.

Maccioni R.B. and Cambiazo V. (1995) Role of microtubule-associated proteins in the control of microtubule assembly. *Physiol Rev* **75**, 835–864.

Mackay D.J.G., Nobes C.D., and Hall A. (1995) The Rho's progress: A potential role during neuritogenesis for the Rho family of GTPases. *Trends Neurosci* **18**, 496–501.

Maness P.F. (1992) Nonreceptor protein tyrosine kinases associated with neuronal development. *Dev Neurosci* **14**, 257–270.

Massé T. and Kelly P.T. (1997) Overexpression of calcium/calmodulin dependent protein kinase II in PC12 cells alters cell growth, morphology, and nerve growth factor- Induced differentiation. *J Neurosci* **17**: 924–931.

Meiri K.F. and Gordon-Weeks P.R. (1990) GAP-43 in growth cones is associated with areas of membrane that are tightly bound to substrate and is a component of a membrane skeleton subcellular fraction. *J Neurosci* **10**, 256–266.

Meiri K.F., Hammang J.P., Dent E.W., and Baetge E.E. (1996) Mutagenesis of ser41 to ala inhibits the association of GAP-43 with the membrane skeleton of GAP-43-deficient PC12B cells: effects on cell adhesion and the composition of neurite cytoskeleton and membrane. *J Neurobiol* **29**, 213–232.

Milligan G., Parenti M., and Magee A.I. (1995) The dynamic role of palmitoylation in signal transduction. *Trends Biochem Sci* **20**, 181–187.

Morton A.J. and Buss T.N. (1992) Accelerated differentiation in response to retinoic acid after retrovirally mediated gene transfer of gap-43 into mouse neuroblastoma cells. *Eur J Neurosci* **4**, 910–916.

Neel V.A. and Young M.W. (1994) *Igloo*, a GAP-43-related gene expressed in the developing nervous system of *Drosophila*. *Development* **120**, 2235–2243.

Nielander H.B., Schrama L.H., Van Rozen A.J., Kasperaitis M., Oestreicher A.B., Gispen W.H., and Schotman P. (1990) Mutation of serine 41 in the neuron-specific protein B-50 (GAP-43) prohibits phosphorylation by protein kinase C. *J Neurochem* **55**, 1442–1445.

Nielander H.B., French P., Oestreicher A.B., Gispen W.H., and Schotman P. (1993) Spontaneous morphological changes by overexpression of the growth-associated protein B-50/GAP-43 in a PC12 cell line. *Neurosci Lett* **162**, 46–50.

Nobes C.D. and Hall A. (1995) Rho, Rac, and Cdc42 GTPases regulate the assembly of multimolecular focal complexes associated with actin stress fibers, lamellipodia, and filopodia. *Cell* **81**, 53–62.

O'Connor T.P. and Bentley D. (1993) Accumulation of actin in subsets of pioneer growth cone filopodia in response to neural and epithelial guidance cues insitu. *J Cell Biol* **123,** 935–948.

O'Connor T.P., Duerr J.S., and Bentley D. (1990) Pioneer growth cone steering decisions mediated by single filopodial contacts in situ. *J Neurosci* **10,** 3935–3946.

Oestreicher A.B., De Graan P.N.E., Gispen, W.H., Verhaagen, J and Schrama, L.H. (1997) B-50, the growth-associated protein-43: modulation of cell morphology and communication in the nervous system. *Prog Neurobiol*, **53,** 627–686

Pawson T. (1995) Protein modules and signalling networks. *Nature* **373,** 573–580.

Pollard T.D. (1990) Actin. *Curr Opin Cell Biol* **2,** 33–40.

Ramón y Cajal S. (1966) *Recollections of my life [by] Santiago Ramón y Cajal, transl. [from the Spanish] by E. Horne Craigie, with the assistance of Juan Cano.* Cambridge, Massachusetts: M.I.T. Press.

Richardson A. and Parsons J.T. (1995) Signal transduction through integrins: A central role for focal adhesion kinase. *BioEssays* **17,** 229–236.

Rosen L.B., Ginty D.D., Weber M.J., and Greenberg M.E. (1994) Membrane depolarization and calcium influx stimulate MEK and MAP kinase via activation of Ras. *Neuron* **12,** 1207–1221.

Schlessinger J. and Ullrich A. (1992) Growth factor signaling by receptor tyrosine kinases. *Neuron* **9,** 383–391.

Schwab M.E. (1990) Myelin-associated inhibitors of neurite growth and regeneration in the CNS. *Trends Neurosci* **13,** 452–456.

Shea T.B. (1994) Delivery of anti-GAP-43 antibodies into neuroblastoma cells reduces growth cone size *Biochem Biophys Res Commun* **203,** 459–464.

Shea T.B. (1995) Inhibition of neuronal surface proteases decreases the requirement for GAP-43 in neurite outgrowth. *Dev Brain Res* **87,** 87–90.

Shea T.B. and Benowitz L.I. (1995) Inhibition of neurite outgrowth following intracellular delivery of anti-GAP-43 antibodies depends upon culture conditions and method of neurite induction. *J Neurosci Res* **41,** 347–354.

Shea T.B., Perrone-Bizzozero N.I., Beermann M.L., and Benowitz L.I. (1991) Phospholipid-mediated delivery of anti-GAP-43 antibodies into neuroblastoma cells prevents neuritogenesis. *J Neurosci* **11,** 1685–1690.

Slemmon J.R. and Martzen M.R. (1994) Neuromodulin (GAP-43) can regulate a calmodulin-dependent target *in vitro*. *Biochemistry* **33,** 5653–5660.

Stevens G.R., Zhang C., Berg M.M., Lambert M.P., Barber K., Cantallops I., Routtenberg A., and Klein W.L. (1996) CNS neuronal focal adhesion kinase forms clusters that co-localize with vinculin. *J Neurosci Res* **46,** 445–455.

Strittmatter S.M., Valenzuela D., Vartanian T., Sudo Y., Zuber M.X., and Fishman M.C. (1992) Growth cone transduction: G_0 and GAP-43. *J Cell Sci* **15 Sup,** 27–33.

Strittmatter S.M., Cannon S.C., Ross E.M., Higashijima T., and Fishman M.C. (1993) GAP-43 augments G-protein-coupled receptor transduction in Xenopus laevis oocytes. *Proc Natl Acad Sci USA* **90,** 5327–5331.

Strittmatter S.M., Fankhauser C., Huang P.L., Mashimo H., and Fishman M.C. (1995) Neuronal pathfinding is abnormal in mice lacking the neuronal growth cone protein GAP-43. *Cell* **80,** 445–452.

Strittmatter S.M., Igarashi M., and Fishman M.C. (1994) GAP-43 amino terminal peptides modulate growth cone morphology and neurite outgrowth. *J Neurosci* **14,** 5503–5513.

Strittmatter S.M., Valenzuela D., and Fishman M.C. (1994) An amino-terminal domain of the growth-associated protein GAP-43 mediates its effects on filopodial formation and cell spreading. *J Cell Sci* **107,** 195–204.

Strittmatter S.M., Valenzuela D., Kennedy T.E., Neer E.J., and Fishman M.C. (1990) G0 is a major growth cone protein subject to regulation by GAP-43. *Nature* **344,** 836–841.

Strittmatter S.M., Valenzuela D., Sudo Y., Linder M.E., and Fishman M.C. (1991) An intracellular guanine nucleotide release protein for G_0. *J Biol Chem* **266,** 22465–22471.

Strittmatter S.M., Vartanian T., and Fishman M.C. (1992) GAP-43 as a plasticity protein in neuronal form and repair. *J Neurobiol* **23,** 507–520.

Sudo Y., Valenzuela D., Becksickinger A.G., Fishman M.C., and Strittmatter S.M. (1992) Palmitoylation alters protein activity - blockade of G_0 stimulation by GAP-43. *EMBO J* **11,** 2095–2102.

Suidan H.S., Stone S.R., Hemmings B.A., and Monard D. (1992) Thrombin causes neurite retraction in neuronal cells through activation of cell surface receptors. *Neuron* **8,** 363–375.

Sydor A.M., Su A.L., Wang F.S., Xu A., and Jay D.G. (1996) Talin and vinculin play distinct roles in filopodial motility in the neuronal growth cone. *J Cell Biol* **134,** 1197–1207.

Teichmann-Weinberg A., Littauer U.Z., and Ginzburg I. (1988) The inhibition of neurite outgrowth in PC12 cells by tubulin antisense oligodeoxyribonucleotides. *Gene* **72,** 297–307.

Théodore L., Derossi D. Chassaing G., Llirbat B., Kubes M., Jordan P., Cheinweiss H., Godement P., and Prochiantz A. (1995) Intraneuronal delivery of protein kinase C pseudosubstrate leads to growth cone collapse. *J Neurosci* **15**: 7158–7167.

Van Dongen C.J., Zwiers H., De Graan P.N.E., and Gispen W.H. (1985) Modulation of the activity of purified phosphatidylinositol 4-phosphate kinase by phosphorylated and dephosphorylated B-50 protein. *Biochem Biophys Res Commun* **128**, 1219–1227.

Van Hooff C.O.M., Holthuis J.C.M., Oestreicher A.B., Boonstra J., De Graan P.N.E., and Gispen W.H. (1989) Nerve growth factor-induced changes in the intracellular localization of the protein kinase C substrate B-50 in pheochromocytoma PC12 cells. *J Cell Biol* **108**, 1115–1125.

Van Hooff C.O.M., De Graan P.N.E., Oestreicher A.B., and Gispen W.H. (1988) B-50 phosphorylation and polyphosphoinositide metabolism in nerve growth cone membranes. *J Neurosci* **8**, 1789–1795.

VanBerkum M.F.A. and Goodman C.S. (1995) Targeted disruption of Ca^{2+}-calmodulin signaling in Drosophila growth cones leads to stalls in axon extension and errors in axon guidance. *Neuron* **14**, 43–56.

Varnum-Finney B. and Reichardt L.F. (1994) Vinculin-deficient PC12 cell lines extend unstable lamellipodia and filopodia and have a reduced rate of neurite outgrowth. *J Cell Biol* **127**, 1071–1084.

Vaughn D.E. and Bjorkman P.J. (1996) The (Greek) key to structures of neural adhesion molecules. *Neuron* **16**, 261–273.

Verhaagen J., Hermens W.T.J.M.C., Oestreicher A.B., Gispen W.H., Rabkin S.D., Pfaff D.W., and Kaplitt M.G. (1994) Expression of the growth-associated protein B-50/GAP43 via a defective herpes-simplex virus vector results in profound morphological changes in non-neuronal cells. *Mol Brain Res* **26**, 26–36.

Widmer F. and Caroni P. (1993) Phosphorylation-site mutagenesis of the growth-associated protein GAP-43 modulates its effects on cell spreading and morphology. *J Cell Biol* **120**, 503–512.

Williams E.J., Furness J., Walsh F.S., and Doherty P. (1994) Activation of the FGF receptor underlies neurite outgrowth stimulated by L1, N-CAM, and N-cadherin. *Neuron* **13**, 583–594.

Wu D.Y., Wang L.C., Mason C.A., and Goldberg D.J. (1996) Association of β1 integrin with phosphotyrosine in growth cone filopodia. *J Neurosci* **16**, 1470–1478.

Yankner B.A., Benowitz L.I., Villa-Komaroff L., and Neve R.L. (1990) Transfection of PC12 cells with the human GAP-43 gene: effects on neurite outgrowth and regeneration. *Mol Brain Res* **7**, 39–44.

Zinn K. (1993) Drosophila protein tyrosine phosphatases. *Semin Cell Biol* **4**, 397–401.

Zuber M.X., Goodman D.W., Karns L.R., and Fishman M.C. (1989) The neuronal growth-associated protein GAP-43 induces filopodia in non-neuronal cells. *Science* **244**, 1193–1195.

Zuber M.X., Strittmatter S.M., and Fishman M.C. (1989) A membrane-targeting signal in the amino terminus of the neuronal protein GAP-43. *Nature* **341**, 345–348.

7

AVIAN HIPPOCAMPUS AS A MODEL TO STUDY SPATIAL ORIENTATION-RELATED SYNAPTIC PLASTICITY

Andrzej Wieraszko*

Department of Biology/Program in Neuroscience
The College of Staten Island/CUNY
2800 Victory Boulevard
Staten Island, New York, New York 10314

1. INTRODUCTION

One of the major goals facing current research in cognitive neuroscience is to identify the neuronal mechanisms that mediate complex cognitive processing (Polser et al.,1991). Ultimately, the goal is to understand the neural basis of human cognition. Human cognition cannot be systematically investigated since, for ethical reasons, one can only manipulate the human nervous system in a limited number of ways. However, several cognitive processes, such as many forms of learning and memory, are common to many vertebrate species including humans and the fundamental mechanisms mediating these phenomena can potentially be studied in a variety of different species (Spetch et al., 1993). Although the molecular processes responsible for memory in the brain are not yet clear, it is reasonable to assume that they are strongly related to synaptic plasticity. For the purpose of this review synaptic plasticity will be defined as changes in the efficiency of synaptic connections following periods of enhanced, or reduced activity. Such changes were thought to occur in the peripheral nervous system and in spinal cord (Larrabee et al., 1938; DelCastillo et al.,1954; Eccles et al., 1951; Magleby,1973; Mallart and Martin,1967), however the duration of this kind of facilitation was too short to be considered as a model for memory in higher CNS regions. The situation changed in the early seventies when Bliss and Lomo described the phenomenon of Long-Term Potentiation (LTP) (Bliss and Lomo, 1973). They showed that brief, high frequency activation of one of the hippocampal pathways called the perforant path could evoke a stable increase in the sy-

* Tel: (718) 982-3941 Fax: (718) 982-3944 E-mail: Wieraszko@Postbox.CSI.CUNY.Edu

Molecular and Cellular Mechanisms of Neuronal Plasticity, edited by Ehrlich. Plenum Press, New York, 1998.

naptic efficiency of the stimulated synapses. Since its discovery LTP has been observed in several brain areas (Racine and Kairiss, 1987), but the most prominent LTP had been recorded so far in mammalian hippocampus and the vast majority of the research devoted to the synaptic plasticity focused on this structure. At present LTP is highly favored candidate for a cellular mnemonic device in the vertebrate brain (Madison et al., 1991; Morris et al., 1991) and an increasing body of the evidence indicates that it may be specifically related to the type of learning which involves spatial orientation (Morris et al., 1986; Morris et al., 1991).

2. MECHANISMS OF LTP

According to current views (Larson et al., 1991) the process of LTP can be divided into three phases: triggering, expression and maintenance. However, while "triggering" is, in general described as an initial sequence of events which induce LTP, "expression" and "maintenance" are rarely distinguished and defining them precisely remains controversial (Larson et al., 1991). As some mechanisms involved in triggering of LTP can also be responsible for its maintenance, the division of LTP into these two phases is made only for the purpose of the clarity of our discussion. The triggering of LTP in CA1 of mammalian hippocampus critically depends on an increase in the intracellular Ca^{+2} concentration (Lynch et al., 1983). That increase can be achieved in two ways: 1) influx of Ca^{+2} from the outside of the cell, or 2) by the liberation of Ca^{+2} from the internal stores. Considering the first possibility, at least two forms of LTP have been identified. LTP, observed for the first time in the hippocampal perforant path (Bliss and Lomo,1973) depends on activation of the N-methyl-D-aspartate (NMDA) subclass of glutamate receptors (Collingridge, 1989). The ionophore operated by the NMDA receptor is blocked by Mg^{+2} in a voltage-dependent manner (Meyer and Westbrook, 1987). The Mg^{+2} block can be removed by either tetanic stimulation (Hanse and Gustafsson, 1992) which results in LTP, or perfusion of the slices in a Mg^{+2}-free medium (Wieraszko and Seyfried, 1993; Avoli et al., 1988). The open NMDA channel permits Na^{+} and Ca^{+2} to enter the cell (Meyer and Westbrook, 1987). The NMDA receptor-dependent form of LTP is generally considered, as operating according to the Hebbian principle: two inputs to the same cell should be activated in order to obtain permanent synaptic facilitation. One of obvious situation requiring such a mechanism is that of spatial learning. The interest in the NMDA-dependent form of LTP has to some extent overshadowed the fact that in several synapses the induction of LTP does not depend on the NMDA-receptor (Johnston et al., 1992). This form of LTP has been described at the mossy fiber-CA3 pyramidal cell synapses (Kauer and Nicoll, 1988; Harris and Cotman, 1986; Katsuki et al., 1991) and subsequently observed in the kitten visual cortex (Komatsu et al., 1991), hippocampal lateral perforant path (Bramham et al., 1991) and in hippocampal Schaffer collaterals after administration of specific, very high-frequency tetanic stimulation (Grover and Teyler, 1995). Recently we have described an NMDA-receptor independent form of LTP in the non-homing pigeon hippocampus (Wieraszko and Ball, 1993). It has been proposed that metabotropic, glutamate receptors (Isumu et al., 1990; Sugiyama et al., 1990; Aronica et al., 1991) or other glutamate, NMDA-independent, voltage-regulated Ca^{+2} channels, may be involved in this form of LTP in mammalian brain (Aniksztejn and Ben-Ari, 1991; Regehr et al., 1989; Chen and Huang, 1992; Isumu et al., 1990; Sugiyama et al., 1990). LTP triggered by elevation of extracellular Ca^{+2} (Turner and Baimbridge, 1982), or by a transient block of K^{+} channels (Aniksztejn and Ben-Ari Y, 1991), while not blocked by NMDA-receptor antagonists, was prevented by antagonists of voltage-dependent Ca^{+2} channels. It has been recently confirmed that voltage-dependent Ca^{+2} channels can, besides NMDA-receptor coupled channels, provide the necessary Ca^{+2} influx required for LTP in the mammalian hippocampus (Kullmann et al., 1992). The second way to elevate intracellular Ca^{+2} concentration is by triphosphoinositol (IP3) induced mobilization of Ca^{+2} from the internal stores (Berridge and Irvine, 1989). IP3 is an intermediate of the phosphatidylinositol signalling pathway which is involved in the regulation of synaptic plasticity (Berridge and Irvine,1989). The initial step in this pathway involves receptor-mediated activation of phospholipase C resulting in hydrolysis of phosphatidylinositol 4,5-biphosphate (PIP2) to IP3 and diacylglycerol (DG). The phosphoinositol pathway is influenced by glutamate (Nicoletti et al., 1986). Moreover, phosphoinositol metabolism (Bar et al.,1984; Lynch et al., 1988) and the sensitivity of phosphoinositol pathway to glutamate (Aronica et al., 1991) changed significantly following the induction of LTP. In several systems the response of the phosphoinositol pathway to glutamate depends on activation of the NMDA receptors (Aronica et al., 1991; Sladeczek et al.,1985; Wroblewski et al., 1986; Nicoletti et al., 1987). Most relevant to our considerations, however, are data showing potentiation of phosphoinositol pathway in the hippocampal slices obtained from rats trained in a spatial orientation test (Nicoletti et al., 1988). The effect was specific for spatial orientation, as training in the avoidance test had no effect (Nicoletti et al., 1988). One of the consequences of intracellular Ca^{+2} elevation is an intracellular protein phosphorylation, a potent mechanism used to regulate neuronal activity, which is also involved in LTP (Pasinelli et al., 1995). The involvement of protein phosphorylation in mammalian LTP has been recently confirmed in several laboratories (Akers and Routtenberg, 1985; Muller et al., 1988a; Reyman et al., 1990). A general conclusion emerges that while antagonists of the key enzyme suggested to participate in LTP, protein kinase C prevent induction of LTP, they do not influence already established LTP (Reyman et al., 1990; Muller et al., 1992; Tsien et al., 1990). Kinase C is normally present in the cytoplasm of presynaptic terminals (Kase et al., 1987), implying a presynaptic action in LTP (Davies et al., 1989 ; Recasens, 1995). Such LTP inducing factors, like electrical stimulation (Colley and Routtenberg, 1993), higher Ca^{+2} concentration (Turner et al., 1982), and PIP2 breakdown (Berridge and Irvine1989) can activate PKC by translocating that enzyme from the cytoplasm to a membrane fraction (Recasens, 1995). The products of PIP2 hydrolysis diacylglycerol (DG) and the already mentioned IP3, can both influence PKC activity. DG acts through direct interaction with the enzyme, while IP3 acts indirectly by increasing intracellular calcium concentration (Berridge and Irvine, 1989). Although sometimes contradictory results demonstrating involvement of protein kinase C in triggering and maintenance of LTP were reported (Malinow et al.,1988; Hu et al., 1987; Klann et al., 1993) activation of this enzyme is the most consistently observed biochemical event correlated with LTP.

At present two very different types of hypothesis concerning the long-lasting changes underlying LTP exist. The first hypothesis, involving a presynaptic mechanism favors an increase in transmitter release (Dolphin et al., 1982; Malinow and Tsien, 1990; Bliss et al., 1991; Ghijsen and LopesDaSilva, 1991; Malgaroli et al., 1992). Results from different laboratories showed a persistent increase in the release of labeled (Dolphin et al., 1982) and endogenous (Ghijsen and LopesDaSilva, 1991) glutamate following induction of LTP in hippocampal pathways. The presynaptic mechanism of LTP may also include a reduction in efficiency of high affinity, presynaptic glutamate uptake (Wieraszko 1983; Goh et al., 1986). As a result, more glutamate will remain in the synaptic cleft (in addition to a greater release discussed above) resulting in greater availability to postsynaptic receptors which could produce a potentiated trans-synaptic response. In the last few years Bliss and collaborators favored the idea that arachidonic acid is a retrograde messenger in LTP (Klann et al., 1993; Bliss et al., 1991; Williams et al., 1991). Arachidonic acid, released from the postsynaptic neuron during the induction of LTP would diffuse to presynaptic part initiating changes leading to permanent increase in neurotransmitter release observed in LTP (Dolphin et al., 1982; Ghijsen and LopesDaSilva, 1991). Although the role of arachidonic acid, as a retrograde messenger in LTP is attractive, new evidence indicate, that another molecule, nitric oxide (NO) may also be involved in LTP as a retrograde messenger (Haley and Schuman, 1994). The second hypothesis suggests an increase in postsynaptic response (Worley et al., 1986). An attractive hypothesis proposed by Baudry and Lynch (Lynch and Baudry, 1984) suggesting an increase in the number of glutamate receptors was not confirmed by others (Reyman et al., 1990; Lynch et al., 1985), although changes in the sensitivity of postsynaptic, glutamergic receptors do occur. Slowly developing (minutes) increase in the sensitivity of AMPA glutamate receptors was reported (Kauer et al., 1988; Muller and Lynch, 1988). However, as discussed by Collingridge and others (Bashir et al., 1991) the issue of changes in the postsynaptic receptors during LTP is still far from clear (Baudry and Davis,1991). Quite recently, two important papers with contradictory results appeared (Manabe and Nicoll, 1994; Stevens and Wang, 1994). Using new approaches, the authors reached opposing conclusions and pre- versus postsynaptic locus of LTP maintenance remains unsolved problem. All these results strongly indicate that LTP is not a unitary phenomenon, but a family of mechanisms regulating synaptic efficiency. These mechanisms may be expressed together or separately in neuronal pathways and their involvement in specific types of memory may differ.

3. HIPPOCAMPUS AND SPATIAL MEMORY

The hippocampus is regarded as an integral component of a number of memory systems (Sherry and Schacter 1987). Behavioral investigation of hippocampal structure and cognitive function tend to emphasize either spatial memory acquisition (Nadel, 1991; Squire and Zola-Morgan,1991), or the role of non-spatial aspects of memory, such as declarative memory in humans (Squire and Zola-Morgan,1991). Although there is a substantial amount of variation in hippocampal anatomy, physiology and neurochemistry between different species, the role of the hippocampus in spatial orientation applies across all vertebrate species (Nadel, 1991; Sherry and Schacter 1987; Bingman, 1992; O'Keefe and Nadel, 1978; Rolls,1991). Hence, comparative investigations of the biochemical and electrophysiological properties of the hippocampus among vertebrate species can provide fundamental insight into the neural basis of spatial orientation and memory in different animals including humans (Eichenbaum, 1992; Squire and Backercave,1991). From an evolutionary view of cognitive function (Bingman, 1992) the results from comparative studies point to spatial memory acquisition as a basic function of the hippocampus and one likely precursor of other hippocampal function. Several researchers share the view that at least some forms of LTP could participate in the formation of spatial memory. It is well documented that damage to the hippocampus (Olton and Papas, 1979; Olton et al., 1979; Olton, 1983), or pharmacological inhibition of LTP disrupts new spatial learning in rats and other mammals (Morris et al., 1991). The rat trained to use spatial clues to find a platform hidden under water, failed the test if its hippocampi were perfused with the drug blocking the induction of LTP. The effect was specific for spatial orientation, as the properties of visual discrimination remained intact (Morris et al., 1991). The relation between LTP and spatial orientation was recently confirmed by the results of the experiments with mutant mice lacking the gene for the alpha-isoform of the calcium-calmodulin-dependent protein kinase II, one of several kinases, which are probably involved in LTP (see below). These mice demonstrated impaired spatial learning (Silva et al., 1992a). Additionally, LTP was significantly reduced in the hippocampal slices obtained from them, even though the synaptic function was normal (Silva et al., 1992b). The results however, were not conclusive, as the genetic modifications influenced these animals throughout their life including embryonic development. Therefore, researchers could not be sure that memory deficiency in the altered mice were not the product of development deficiencies, or changes outside the hippocampus. However, new more precise experiments performed in the same laboratory gave further support for the idea of the involvement of LTP in the mechanism of spatial orientation (Rotenberg et al., 1996). Through genetic manipulations a strain of mice was produced which expressed Ca^{+2}-independent (permanently active) form of CaMKII. Although these animals showed normal LTP triggered by 100 Hz stimulation, another form of LTP, inducible by lower frequency (5–10 Hz range) was missing (Rotenberg et al., 1996). This deficit was correlated with severe impairment of spatial orientation and loss of the function of the hippocampal place cells (Rotenberg et al., 1996). The place cells are individual hippocampal neurons which fire in response to the placement of the animal in a familiar environment (O'Keefe and Dostrovsky, 1971). The new place cells may be formed in minutes in new environment (Wilson and McNaughton, 1993) and are stable for weeks or even months (Bostock et al., 1991). These data provide an important functional correlation between specific form of LTP and function of the place cells. The notion about the involvement of LTP in spatial orientation was further supported by the experiments performed on mice with mutations in gene responsible for tyrosine kinase, another enzyme involved in LTP (Grant et al., 1992; O'dell et al., 1991). These animals demonstrated reduced LTP and impaired ability to learn in spatial orientation test (Grant et al., 1992). Other forms of synaptic plasticity represented by paired-pulse facilitation and postetanic potentiation remained intact (Grant et al., 1992). Furthermore, the subcellular redistribution of hippocampal kinase C was observed following the spatial-orientation training procedure (VanderZee et al., 1992; Olds et al., 1990) and poor spatial orientation was correlated with reduced kinase C activity in the hippocampus of DBA/2Ibg strain of mice (Wehner et al., 1990). Although described experiments support the idea that a number of kinases, including kinase C (Pasinelli et al., 1995) play a role in spatial learning, the evidence supporting the role of LTP in learning and memory, especially in spatial orientation is still largely indirect. More direct evidence for participation of NMDA receptors in spatial learning was provided recently by Tonegava's laboratory. Using genetic manipulation technique (Tsien et al., 1996a) he and his collaborators deleted postdevelopmentally the gene for NMDA receptor specifically from CA1 hippocampal pyramidal neurons. These animals did not express any form of LTP and were unable to learn a spatial orientation task (Tsien et al., 1996b; McHugh et al., 1996), although nonspatial learning remained unimpaired. These experiments specifically point on the role of hippocampal, NMDA-dependent form of LTP in spatial orientation.

4. SYNAPTIC PLASTICITY IN THE AVIAN HIPPOCAMPUS

Recent advance in molecular biology and electrophysiology created a very fruitful environment to further our knowledge about possible relation between LTP and spatial memory. However, the vast majority of this research uses traditional approach. The function of the structure modified through pharmacological, electrophysiological or genetic manipulations is compared with the function of the control, unimpaired structure. This type of approach may be especially detrimental for the research on spatial memory, which evolved and specialized through the interaction of the nervous system with specific environment. Research on spatial memory carried on animals bred for generations in the laboratory can pose a basic question: is the mechanism of spatial memory observed in laboratory animals related to the memory which evolved in the natural environment? To address this problem Shettleworth introduced a term of "inverse neuropsychology" which would stress characterization and comparison of the memory function enhanced by specific behavior with normal function (Shettleworth, 1955). This approach requires the comparison of two closely related species which occupy two different ecological niches. Birds can provide an excellent material for this type of research, as some of them developed an excellent spatial orientation.

Several data indicate that avian hippocampus participates in two types of spatial orientation-related behavior: food storing behavior (Krebs et al., 1989; Clayton and Krebs, 1995) and homing behavior (Bingman et al., 1987; Bingman et al., 1988). Hippocampal lesions during field experiments interfere with spatial learning and navigational map acquisition (Bingman et al., 1988; Bingman et al., 1989). The volume of the avian hippocampus is related to the life style of the animal. Differences in the size of the hippocampal region between food storing birds such as the corvids are significant and are attributable to natural selection. For example, significant differences exist between Clark's nutcracker, which lives at high elevations and caches tens of thousands of seeds in the Fall for Winter consumption and the very similar scrub jay of the Southwest which never has to depend on stored food (Sherry et al., 1989; Krebs et al., 1989; Krebs, 1990; Sherry et al., 1993). Recently reported results demonstrated significant enlargement of the hippocampus in birds with highly specialized behavior related to spatial orientation like food storing in chickadees and related species (Sherry et al., 1989; Krebs et al., 1989; Clayton, 1996; Clayton, 1995a; Clayton and Krebs, 1994). The general conclusion emerges that while experience of food storing and retrieving results in an increase in hippocampal volume, absence of experience leads to attrition of hippocampal volume (Clayton, 1995b; Clayton, 1996; Clayton and Krebs 1994). These volumetric changes occur rapidly (days) and are accompanied by an increase and decrease in the cell number, respectively (Clayton, 1995; Clayton, 1996; Clayton and Krebs, 1994). The effect is observed in the hippocampus only and suggests that while experience can trigger a recruitment of neurons, lack of experience may cause their loss (Clayton and Krebs, 1994). Moreover, growth of the hippocampus, triggered by some aspects of memory for retrieving food appeared in food-storing birds only (Clayton, 1995). It implies that an increase in hippocampal volume is a specific, anatomical specialization of the brain of food-storing birds (Clayton, 1995). This morphological specialization is genetically regulated, as it has been observed in young food storers even before onset of storing behavior (Healy et al., 1991). This anatomical plasticity is correlated with biochemical specialization documented by the presence of subpopulation of large, calbindin-like immunopositive neurons in the hippocampus of food-storing birds (Montagnese et al., 1993). Calbindin, a calcium binding protein, is supposed to regulate intracellular Ca^{2+} concentration during changes associated with memory (Persechini et al., 1989). One can speculate that larger population of calbindin-like immunopositive cells can enhance learning ability of these birds. The memory in food-storing birds can be also enhanced by higher concentration of nitric oxide synthase, an enzyme responsible for biosynthesis of nitric oxide (Szekely and Krebs, 1993). There is a strong evidence for involvement of NO in mammalian LTP (Haley and Schulman, 1994) and higher level of nitric oxide synthase in food-storing birds can suggest specific role of NO in spatial orientation. Specific plasticity of the avian hippocampus is not limited to food-storing experience and can be also observed in homing pigeons (Rehkamper et al., 1988). One basic difference between homing and non-homing pigeons stems from their different ability to learn spatial orientation tasks. While a homing pigeon can be easily trained to return to its loft, non-homing pigeons are unable to master this task. Rekhamper and collaborators (Rekhamper et al.,1988) have shown that this behavioral difference is correlated with a brain difference. Homing pigeons have larger hippocampus than non-homing pigeons. The artificial selection of homing/racing ability has led to a larger hippocampus in homing pigeon breeds than in non-homing breeds (Rehkamper, et al., 1988). All these results support the hypothesis that specific, spatial-orientation related abilities/experience may lead to modification of morphology and/or physiology of avian hippocampus. Indeed, in support for this hypothesis, we reported NMDA receptor-dependent and independent forms of LTP in the hippocampus of homing and non-homing pigeons, respectively (Shapiro and Wieraszko, 1996). As a behavioral correlate to these findings Riters and Bingman (Riters and Bingman, 1994), using homing breed, have shown in open field tests that MK-801 (an inhibitor of NMDA receptor-operated channel) interfered with homing behavior and spatial memory, although it did not effect position/color discrimination tasks in an operant chamber. Laboratory tests evaluating the influence of NMDA receptor antagonists on the behavior of homing and non-homing pigeons gave inconclusive results (Meehan, 1996; Meehan and Wieraszko, 1996). The interpretation of these results was hampered by the lack of differences in performance between control homing and non-homing pigeons. It is possible that the test used in this study was too simple to differentiate between spatial ability of these two breeds of pigeons. Possibly, as demonstrated by Riters and Bingman (Riters and Bingman,1994), the flight is required to expose the differences in the pigeon spatial ability. It seems reasonable to assume, that the enlargement of the hippocampus in homing pigeons is correlated with some physiological adaptations, perhaps reflected in the properties of synaptic plasticity related to spatial orientation. One of the basic questions in studying LTP in the mammalian hippocampus is a hope of understanding the molecular mechanism of memory. Considering the suggested role of LTP in the spatial memory (Morris et al., 1991), and the role of the hippocampus in mammalian (Morris et al., 1982) and avian spatial orientation (Bingman et al., 1987; Bingman et al., 1988; Macphail and Good, 1991) the study of the avian hippocampus may be very helpful, as anatomical (Casini et al., 1986; Krayniak and Siegel, 1978), ontogenetic (Kallen, 1962) and embryological (Benowitz, 1980) data demonstrate that the hippocampal complex in birds is homologous to the hippocampal formation in mammals. Additionally, hippocampal lesions result in a similar impairment in orientation in birds (Bingman et al., 1987; Bingman 1988) and rodents (Morris et al., 1982; Olton, 1983). This further supports a hypothesis of homology between avian and mammalian hippocampus and implies a similarity of function. Thus studying LTP in the avian hippocampus can give us additional information about a possible role of this process in the spatial orientation in birds and mammals.

4.1. Preparation of Avian Hippocampal Slices

One of the most fruitful approach in studying synaptic plasticity in the mammalian hippocampus was the technique of hippocampal slices. Mammalian hippocampus is a highly laminated structure which preserves its major connections when sliced into transverse sections (Teyler 1980). The stimulating and recording electrodes may be than relatively easy localized on specific pathways and cell layers under dissecting microscope. However morphological markers clearly visible in mammalian hippocampus are missing in the avian hippocampus. Moreover the localization and the size of avian hippocampus is

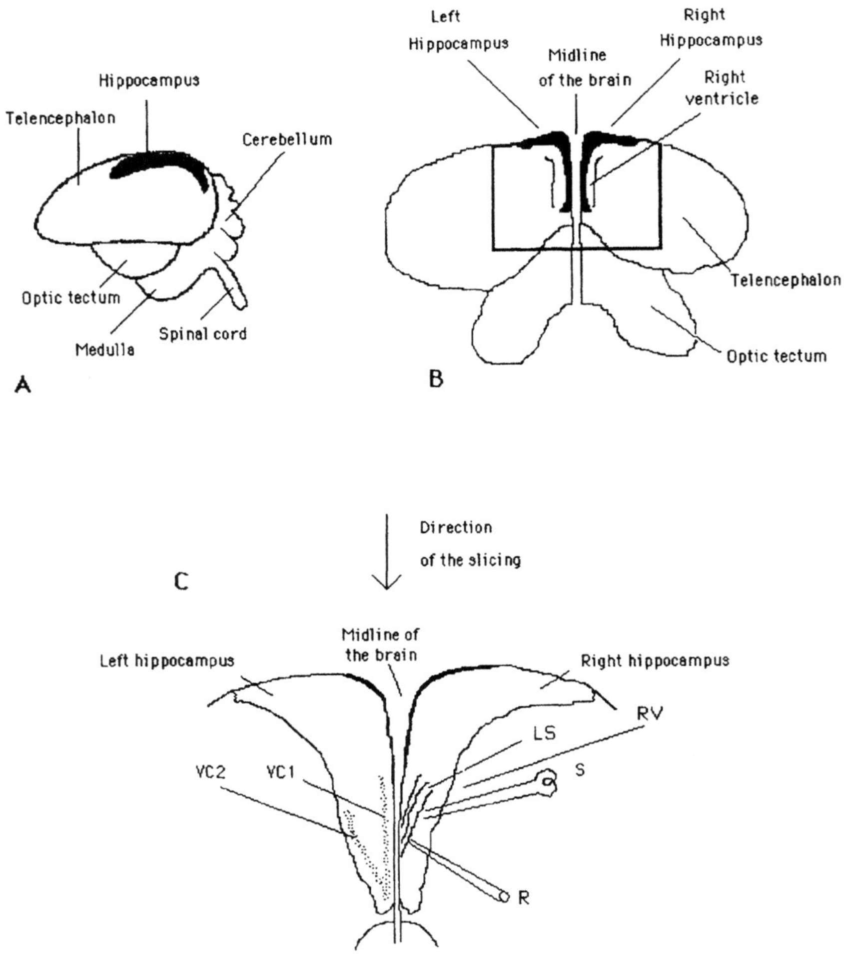

Figure 1. Diagram demonstrating the location of the hippocampus in the pigeon brain; A - sagittal section of the pigeon brain; note the characteristic location of the hippocampus on the surface of the brain; B - transverse section of the pigeon brain. Thick, straight lines indicate the cuts made to trim off posterior, anterior and lateral parts of the brain to leave the block of the tissue containing intact hippocampus. This part was glued to the stage of the vibrotome and sliced; C - hippocampal slice dissected out from the slice of the brain. Left part indicates the location of the cell layers (VC1 and VC2) as described by Krebs et al., (1991). Right part shows laminar structures (LS) visible in Fig 2A and the location of the stimulating (S) and recording (R) electrodes in relation to LS; RV - right ventricle. Reproduced from Shapiro and Wieraszko, 1996; with kind permission from Elsevier Science - NL, Sara Burgerhartstraat 25, 1055 KV Amsterdam, The Netherlands.

different than mammalian hippocampus. Therefore we modified the procedure employed by us previously for mouse hippocampal slice preparation (Wieraszko and Seyfried, 1993) to prepare avian hippocampal slices. Following removal, the brain was trimmed (5 mm off the left, right and anterior parts) and the cerebellum was cut off. The brain was glued to the stage of the vibrotome and submerged in ice-cold Ringer's solution. As the avian hippocampus is too small to be dissected out (Fig 1), the whole brain had to be sliced. The slices (400 μ thick) were placed in the incubation chamber and transferred to the recording chamber 1 hr later. In some cases the hippocampal complex could be separated from the rest of the slice, but no difference was observed in electrophysiological recordings whether the hippocampus was separated or not. Our technique to slice the whole brain first and to dissect the hippocampal area later yielded a very high rate of slices demonstrating evoked population spikes. These population spikes could be recorded several hours following tissue preparation. It indicates that survival rate of mammalian and avian hippocampal slices is similar, although no systematic comparisons have been made.

4.2. Synaptic Plasticity in the Hippocampus of Song Sparrow

We initialized our research on the avian hippocampal slices few years ago choosing a common American North American passerine (songbird) species, the song sparrow (Melospiza melodia). The distance between the stimulating and recording electrodes localized on the avian hippocampus never exceed 2–3 mm. A typical potential recorded from the song sparrow hippocampal slices was much smaller and more irregular than one recorded from the mammalian hippocampus, although the shape was very similar (Fig 2A). This could be due to the lack of laminar organization of the song sparrow's hippocampus and smaller number of synchronously activated neurons. The potential often consisted of two components, of which only the second one with longer latency was calcium-dependent, indicating its synaptic origin (Fig 2A). The song sparrow's hippocampus shows some forms of synaptic plasticity typical for mammalian hippocampus. One of these forms was paired-pulse facilitation induced by two pulses delivered in rapid succession. At shorter time intervals between two pulses (10–20ms) we observed an inhibition of the second response (Fig 2Ba). This depressive effect was latency-dependent, being the greatest at 10 ms and becoming smaller as the latency increased. The inhibition of the second response with a latency below 20 ms gives some clues concerning avian hippocampus anatomy. A similar pattern was observed in the rat hippocampus (Creager et al., 1980). Therefore, one can suggest that the anatomical connections that underlay the recurrent inhibitory loop in the mammalian hippocampus (Lee and Schubert, 1982) are also present in the avian hippocampus. Longer intervals between two pulses (20–40 ms) resulted in the facilitation of the second response (Fig 2Bb). At intervals over 50 ms no facilitation of the second response was observed (Fig 2Bc). The synaptic plasticity in the sparrow's hippocampus was not limited to paired-pulse facilitation. 20 Hz stimulation produced prolonged, although decremental facilitation of extracellular potentials in the sparrow's hippocampus (Fig 2C). Interestingly, the high-frequency stimulation of 100 Hz which was the most successful in inducing LTP in the mammalian hippocampus, was ineffective in sparrow's hippocampus.

4.3. Synaptic Plasticity in the Hippocampus of Homing and Non-Homing Pigeon

Although the very characteristic laminar structure observed under the light microscope in the mammalian hippocampus is not visible in the pigeon hippocampus, we were

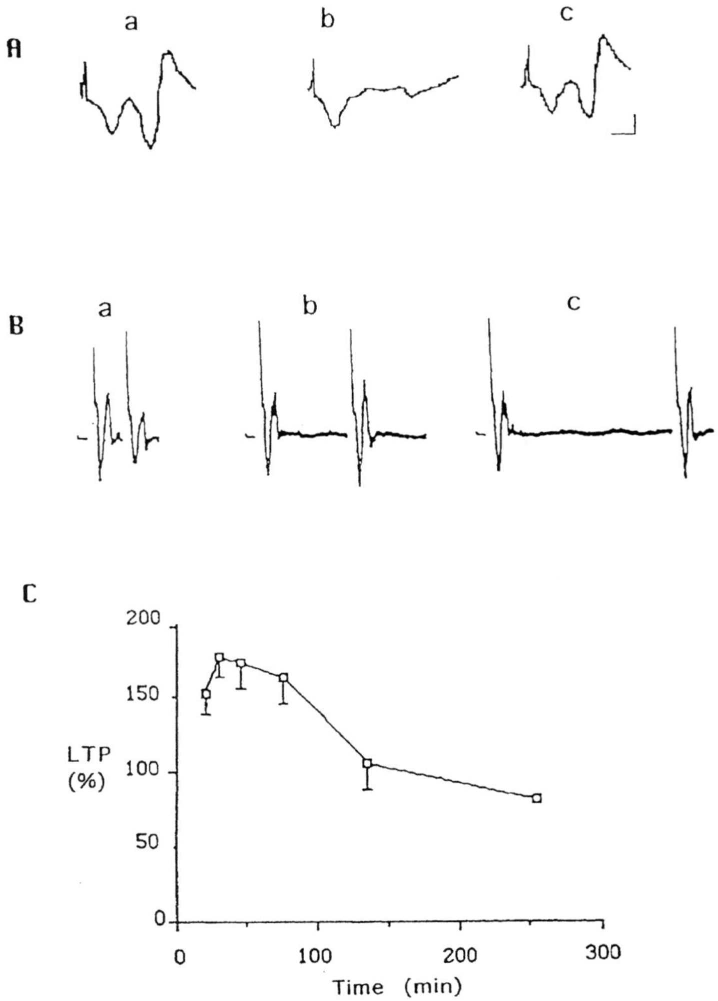

Figure 2. Evoked potentials recorded from the hippocampus of song sparrow; A - Calcium dependency of the evoked potentials: a - control potentials recorded in the presence of 3.1 mM Ca^{2+}; b - the potential recorded 20 minutes after initiating of the perfusion with Ca^{2+}-free Ringer solution; c - the potential recorded 30 minutes after addition of $CaCl_2$ to achieve original Ca^{2+} concentration. Note that the record consists of two potentials of which only the second one demonstrates Ca^{2+} dependency. The first potential represents compound action potential and is Ca^{2+}-independent. Calibration: 2 mS, 0.2 mV. B - Paired-pulse facilitation; the hippocampus was stimulated with twin pulses with 10ms (a), 30 ms (b) and 70 ms (c) latency. Calibration as in A. C - Prolonged facilitation of the potentials following three 1-s trains of pulses of 20 Hz, every 10 s. The potential, elevated immediately after high-frequency stimulation, declined steadily to reach the control, prestimulus value after about 2 hr. This figure represents the data collected from 9 slices (the last time point was calculated from two experiments only). The results are expressed as mean ± SEM. Reproduced from Wieraszko and Ball, 1993, with kind permission from Elsevier Science - NL, Sara Burgerhartstraat 25, 1055 KV Amsterdam, The Netherlands.

able to observe fiber-like tracts in the hippocampus of homing and non-homing pigeon in the location indicated on Fig 1C. These fiber tracts, which could be observed under the dissecting scope without any staining appear as whitish lines emanating from the medial areas of both left and right hippocampus (Fig 3). These structures, observed in hippocampus of homing and non-homing pigeons, correspond in location to the medial fiber tract of the homing pigeon hippocampus which was described on the basis of immunohistochemical staining for a number of neurotransmitters and neuropeptides (Erichsen et al., 1991; Krebs et al., 1991). The stimulating and recording electrodes were placed within areas of visible fiber path (Fig 1 C). As this fiber path was observed in the hippocampus of homing and non-homing pigeon, we could place the recording (Fig 1C; R) and stimulating (Fig 1C; S) electrodes in corresponding locations in different slices coming from different breeds of pigeons. The division of the avian hippocampus described recently (Erichsen et al., 1991; Krebs et al., 1991) indicates that the electrodes in our experiments were localized in ventromedial regions of the hippocampus ("VM" according to Erichsen et al., 1991; Krebs et al., 1991). It was demonstrated with Nissl staining that the concentration of cells in this area is arranged in the shape of the letter "V" (see VC1 and VC2 in Fig 1 C). Immunostaining techniques revealed the presence of several transmitter-related enzymes, including glutamate decarboxylase, scattered through hippocampal area. The fiber-like structures observed by us in the vicinity of "V" shaped cell layer may represent bundles of axons targeting cells arranged in vertical arm of the letter "V" (see V1 in Fig 1C). The visibility of these fibers under the dissecting microscope allows placement of the recording and stimulating electrodes in a reproducible manner in the hippocampal slices obtained from different pigeons. No differences were noted in the arrangement of these fibers in the hippocampal slices obtained from homing and non-homing pigeon. The population spikes evoked by stimulation of these axons were attenuated by CNQX which indicates that glutamate may be a neurotransmitter in this system (Fig 4). In slices prepared from homing pigeons, we also recorded intracellularly. The purpose of these intracellular experiments was to better characterize the physiological basis of the extracellular evoked responses. Following successful placement of the extracellular recording electrode, which resulted in the recording of extracellular field potentials (population spikes) in response to electrical stimulation, the intracellular recording electrode was positioned in the immediate vicinity of the extracellular recording electrode, and slowly advanced into the slice to impale cells. Impalements were monitored by observing the changes in potential on the oscilloscope (Fig 5). Impalement were considered successful if they recorded a rapid shift in potential of at least -40 mV, which could be held for at least 5 min. During these experiments, 9 cells met our criteria for successful intracellular recording, a resting membrane potential of at least -40V which could be recorded for at least 5 min. In these 9 cells the average resting membrane potential was -57 mV ± 12 mV, SD. During all of these recordings we looked for intracellular potentials in response to the extracellular stimulation. In five of the nine cells we recorded excitatory synaptic potentials which followed the extracellular stimulus at a short and constant latency. In two of these cells, the evoked excitatory potential was large enough to trigger an action potential. Figure 5 shows an example of a simultaneous intracellular and extracellular recording. We also observed spontaneous inhibitory synaptic potentials in three out of 9 cells. At the end of these experiments hyperpolarizing current pulses were delivered through the bridge circuit to eject the Lucifer Yellow and to stain the cell. During the entire session of intracellular recordings the extracellular recording and stimulation protocols were continuous. A map of the slice was drawn, and the position of the electrodes marked on the map. At the end of the recording session the slices were fixed in 4% paraformaldehyde in pigeon Ringer, and ob-

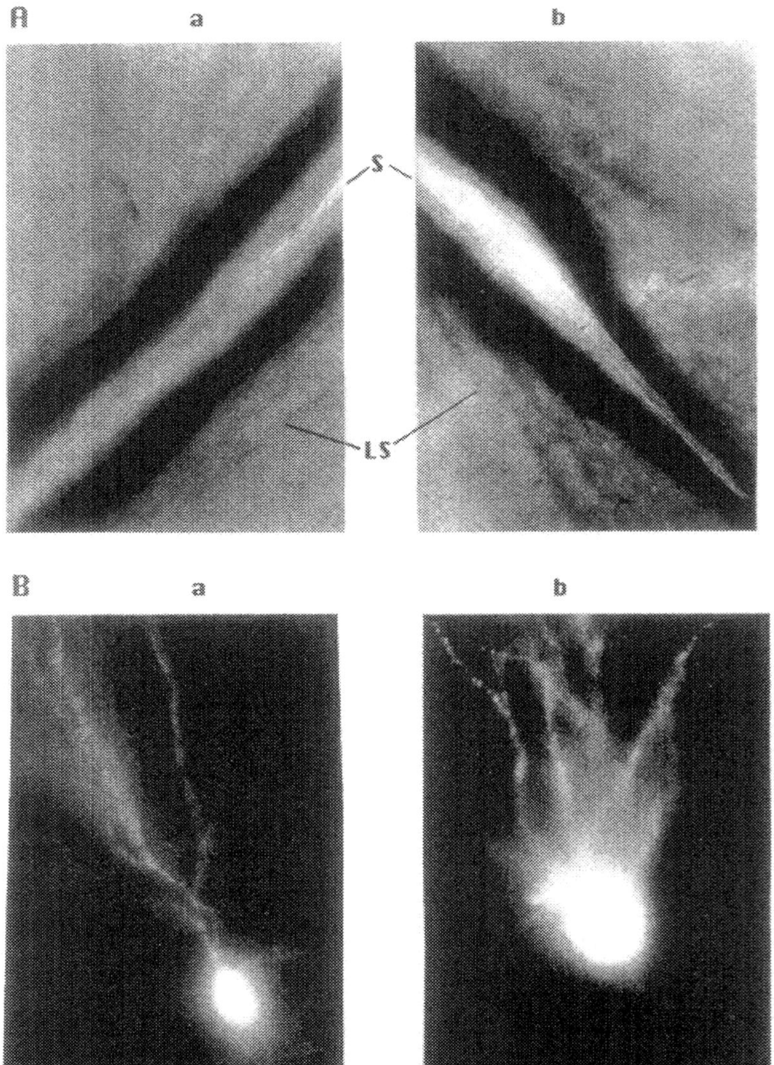

Figure 3. Photographs showing fibers and cells observed in the pigeon hippocampal slice. A - the picture of unstained fragment of homing (a) and non-homing (b) pigeon hippocampal slice as seen in the recording chamber under dissecting microscope (mag. 40x). Left and right hippocampi were not separated in this preparations. Note the laminar structures (LS) running in dorso-lateral direction clearly visible in left and right hippocampi of homing and non-homing pigeon. They focus toward the ventral direction and they were used as markers for reproducible location of the stimulating and recording electrodes. S - space between both hippocampi. B - fluorescent picture of homing pigeon hippocampal neurons filled with Lucifer Yellow (x 600); a - the cell with multiple dendrites concentrated at one region of the cell soma. The electrical response of this cell to extracellular stimulation is shown in Fig 5; b -the cell with the dendrites radially arranged around the cell soma.

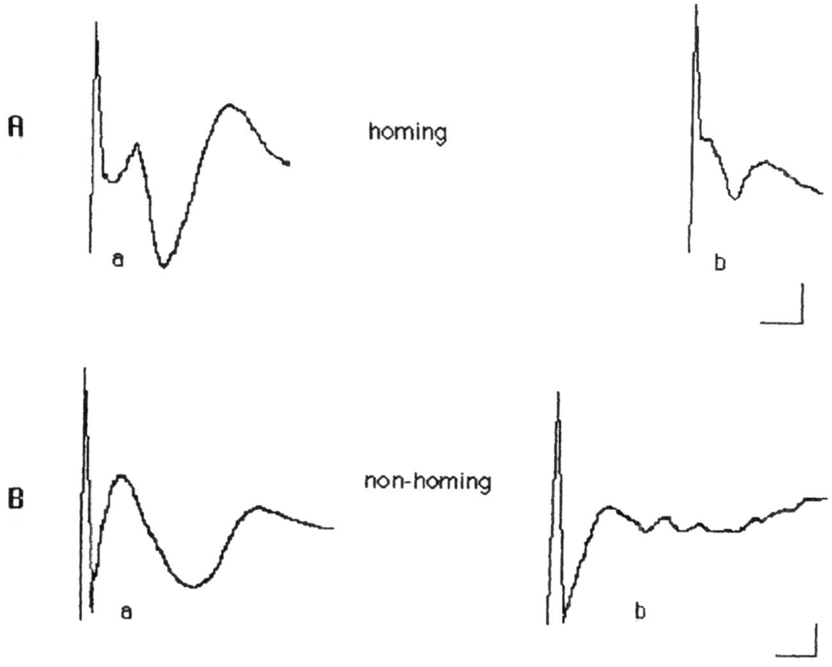

Figure 4. Inhibition of evoked population spikes recorded extracellularly from the hippocampus of homing (A) and non-homing (B) pigeons by CNQX; a and b represent the population spikes recorded 5 min before and 20 minutes after addition of 10 μM CNQX, respectively. Calibration: 2 mS; 0.2 mV. Reprinted from Shapiro and Wieraszko, 1996; with kind permission from Elsevier Science -- NL, Sara Burgerhartstraat 25, 1055 KV Amsterdam, The Netherlands.

Figure 5. Evoked responses recorded intracellularly (top trace) and extracellularly (lower trace) in the hippocampus of homing pigeon. The resting membrane potential of this cell was -50 mV. The intracellular recording and the large evoked response from this cell was held for over 50 min. The intracellular response shows a large excitatory synaptic potential which lasts for the duration of the figure with an action potential riding on it. The intracellular evoked response is well-correlated in time with the extracellularly recorded evoked field response (average of 10 responses). The scale bars represent 3 ms (time scale), 15 mV (top trace), 0.5 mV (lower trace). The morphology of this cell is shown in the fluorescence photomicrograph Figure 3. Reproduced from Shapiro and Wieraszko, 1996; with kind permission from Elsevier Science - NL, Sara Burgerhartstraat 25, 1055 KV Amsterdam, The Netherlands.

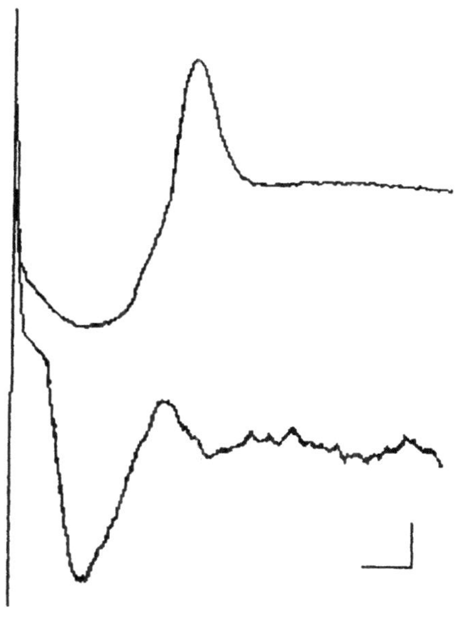

served and photographed using fluorescent microscopy. Our results showed that we could find cells which responded to extracellular stimulation with excitatory synaptic responses or with evoked action potentials. Not all cells responded to the stimulation. This result may be due to the possibility that the fiber pathways we stimulated are extremely restricted in their area of innervation. Alternatively, the efficacy of the synapses may be low, and potentiation may result in an increased number of cells responding to extracellular stimulation. We examined the morphology of cells filled with the fluorescent dye, Lucifer Yellow in four slices (6 cells). Figure 3 shows an example of a cell stained with Lucifer Yellow. The dendritic spread of the cell, as well as dendritic varicosities can be observed. Observation of the staining patterns of cells showed two types of morphology. Some cells demonstrated dendrites which were radially arranged around the cell soma (Fig 3Ba). Other cells appeared to be polarized, and had multiple dendrites concentrated at one region of the cell soma (Fig 3Bb). Both types of cells have been previously described in the hippocampus of homing pigeons (Erichsen et al., 1991; Krebs et al., 1991) with immunohistochemical staining techniques. One of the major goal of our research of pigeon hippocampus was to determine whether LTP-like phenomenon can be observed in this structure. While the attempt to induce LTP in the hippocampus of homing pigeon failed in the presence of competitive NMDA receptor antagonist APV (Fig 6C), it did not block induction of LTP in the hippocampus of non-homing pigeon (Fig 6D). This observation indicates that properties of LTP might be different in these two breeds of the pigeons. Further support for this conclusion comes from the experiments in which pigeon hippocampal slices were perfused with Mg^{2+}-free medium. While this treatment significantly facilitated population spike recorded from the slices obtained from homing pigeon (Fig 6A), it had no effect on the population spikes recorded from hippocampal slices obtained from non-homing pigeon (Fig 6B). The increase in the size of the population spikes recorded from homing pigeon hippocampus after perfusion with the Mg^{2+}-free solution was similar to that reported previously in rat (Avoli et al., 1988; Mody and Heinemann, 1987) and mouse (Wieraszko and Seyfried, 1993) hippocampal slices. Whether the increased potential resulting from low Mg^{2+} is identical to that resulting from LTP (Avoli et al., 1988; Mody and Heinemann, 1987; Wieraszko and Seyfried, 1993) is not yet clear, although both LTP and low-Mg^{2+}- induced facilitation are blocked by antagonists of NMDA receptor in the mouse hippocampus (Wieraszko and Seyfried, 1993). Besides activating the postsynaptic NMDA receptor, reduced extracellular Mg^{2+} may change the excitability of the stimulated fibers and enhance presynaptic influx of Ca^{2+} (Kuno and Takahashi, 1986). Moreover, a low-Mg^{2+} activated presynaptic NMDA receptor exists that may participate in amplification of the population spike through enhanced neurotransmitter release (Chapman and Bowker, 1987). However, regardless of the interpretation, these results indicate that the involvement of the NMDA receptors in the synaptic plasticity is different in the hippocampus of homing and non-homing pigeons. NMDA receptors were recently localized by Casto and Ball in the hippocampus of Zebra Finch and Starling (Ball and Casto, 1991). Assuming that localization of these receptors is similar in the pigeon hippocampus, the existence of NMDA-independent form of LTP in the hippocampus of non-homing pigeon seems to be unexpected. However, similar discrepancy in the properties of LTP had been observed in the rat hippocampal lateral and medial perforant path (Bramham et al., 1991). NMDA receptors had been localized in the synaptic zones of both of these pathways (Monagan et al., 1984). However, LTP induced in the lateral perforant path, in contrast to LTP induced in the medial perforant path was not blocked by NMDA receptor antagonists (Bramham et al., 1991). It has been suggested (Bramham et al., 1991) that NMDA receptors are localized extrasynaptically in the terminal zone of lateral perforant

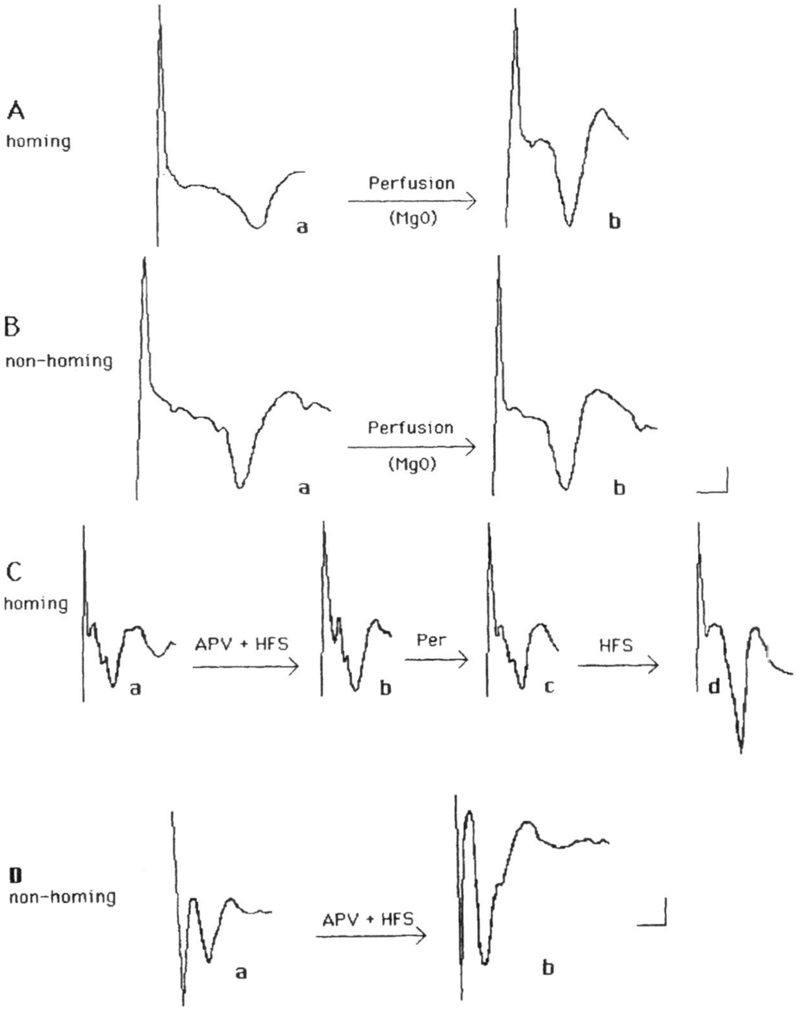

Figure 6. Factors influencing evoked population spikes recorded from pigeon hippocampus. The influence of reduced Mg^{2+} concentration on the population spikes recorded from the hippocampus of homing (A) and non-homing (B) pigeon hippocampus. Population spikes marked by a and b were recorded 5 minutes before and 30 minutes after start of the perfusion with Mg^{2+} -free medium, respectively. While the potential recorded from the hippocampus of homing pigeon was elevated by decrease in Mg^{2+} concentration (A), the potential recorded from the hippocampus of non-homing pigeon remained unchanged in these conditions (B). C- The influence of APV (60 μM) on the electrically-induced LTP in the hippocampus of homing (C) and non-homing (D) pigeon. Application of HFS to the hippocampus of homing pigeon incubated with APV had no influence on the magnitude of the evoked potential (compare C: a and b). However, following 20 min perfusion and washing APV away LTP could be induced by HFS in the same slice; c and d represent the population spikes after perfusion recorded 5 min before and 30 min after application of HFS, respectively. D - induction of LTP in the hippocampus of non-homing pigeon in the presence of 60 μM APV; a and b represent the population spikes recorded 5 min before and 30 min after application of HFS, respectively. Calibration A and B - 2 mS, 0.3 mV; C and D - 2 mS, 0.1 mV. Reproduced from Shapiro and Wieraszko, 1996; with kind permission from Elsevier Science - NL, Sara Burgerhartstraat 25, 1055 KV Amsterdam, The Netherlands.

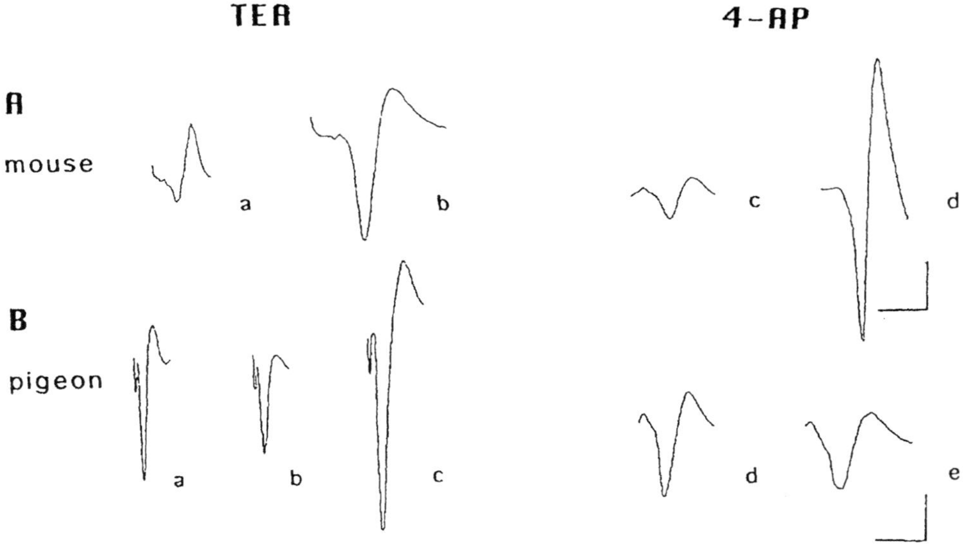

Figure 7. The influence of K+ channels blockers, tetraethylammonium (TEA) and 4-aminopyridine (4-AP) on the evoked potentials recorded from mouse (A) and pigeon (B) hippocampal slices. Bb represents one of 5 experiments in which the average magnitude of the potential was reduced by 8.6% (n.s., n = 5) 40 min after addition of TEA. Be represents one of 5 experiments in which the average amplitude of the potential was reduced by 28.2% (n.s., n=5) 40 minutes after application of 4-AP. The potentials recorded from mouse hippocampus before addition of TEA and 4-AP and 40 minutes after addition of TEA and 4-AP are marked as Aa, Ac and Ab, Ad, respectively. The potentials recorder from pigeon hippocampus before addition of TEA and 4-AP and 40 minutes after addition of TEA and 4-AP are marked as Ba, Bd and Bb, Be, respectively. Note that while both TEA (10 mM) and 4-AP (1μM) amplified potentials recorded from the mouse hippocampus, they attenuated potentials recorded from the pigeon hippocampus. Bc represents the potential triggered by 5 fold stronger stimulus. Calibration: A - 5 mS, 1 mV; B - 5 mS, 0.2 mV.

path. This subcellular localization would eliminate these receptors from participation in LTP. By analogy, NMDA receptor in the hippocampus of the homing pigeon could be located at the synaptic region, while in non-homing pigeon these receptors may be located extrasynaptically. The different location of these receptors could be a result of selective breeding of many generations. While the homing pigeon has been selected for its distinctive ability to find its home loft from distant locations, non-homing breeds are selected mainly for their appearance, size, and breeding ability. It is possible that the NMDA receptor-dependent form of LTP evolved under selective pressure directed towards obtaining birds with superior spatial orientation to secure fast return from distant release points. We do not know, if these two forms of LTP, observed by us in the hippocampus of homing and non-homing pigeons differ by the triggering mechanism only (NMDA receptor dependent versus NMDA receptor independent) or if the whole cascade of biochemical reactions involved in LTP is different in these two breeds of pigeons.

Differences in the modulation of the synaptic plasticity between mammalian and avian hippocampus are not limited to the function of the NMDA receptor. We have recently observed interesting properties of potassium channels in the pigeon hippocampus. The blockers of K^+ channels, tetraethylammonium (TEA) and 4-aminopyridine (4AP) amplify evoked responses recorded from the mouse hippocampal slices (Fig 7A) most likely by enhancing release of neurotransmitter (Thesleff, 1980). The amplification evoked by TEA is permanent and resembles electrically induced LTP (Aniksztejn and Ben-Ari, 1991). Surprisingly, neither TEA nor 4AP amplified evoked potential recorder from the pigeon hippocampus. To the contrary, both reduced the size of pigeon hippocampal potentials (Fig 7B). The attenuation induced by TEA was not due to unspecific "killing" effect, as the 5-fold increase in the strength of the stimulation significantly amplified evoked potential (Fig 7Bc). These preliminary experiments demonstrate the basic difference in mammalian and avian K^+ channels and imply novel mechanisms involved in the regulation of neurotransmitter release in the pigeon hippocampus.

5. CONCLUSIONS

There is currently enough experimental evidence to support the idea that the avian brain provides an excellent model to study spatial-orientation related synaptic plasticity. Although the focus of this review was on spatial-orientation related modifications in the avian hippocampus, the investigation of the avian song control system demonstrated that other structures, besides hippocampus express changes induced by a specific type of behavior (Nordeen and Nordeen 1990). Avian auditory experience regulates the size of song nuclei (Bernard et al., 1996; Brenowitz et al., 1991; Arai et al., 1989; Nottebohm, 1981), synaptic morphology (Hill and DeVoogd, 1991; DeVoogd et al., 1985) and incorporation of new neurons in song control nuclei (Nottebohm et al., 1994; Alvarez-Buylla et al., 1988). Recently a phenomenon of Long-Term Depression (LTD), originally described in mammalian cerebellum (Ito et al., 1982) and later observed in mammalian hippocampus (Dudek and Bear, 1992), was also recorded in the auditory pathway of a song bird (Chew et al., 1996). All these briefly mentioned characteristics of the song nuclei support the notion that the avian brain expresses different forms of synaptic plasticity providing unique opportunity to study the association between memory specialization and structural organization of the brain. Modification of the routine techniques used for mammalian hippocampal slices allows to maintain "alive" avian hippocampal slices in vitro and to follow their evoked synaptic activity for several hours. Laminar structures identified in some specific parts of pigeon hippocampus can be used as markers to place the electrodes in homologous locations in different hippocampal slices. Potentials recorded from these structures had glutamergic origin and expressed some forms of synaptic plasticity (paired-pulse facilitation and LTP) typical for mammalian hippocampus. While LTP in the hippocampus of non-homing pigeons (a poor spatial performer) does not require activation of the NMDA receptor, the LTP in the hippocampus of homing pigeon (good spatial performer) is NMDA receptor dependent. Our research suggests that anatomical adaptation, expressed as an enlarged hippocampus in homing pigeons, are also represented by modified mechanisms that mediate the triggering and maintenance of LTP. It is not known whether observed differences in the properties of LTP are genetically regulated or are induced by training procedures. An attractive and testable hypothesis may be suggested stating that these two differentially triggered forms of LTP which probably evolved under selective pressure of breeding, are related to different mechanisms of LTP and correlate with different spatial ability of these two breeds of pigeons.

ACKNOWLEDGMENTS

This work was supported by a 1996 Presidential Research Fellowship and by PSC CUNY grant #668219.

REFERENCES

Akers R.F., and Routtenberg A. (1985) Protein kinase C phosphorylates a 47 Mr protein (F1) directly related to synaptic plasticity. Brain Research 334, 147–151.

Alvarez-Buylla A., Kirn J., and Nottebohm F. (1990) Birth of projection neurons in adult avian brain may be related to perceptual or motor learning. Science 249, 1444–1446.

Anikszteјn L., and Ben-Ari Y. (1991) Novel form of long-term potentiation produced by a K^{-}channel blocker in the hippocampus. Nature 349, 67–69.

Arai O., Taniguchi I., Saito N. (1989) Correlation between the size of song control nuclei and plumage color change in orange bishop birds. Neurosci. Lett. 98, 144–148.

Aronica E., Frey U., Wagner M., Schroeder H., Krug M., Ruthrich H., Catania M.V., Nicoletti F., and Reyman K.G. (1991) Enhanced sensitivity of "metabotropic" glutamate receptors after inducing of Long-term Potentiation in rat hippocampus. J. Neurochem. 57, 376–383.

Avoli M., Drapeu C., and Kostopoulos. G. (1988) Changes in the synaptic transmission evoked in the in "in vitro" hippocampal slice by a brief decrease of Mg^{2+}: a correlate of Long-term potentiation? In H.L. Haas, and G. Buzsaki (Eds). Synaptic plasticity in the hippocampus. Springer-Verlag, Heidelberg, pp.9–12.

Ball G.F., and Casto J.M. (1991) Autoradiographic localization of NMDA receptors in the avian song control system using (3H)MK 801. Soc. Neurosci. Abstr. 17, 1053.

Bar R.P., Wiegant F. Lopes da Silva A., and Gispen F.H. (1984) Tetanic stimulation affects the metabolism of phosphoinositides in hippocampal slices. Brain Research, 321, 381–385.

Bashir Z.I., Alford S.T., Blake J., Frenguelli B.G., Davies S.N., Reyman K.G., and Collingridge G.L. (1991) On the locus of the maintenance of long-term potentiation in area CA1 of rat hippocampus.In: Long-term potentiation. M. Boudry, J. Davies (Eds.), MIT Press, pp. 31- 44.

Baudry M., and Davis J.L., (Eds.,) Long-Term Potentiation. (1991) MIT Press; pp. 3–121.

Benowitz L., (1980)Functional organization of the avian telencephalon. In Ebbesson, S.O. (Ed) Comparative neurology of telencephalon, Academic Press, pp.389–421.

Bernard D.L., Eens M., Ball G.F., (1996) Age- and behavior-related variation in volumes of song control nuclei in male European Starlings. J. Neurobiol. 30, 329–339.

Berridge M.J., and Irvine R.F., (1989) Inositol phosphates and cell signaling. Nature 341, 197–205.

Bingman V.P., Ioale P., Casini G., and Bagnoli P. (1987) Impaired retention of preoperatively acquired spatial reference memory in homing pigeons following hippocampal ablation. Behav. Brain Research 24, 147–156.

Bingman V.P., Ioale P., Casini G., and Bagnoli P. (1988) Unimpaired acquisition of spatial reference memory, but impaired homing performance in hippocampal-ablated pigeons. Behav. Brain Research 27, 179–187.

Bingman V.P., Bagnoli P., Ioale P., and Casini G. (1989) Behavioral and anatomical studies of the avian hippocampus. In The Hippocampus -new vistas. Neurol. Neurobiol. 52, 379–395.

Bingman V.P. (1992) The importance of comparative studies and ecological validity for understanding hippocampal structure and cognitive function. Hippocampus 2, 213–220.

Bliss T.V.P., and Lomo T. (1973) Long-lasting potentiation of synaptic transmission in the dentate area of the anaesthetized rabbit following stimulation of the perforant path. J. Physiol. 232, 331–356.

Bliss T.V.P., Clements M., Errington M.L., Lynch M.A., and Williams J.H. (1991) Presynaptic changes associated with long-term potentiation in the dentate gyrus. in Long-Term Potentiation. (Baudry, M., and Davis, J.L. Eds.) MIT Press. pp. 3–19.

Bostock E., Muller R.U., and Kubie J.L. (1991) Experience-dependent modifications of hippocampal place cell firing, Hippocampus 1, 193–205.

Bramham C.R., Milgram N.W., and Srebro B. (1991) Activation of AP5-sensitive NMDA receptors is not required to induce LTP of synaptic transmission in the lateral perforant path. Eur. J.Neurosci. 3, 1300–1308.

Brenowitz E.A., Nalls B., Wingfield J.C., and Kroodsma D.E. (1991) Seasonal changes in avian song nuclei without seasonal changes in song repertoire. J. Neurosci. 11, 1367–1374.

Casini G., Bingman V.P., and Bagnoli P., (1986) Connections of the pigeon dorsomedial forebrain studied with WGA-HRP and 3H-proline. J. Comp.Neurol. 245, 454–470.

Chapman A.G., and Bowker, H.M. (1987) Inhibition of hippocampal 3H-D-aspartate release by 2-APB, APV and 2-APH. In Excitatory amino acid transmission, Neurology and Neurobiology, 24.(T.P., Hicksand H. McLennan Eds.) A.R. Liss NY. pp. 165–171.
Clayton N.S., and Krebs J.R., (1994) Hippocampal growth and attrition in birds affected by experience. Proc. Natl. Acad. Sci. 91, 7410–7414.
Clayton N.S. (1995a) Comparative studies of food-storing, memory, and the hippocampal formation in Parids. Hippocampus 5, 499–510.
Clayton, N.S., (1995b) Development of memory and the hippocampus: comparison of food-storing and nonstoring birds on a one-trial associative memory task. J. Neurosci. 15, 2796–2807.
Clayton N.S., and Krebs J.R. (1995) Memory in food-storing birds: from behaviour to brain, Curr. Opinion in Neurobiol. 5, 149–154.
Clayton N.S. (1996) Development of food-storing and the hippocampus in juvenile marsh tits (Parus palustris). Beh. Brain Research 74, 153–159.
Chen L., and Huang L. (1992) Protein kinase C reduces Mg^{2+} block of NMDA-receptor channels as a mechanism of modulation. Nature 356, 521–523.
Chew S.J., Vicario D., and Nottebohm F. (1996) Quantal duration of auditory memories. Science 274, 1909–1912.
Colley P. A., and Routtenberg A., (1993) Long-term potentiation as synaptic dialog. Brain Res. Rev. 18, 115–122.
Collingridge H., Synaptic function of N-methyl-D-Aspartate receptors in the hippocampus. (1989) In The hippocampus - New vistas, Neurology and Neurobiology, 52 (V. Chan-Palay and Ch. Kohler eds.) A.R. Liss NY pp. 329–345.
Creager R., Dunwiddie T., and Lynch G. (1980)Paired-pulse facilitation and frequency facilitation in the CA1 region of the in vitro rat hippocampus. J. Physiol. 299, 409–424.
Davies S.N., Lester R.A., Reymann K., and Collingridge G. (1989)Temporally-distinct pre- and postsynaptic mechanisms maintain long-term potentiation. Nature 338, 500–503.
DelCastillo J., and Katz B. (1954) Quantal components of the end-plate potential, J. Physiol. (Lond) 124, 560–573.
DeVoogd T.J., Nixdorf B., and Nottebohm F. (1985) Synaptogenesis and changes in synaptic morphology related to acquisition of new song behavior. Brain Res. 329, 304–308.
Dolphin A.C., Errington M.L., and Bliss T.V.P. (1982) Long-term potentiation of the perforant path in vivo is associated with increased glutamate release. Nature 297, 496–498.
Dudek S.M., and Bear F.M. (1992) Homosynaptic long-term depression in area of CA1 of hippocampus and effects of N-methyl-D-aspartate receptor blockade. Proc. Natl. Acad. Sci. 89, 4363- 4367.
Eccles J.C., and McIntyre A.R., (1951)Plasticity of mammalian monosynaptic reflexes, Nature (Lond) 167, 466–468.
Eichenbaum H., (1992) The hippocampal system and declarative memory in animals. J. Cog. Neurosci. 4, 217–231.
Erichsen J.T., Bingman V.P., and Krebs J.R., (1991) The distribution of neuropeptides in the dorsomedial telencephalon of the pigeon (Columba Livia): a basis for regional subdivision. J. Comp. Neurol. 314, 478–492.
Ghijsen W.E.J. and LopesDaSilva F. (1991) Increase in endogenous amino acid release. In Long-term potentiation.(Baudry, M., and Davis, J.L.Eds.,) MIT Press pp. 45–57.
Goh J.W., Ho-Asjoe M., and Sastry B.R., (1986) Tetanic stimulation induced changes in (3H)glutamate binding and uptake in rat hippocampus. Gen. Pharm. 17, 537–542.
Grant S.G.N., O'Dell T.J., Karl K.A., Stein P.L., Soriano P., and Kandel E.R. (1992)Impaired long- term potentiation, spatial learning, and hippocampal development in fyn mutant mice. Science 258, 1903–1910.
Grover L.M. and Teyler T.J. (1995) Different mechanisms may be required for maintenance of NMDA receptor-dependent and independent forms of long-term potentiation. Synapse 19, 121- 133.
Haley J.E., and Schuman E.M. (1994) Involvement of nitric oxide in synaptic plasticity and learning. Seminars in Neurosci. 6, 11–20.
Hanse E., and Gustafsson B. (1992) Long-term potentiation and field EPSPs in the lateral and medial perforant path in the dentate gyrus in vitro: a comparison. Eur. J. Neurosci. 4, 1191- 1201.
Harris E., and Cotman C. (1986) Long-Term potentiation of guinea pig mossy fiber responses is not blocked by N-methyl-D-aspartate antagonist. Neurosci., Lett. 70, 132–137.
Healy S., Clayton N., and Krebs J., Grey E. (1991) Food storing birds: development of brain and behavior. Abstr. Sat. Symp. 14th ENA Meeting p.10.
Hill K.M., and DeVoogd T.J. (1991) Altered daylenght affects dendritic structure in a song-related brain region in red-winged blackbirds. Behav. Neural. Biol. 56, 240–250.
Hu G.Y., Hvalby O., Walaas S.I., Albert K.A., Skjeflo P., Andersen P., and Greengard P. (1987) Protein kinase C injection onto hippocampal pyramidal cells elicits features of long-term potentiation. Nature 328, 426–429.

Isumu Y., Ito K., Kato K., Uruno K., and Kato H. (1990) The role of quisqualate receptors in the induction of hippocampal long-term potentiation. Medical Hypothesis 33, 89–93.

Ito M., Sakurai M., Tongroach P. (1982) Climbing fibre induced depression of both mossy fibre responsivness and glutamate sensitivity of cerebellar Purkinje cells. J. Physiol. 324, 113- 134.

Johnston D., Williams S., Jaffe D., and Gray R. (1992) NMDA-receptor-independent long-term potentiation. Annu. Rev. Physiol. 54, 489–505.

Kallen, B. (1962) Embryogenesis of brain nuclei in the chick telencephalon. Ergeb. Anat Entwicklungsgesch, 36, 62–82.

Kase H., Iwahashi K., Nakanishi S., Matsuda Y., Yamada K., Takahaschi M., Murakada Ch., Sato A., and Kaneko M. (1987) K-252 compounds, novel and potent inhibitors of protein kinase C and cyclic nucleotide-dependent protein kinase. Biochem. Biophys. Res. Comm. 142, 436–440.

Katsuki H., Kaneko S.,Tajima A., and Satoh M. (1991)Separate mechanisms of long-term potentiation in two input systems to CA3 pyramidal neurons of rat hippocampal slices as revealed by the whole-cell patch-clamp technique. Neurosci. Res. 12, 393–402.

Kauer J.A., and Nicoll R.A. (1988) An APV-resistant non-associative form of long-term potentiation in the rat hippocampus. In Synaptic Plasticity in the hippocampus. (Haas, H. and Buzsaki, G., Eds) pp 65–66.

Kauer J.A.R., Malenka C., and Nicoll R.A. (1988) A persistent postsynaptic modification mediates long-term potentiation in the hippocampus. Neuron 1, 911–917.

Klann E., Chen S.-J., and Sweatt J.D. (1993) Mechanism of protein kinase C activation during the induction and maintenance of long-term potentiation probed using a selective peptide substrate. Proc. Natl. Acad.Sci. 90, 8337–8341.

Komatsu Y., Nakijima S., and Toyama K. (1991) Induction of long-term potentiation without participation of N-methyl-D-Aspartate receptors in kitten visual cortex. J. Neurophysiol. 65, 20–31.

Krayniak P.F. and Siegel A. (1978) Efferent connections of the hippocampus and adjacent regions in the pigeon. Brain Bev. Evo. 15, 372–388.

Krebs J.R., Sherry D.F., Healy S., Perry V.H., and Vaccarino A.L. (1989) Hippocampal specialization of food storing birds, Proc. Nat. Acad. Sci. 86, 1388–1392.

Krebs J.R. (1990). Food storing birds: Adaptive specialization in brain and behaviour? Philosophical Transactions of the Royal Society B 329, 153–160.

Krebs J.R., Erichsen J.T., and Bingman V.P. (1991) The distribution of neurotransmitters and neurotransmitter-related enzymes in the dorsomedial telencephalon of the pigeon (Columba livia). J. Comp. Neurol. 314, 467–477.

Kullmann D.M., Perkel D.J., Manabe T., and Nicoll R.A., (1992) Ca2+ entry via postsynaptic voltage- sensitive Ca2+ channels can transiently potentiate excitatory synaptic transmission in the hippocampus. Neuron 9, 1175–1183.

Kuno M., and Takahashi T., (1986) Effect of calcium and magnesium on transmitter release at Ia synapses of rat spinal motoneurons. J. Physiol. (Lond) 376, 543–553.

Larson L., Ambros-Ingerson J., and Lynch G. (1991) Sites and mechanisms for expression of long- term potentiation, In: Long-term potentiation. (M. Boudry, J. Davies Eds.) pp. 121–139.

Larrabee M.G., and Bronk D.W. (1938) Long-lasting effects of activity on ganglionic transmission, Am. J. Physiol. 123, p. 126.

Lee K., and Schubert P. (1982) Modulation of an inhibitory circuit by adenosine and AMP in the hippocampus. Brain Research 246, 311–314.

Lynch G., Larson J., Kelso, K., Barrionuevo G., and Schottler F. (1983) Intracellular injections of EGTA block induction of hippocampal long-term potentiation. Nature 305, 719–721.

Lynch G., and Baudry M. (1984) The biochemical basis of learning and memory: a new and specific hypothesis. Science 224, 1057–1063.

Lynch M.A., Errington M.L., and Bliss T.V.P. (1985) long-term potentiation of synaptic transmission in the dentate gyrus: increased release of (14C) glutamate without increase in receptor binding. Neurosci. Lett. 62, 123–129.

Lynch M.A., Clements M.P., Errington M.L., and Bliss T.V.P. (1988) Increased hydrolysis of phosphatidylinositol-4,5-biphosphate in long-term potentiation. Neurosci. Lett. 84, 291- 296.

Macphail E.M., and Good M. (1991) Effects of hippocampal ablation on pigeon learning and memory in laboratory task. Abstr. Sat. Symp. 14th ENA Meeting p.12.

Madison D.V., Malenka R.C., and Nicoll R.A. (1991) Mechanisms underlying long-term potentiation of synaptic transmission. Annu.Rev.Neurosci. 14, 379–397.

Magleby K.L., (973) The effects of repetitive stimulation on facilitation of transmitter release at the frog neuromuscular junction, J. Physiol. (Lond) 234, 327–352.

Malgaroli A., Malinow R., Schulman H., and Tsien R.W. (1992) Persistent signalling and changes in presynaptic function in long-term potentiation. In: Interactions among cell signalling systems (Ciba Fund. Symp. 164) pp. 176–196.

Malinow R., Madison D.V., and Tsien R.W. (1988) Persistent protein kinase activity underlying long- term potentiation. Nature 335, 820–824.

Malinow R. and Tsien R.W. (1990) Presynaptic enhancement shown by whole cell recordings of long- term potentiation in hippocampal slice. Nature 346, 177–180.

Mallart A., and Martin A.R. (1967) An analysis of facilitation of transmitter release at the neuromuscular junction of the frog. J. Physiol. (Lond) 193, 679–694.

Manabe T., and Nicoll R.A. (1994) Long-term potentiation: evidence against an increase in transmitter release probability in the CA1 region of the hippocampus. Science 265, 1888–1892.

McHugh T., Blum K.I., Tsien J.Z., Tonegawa S., and Wilson M.A., (1996) Impaired hippocampal representation of space in CA1-specific NMDAR1 knockout mice. Cell 87, 1339–1349.

Meehan E. (1996) Effects of MK-801 on spatial memory in homing and non-homing pigeon breeds. Beh. Neurosci. 110, 1–6.

Meehan E., and Wieraszko A. (1996) Effects of MK-801 on spatial memory in homing and non-homing pigeons breeds. Abstr. Soc. Neurosci. 22, 143.

Meyer M.L., and Westbrook G.L. (1987) Permeation and block of N-methyl-D-aspartic acid receptor channels by divalent cations in mouse cultured central neurons. J. Physiol. 394, 501- 527.

Mody I., and Heinemann H. (1987) NMDA receptors of dentate gyrus granule cells participate in synaptic transmission following kindling. Nature 326, 701–704.

Monagan D.T., Yao D., Olverman H.J., Watkins J.C. and Cotman C.W. (1984) Autoradiography of D- 2(3H) amino-5-phosphopentoate binding sites in rat brain. Neurosci. Lett. 52, 253- 258.

Montagnese C.M., Krebs J.R., Szekely A.D., and Csillag A. (1993) A subpopulation of large calbindin- like immunopositive neurones is present in the hippocampal formation in food-storing but not in non-storing species of bird. Brain Research 614, 291–300.

Morris R., Garrud P., Rawlins J., and O'Keefe J. (1982) Place navigation impaired in rats with hippocampal lesions. Nature 297, 681–683.

Morris R.G.M., Anderson E., Lynch G., and Baudry M. (1986) Selective impairment of learning and blockade of long term potentiation by a N-methyl-D-aspartate receptor antagonist, AP5. Nature 319, 775–776.

Morris R., Davis, S., Buther S. (1991) Hippocampal synaptic plasticity and N-methyl-D-aspartate receptors: a role in information storage? In Long-Term Potentiation. (Baudry, M., Davis, J.L., Eds.)MIT Press, pp. 267–300.

Muller D., Lynch G. (1988) Long-term potentiation differentially affects two components of synaptic responses in hippocampus. Proc.Nat.Acad.Sci. 85, 9346–9350.

Muller D., Joly M., Lynch G., (1988) Contribution of quisqualate and NMDA receptors to the induction and expression of LTP. Science 242, 1694–1700.

Muller D., Bittar P., Boddeke H. (1992) Induction of stable long-term potentiation in the presence of the protein kinase C antagonist staurosporine. Neurosci. Lett. 135, 18–22.

Nadel L. (1991) The hippocampus and space revisited. Hippocampus 1, 221–229.

Nicoletti F., Meek J.L., Iadarola M.J., Chuang D.M., Roth B.L., and Costa E. (1986) Coupling of inositol phospholipid metabolism with excitatory amino acid recognition sites in rat hippocampus. J. Neurochem. 46, 40–46.

Nicoletti F., Wroblewski J.T., and Costa E. (1987) Magnesium ions inhibit the stimulation of inositol phospholipids hydrolysis by endogenous excitatory amino acids in primary cultures of cerebellar granule cells. J. Neurochem. 48, 967–973.

Nicoletti F., Valerio C., Pellegrino C., Drago F., Scapagnini U., Canonico P.L. (1988) Spatial learning potentiates the stimulation of phosphoinositide hydrolysis by excitatory amino acids in rat hippocampal slices. J. Neurochem. 51, 725–729.

Nordeen E.J., and Nordeen K.W. (1990) Neurogenesis and sensitive periods in avian song learning. TINS 13, 31–36.

Nottenbohm F., O'Loughlin B., Gould K., Yohay K., and Alvarez-Buylla A. (1994) The life span of new neurons in a song control nucleus of the adult canary brain depends on time of year when cells are born. Proc. Natl. Acad. Sci. 91, 7849–7853.

Nottenbohm F. (1981) A brain for all seasons: cyclical anatomical changes in song control nuclei of the canary brain. Science 214, 1368–1370.

O'dell T.J., Kandel E.R., Grant S.G.N. (1991) Long-term potentiation in the hippocampus is blocked by tyrosine kinase inhibitors. Nature 353, 558–561.

O'Keefe J., and Dostrovsky, J. (1971) The hippocampus as a spatial map. Preliminary evidence from unit activity in the freely-moving rats. Brain Research 34, 171–175.

O'Keefe J., and Nadel, L. (1978) The hippocampus as a cognitive map. Clarendon Press. Oxford.

Olds J.L., Golski S., McPhie D.L., Olton D., Miskin M., and Alkon D.L. (1990) Discrimination learning alters the distribution of protein kinase C in the hippocampus of rats. J. Neurochem. 10, 3707–3713.

Olton D.S., and Papas B.C. (1979). Spatial memory and hippocampal function. Neuropsychologia 17, 669–682.

Olton D.S., Becker J.T., and Handelmann G.E. (1979) Hippocampus, time, and memory. Behavioral and Brain Sciences 2, 313–365.

Olton D., (1983) Memory function and the hippocampus. In Neurobiology of the hippocampus (Seifert W. Ed.,) Academic Press.

Pasinelli P., Ramakers G.M.J., Urban I.J.A., Hens J.J.H., Oestreicher A.B., deGraan P.N.E., and Gispen W.H. (1995) Long-term potentiation and synaptic phosphorylation. Beh. Brain Research 66, 53–59.

Persechini A., Moncrief N.D., Kretsinger R.H. (1989) The EF-hand family of calcium-modulated proteins. TINS 12, 462–467.

Polser M.R., Nadel L., Schacter L. (1991) Cognitive neuroscience analyses of memory: a historical perspective. J. Cog. Neurosci. 3, 95–116.

Racine R.J. and Kairiss E.W. (1987) In Neuroplasticity, Learning and Memory, Alan Liss, Inc., pp. 173–197.

Recasens M., (1995) Putative molecular mechanisms underlying long term potentiation (LTP): the key role of excitatory amino acid receptors. Therapie 50, 19–34.

Regehr W.G., Connor A., and Tank D.W. (1989) Optical imaging of calcium accumulation in hippocampal pyramidal cells during synaptic activation. Nature 341, 533–536.

Rehkamper G., Haase E., and Frahm H.D. (1988) Allometric comparison of brain weight and brain structure volumes in different breeds of the domestic pigeon, Columba livia. Brain Behav.Evol. 31, 141–149.

Riters, L.V., and Bingman V.P. (1994) The NMDA-receptor antagonist MK-801 blocks navigational learning in homing pigeons. Beh. Neural Biol. 62, 50–59.

Reyman K.G., Davies S.N., Matthies H., Kase H., and Collingridge G.L. (1990) Activation of a K-252b-sensitive protein kinase is necessary for a postsynaptic phase of long-term potentiation in area CA1 of rat hippocampus. Eur. J. Neurosci. 2, 481–486.

Rolls E.T. (1991)Functions of the primate hippocampus in spatial and nonspatial memory. Hippocampus 1, 258–261.

Rotenberg A., Mayford,M., Hawkins R.D., Kandel E.R., and Muller R.U. (1996) Mice expressing activated CaMKII lack low frequency LTP and do not form stable place cells in the CA1 region of the hippocampus. Cell 87, 1351–1361.

Shapiro E., and Wieraszko A. (1996) Comparative, in vitro, studies of hippocampal tissue from homing and nonhoming pigeon. Brain Research 725, 199–206.

Sherry D.F., and Schacter D.L. (1987) The evolution of multiple memory systems. Psychological Rev. 94, 439–454.

Sherry D.F. Vaccarino A.L., Buckenham K., and Herz R.S. (1989)The hippocampal complex of food storing birds, Brain Behav. Evol. 34, 308–317.

Sherry D.F., M.R.L. Forbes M.R.L., Khurgel M., and Ivy G.O. (1993) Females have a larger hippocampus than males in the brood-parasitic brown-headed cowbird. Proc. Natl. Acad. Sci. USA 90, 7839–7843.

Shettleworth S.J., (1995) Comparative studies of memory in food storing birds. From field to the Skinner box. In Behavioral brain research in naturalistic and seminaturalistic settings. NATO ASI series. (E. Alleva, A. Fasolo, H.P. Lipp, L. Nadel, L. Ricceri, B. Dordrecht eds.) Kluver Academic Publisher, 159–192.

Silva A.J., Paylor R., Wehner J.M., and Tonegawa S. (1992a) Impaired spatial learning in alpha- calcium-calmodulin kinase II mutant mice. Science 257, 206–211.

Silva A.J., Stevens Ch., Tonegawa S., and Wang Y. (1992b) Deficient hippocampal long-term potentiation in alpha-calcium-calmodulin kinase II mutant mice. Science 257, 201- 206.

Sladeczek F., Pin J., Recasens M., Bockaert J., and Weiss S. (1985) Glutamate stimulates inositol phosphate formation in striatal neurons. Nature 317, 717–719.

Spetch M.L., and Mondloch M.V. (1993) Control of pigeons spatial search by graphic landmarks in a touch screen task. J. Exp. Psychol. Animal Beh. Proc. 19, 353–372.

Stevens C.F., and Wang Y. (1994) Changes in reliability of synaptic function as a mechanism for plasticity. Nature 371, 701–706.

Squire L.R., and Zola-Morgan S. (1991) The medial temporal lobe system. Psych. Rev. 94, 1380- 1386.

Squire L.R., and Backercave C. (1991) The hippocampus, memory and space. Hippocampus 1, 258- 261.

Szekely A.D., and Krebs J.R. (1993) More hippocampal nitric oxide synthase in food-storing than in non-food storing birds. Abstr. Soc. Neurosci. 19, p.1448.

Sugiyama H., Ito I., and Okada D. (1990) Roles of metabotropic glutamate receptors in the long-term potentiation of hippocampal mossy fiber synapses. In Excitatory Amino acids and neuronal Plasticity (Ben-Ari. Y. Ed.) pp. 387–394.

Teyler T.J. (1980) Brain slice preparation: hippocampus. Brain Res. Bull. 5, 391–403.

Thesleff S. (1980) Aminopyridines and synaptic transmission. Neuroscience 5, 1413–1419.

Tsien J.Z., Chen D.F., Gerber D., Tom C., Mercer E.H., Anderson D.J., Mayford M., Kandel E.R., and Tonegawa, S. (1996a) Subregion- and cell type-restricted gene knockout in mouse brain. Cell 87, 1317–1326.

Tsien J.Z., Huerta P.T., and Tonegawa S. (1996b) The essential role of hippocampal CA1 NMDA receptor-dependent synaptic plasticity in spatial memory. Cell 87, 1327–1338;

Tsien R., Schulman and H., Malinow R. (1990) Peptide inhibitors of PKC and CaMK block induction, but not expression of long-term potentiation. In The Biology and Medicine of Signal Transduction. (Y. Nishizuka ed.) pp. 101–107.

Turner R.W., Baimbridge K.G., and Miller J.J. (1982) Calcium-induced long-term potentiation in the hippocampus. Neuroscience 7, 1411–1416.

Wehner J.M., Sleight S., and Upchruch M. (1990) Hippocampal protein kinase C activity is reduced in poor spatial learners. Brain Research 523, 181–187.

Wieraszko A. (1983) Stimulation and potentiation-dependent D-aspartate uptake by hippocampal slices following Schaffer collaterals and perforant path stimulation. Brain Research 259, 324- 326.

Wieraszko A., and Seyfried T. (1993) Influence of audiogenic seizures on synaptic facilitation in mouse hippocampal slices is mediated by N-Methyl-D-Aspartate receptor. Epilepsia 34, 979–984.

Wieraszko A., and Ball G. (1993) Long-term potentiation in the avian hippocampus does not require activation of the N-methyl-D-aspartate (NMDA) receptor. Synapse 13, 173–178.

Williams J.H., Clements, M.P., Errington M.L., Voss K.L., Lynch M., Bliss T.V.P. (1991) Retrograde messenger in long-term potentiation: a role for arachidonic acid? In Excitatory amino acids (Meldrum B., Moroni F., Costa C., Eds.) Raven Press pp.487–493.

Wilson M.A., and McNaughton B.L. (1993) Dynamics of the hippocampal ensemble code for space. Science 261, 1055–1058.

Worley F., Baraban J.M., and Snyder S.H. (1986) Heterogenous localization of protein kinase C in rat brain: autoradiographic analysis of phorbol ester receptor binding. J. Neurosci. 6, 199- 207

Wroblewski J.T., Nicoletti F., Fadda E., and Costa E. (1986) Excitatory amino acid recognition sites coupled with inositol phospholipid metabolism: developmental changes and interaction with alpha1 adrenoreceptors. Proc. Natl. Acad. Sci. 83, 1931–1935.

VanderZee E.A., Compaun C., deBoer M., Luiten P.G.M. (1992) Changes in PKC immunoreactivity in mouse hippocampus induced by spatial discrimination learning. J. Neurochem. 12, 4808- 4815.

8

METABOTROPIC GLUTAMATE RECEPTORS IN THE PLASTICITY OF EXCITATORY RESPONSES IN THE HIPPOCAMPUS

Clinical Impact

Lisa R. Merlin

Department of Neurology and Pharmacology and Department of Physiology and Pharmacology
State University of New York Health Science Center at Brooklyn
Brooklyn, NewYork 11203

1. INTRODUCTION

Excitatory connections in the hippocampal network have been shown tc exhibit plasticity when subjected to specific modes of activation. These include short term and long term potentiation (STP and LTP, respectively) as well as long term depression (LTD). Ionotropic glutamate receptor (iGluR) activation, especially that mediated via NMDA receptors, is key in regulating many of these processes and is discussed in more detail in other chapters in this text. This chapter will demonstrate the participation of metabotropic glutamate receptors (mGluRs) in the plastic processes of the hippocampal circuit and will emphasize the possible clinical relevance of these processes. I will first provide a review of the receptor classification and the actions of the various mGluR subgroups. I will briefly summarize the work to date demonstrating the different ways in which each of these subgroups participates in synaptic plasticity. I will discuss ways in which mGluR-mediated activities may influence clinical conditions such as epilepsy. And finally, I will present the results of my studies on the role of these receptors in the production of epileptiform activities in the *in vitro* hippocampus, focusing on their unique capacity for autopotentiation, providing a potential mechanism for epileptogenesis.

2. THE METABOTROPIC GLUTAMATE RECEPTORS

The metabotropic glutamate receptors (mGluRs) are a family of second-messenger-linked receptors activated by the excitatory amino acid glutamate. At least eight subtypes have been identified to date; these have been subclassified into three subgroups based on

Molecular and Cellular Mechanisms of Neuronal Plasticity, edited by
Ehrlich. Plenum Press, New York, 1998.

Table 1. Simplified overview of the mGluR subgroups

mGluR subgroup	Second messenger coupling	Predominant localization	Major physiological effect*
Group I	IP3 generation-Ca^{++} mobilization	perisynaptic	Enhanced cellular excitability
Groups II and III	Negatively coupled to adenylate cyclase: reduced cAMP	presynaptic	Reduced synaptic transmission

*see Wong et al., 1999 (in press) for more detailed review

their amino acid sequence similarities, signal transduction mechanisms, and agonist specificities. Group I is comprised of mGluRs 1 and 5. This subgroup activates phosphoinositide hydrolysis, leading to the production of inositol triphosphate (IP3) and activation of protein kinase C (PKC). IP3 generation secondarily increases intracellular free calcium via release from IP3-sensitive stores (Abe et al., 1992; Houamed et al., 1991; Masu et al., 1991). Group II (which includes mGluRs 2 and 3) and group III (mGluRs 4, 6, 7, and 8) are both negatively coupled to adenylate cyclase to inhibit cyclic AMP formation (Tanabe et al., 1992; Duvoisin et al., 1995). (1S, 3R)-1-aminocyclopentane-1,3-dicarboxylic acid (ACPD) activates both groups I and II (Schoepp et al., 1991; Tanabe et al., 1992), while (+)-α-methyl-4-carboxyphenylglycine (MCPG) is an antagonist at the same two subgroups (Eaton et al., 1993; Hayashi et al., 1994; Jane et al., 1993) (see Tables 1 and 2 for overview).

Prior to the development of selective mGluR agonists, ACPD was the agent of choice for the examination of the effects of mGluR activation, and it remains the most extensively used mGluR agonist. Thus, many of the numerous mGluR-induced effects which have been reported have not yet been attributed to any particular subgroup. Since ACPD has no activity at group III mGluRs, any action elicited by ACPD is most likely mediated either by activation of group I or group II mGluRs, or may be a combined effect of dual activation. Most of this chapter will therefore focus on actions shown to be mediated by group I and/or group II mGluRs.

There are data revealing that ACPD has group II selectivity when used in low concentrations, and activates group I mGluRs only at higher concentrations (Cartmell et al., 1993). Effects requiring elevated concentrations of ACPD are therefore often presumed to be mediated by group I mGluRs. More selective agonists and antagonists have recently been developed; studies utilizing these agents are useful to distinguish with more certainty effects mediated by group I mGluRs from those mediated by group II.

Table 2. mGluR subgroup specificities of pharmaceutical agents*

Subgroup	Major agonists	Major antagonists
Group I	(1S,3R)ACPD, tADA, (S)3HPG, (S)DHPG	(+)MCPG, (S)4CPG, (R,S)AIDA
Group II	(1S,3R)ACPD, L-CCG-I, DCG-IV, (S)4CPG	(+)MCPG, EGLU, MCCG
Group III	L-AP4, L-SOP	MSOP, CPPG

(S)3HPG: 3-hydroxyphenylglycine; tADA: *trans*-azetidine-2,4-dicarboxylic acid; (S)DHPG: 3,5-dihydroxyphenylglycine; (R,S)AIDA: 1-aminoindan-1,5-dicarboxylic acid; L-CCG-I: (2S,3S,4S)-α-(carboxycyclopropyl)-glycine;DCG-IV:(2S,2'R,3'R)-2-(2',3'-dicarboxycyclopropyl)glycine;(S)4CPG:4-carboxyphenylglycine; EGLU: (2S)-α-ethylglutamic acid; MCCG: α-methyl-L-CCG-I; L-AP4: (+)-2-amino-4-phosphonobutyric acid: L-SOP: L-serine-O-phosphate; MSOP: (R,S)-α-methylserine-O-phosphate; CPPG: (R,S)-α-cyclopropyl-4-phosphonophenylglycine
*see Tocris Cookson catalogue for references

Biochemical studies on intracellular transduction systems linked to mGluRs have been in progress since the late 1980's; these studies have made extensive use of cloned receptors. More recent studies have begun to elucidate the physiological actions and *in vivo* implications of activation of these mGluRs. There have been occasional discrepancies between neurochemical results and anticipated physiological effects of some agents. This most likely stems from differences between biochemical effects on cloned receptors subtypes and those on the endogenously-occuring mixtures of receptors found in cortical slice systems. Electrophysiological effects have generally shown better correlation with results on cloned receptors (Watkins and Collingridge, 1994).

For the purpose of simplification, I will assign the various recognized electrophysiological actions of the mGluRs to the responsible mGluR subgroup whenever possible. However, one should keep in mind that there may well be subtle but important differences between the actions of the individual members of a given subgroup; if so, variations in receptor subtype localization and distribution in different brain regions may have important functional consequences.

3. LTP AND LTD IN THE HIPPOCAMPUS: ROLE OF mGLURS

A number of review articles have been published recently reviewing the role of mGluRs in hippocampal long-term potentiation (LTP) (see for example Riedel and Reymann, 1996). In brief, it has been shown that mGluR activation participates in the induction of some forms of LTP (Petrozzino and Connor, 1994; Platt et al., 1995; Sagratella et al., 1996), and transient mGluR activation facilitates the subsequent induction and maintenance of LTP, apparently lowering the threshold for tetanic LTP induction (Cohen and Abraham, 1996; Aniksztejn et al., 1992; Riedel et al., 1995). Activation of mGluRs with the exogenous mGluR agonist ACPD can also induce a slow-onset potentiation (Bortolotto and Collingridge, 1993; 1995). Furthermore, potentiation of action potential-driven intracellular calcium increases in the dorsolateral septum has also been elicited with ACPD (Zheng et al., 1996). Consistent with these findings, some studies indicate that the mGluR antagonist MCPG blocks both tetanus-induced and slow-onset potentiation in naïve slices (Bortolotto et al., 1994; Bashir et al., 1993).

Some forms of ACPD-induced LTP appear to be dependent on NMDA receptor activation (Breakwell et al., 1996; O'Connor et al., 1994). However, ACPD elicits an NMDA receptor-independent LTP in the hippocampal CA1 region (Bortolotto and Collingridge, 1993; 1995). Using selective group I mGluR agonist, O'Leary also elicited an NMDA-independent, relatively rapid-onset LTP in the dentate gyrus (O'Leary and O'Connor, 1997).

Most studies indicate that the mGluR-induced LTP is mediated by the group I metabotropic receptors (Conquet et al., 1994; Davis and Laroche, 1996; Manahan-Vaughan et al., 1996; Bordi, 1996). The mechanism for the ACPD-induced potentiation remains unclear and may differ in different brain regions. While ACPD increases depolarization-induced transmitter release in cortical slices, it decreases release in striatal slices (Lombardi et al., 1996). Other possible contributing factors include an upregulation of receptor number, receptor sensitivity, or receptor responsiveness affecting NMDA and/or AMPA receptors, or even the mGluRs themselves. Here, too, there have been conflicting results. Most recent data suggest that group I mGluR activation results in a long-lasting enhancement of AMPA responses in the hippocampus (Bortolotto and Collingridge, 1995, contesting the earlier findings of Aniksztejn et al., 1991). Potentiation of NMDA receptor-mediated responses is fairly well established and is also long-lasting (O'Connor et al., 1994; Fitzjohn et al., 1996), and there is evidence for potentiation of group I mGluR-mediated responses as well (Merlin and Wong, 1997).

In contrast to the group I mGluR-mediated potentiation, activation of group II and group III mGluRs primarily results in a presynaptic reduction of transmitter release (Baskys and Malenka, 1991; Desai et al., 1994; Vignes et al., 1995; Yoshino et al., 1996; Kamiya et al., 1996; Forsythe and Clements, 1990; Gereau and Conn, 1995). On this basis, a role for these receptors in the production of long-term depression (LTD) is not unexpected. An mGluR-mediated long-lasting suppression of inhibitory transmitter release has been reported (Liu et al., 1993); mGluR activation has also induced a long-lasting suppression of excitatory transmission (Garciamunoz et al., 1996).

The role of these presynaptic mGluRs in LTD is supported by studies showing that MCPG, the mGluR antagonist, suppresses hippocampal LTD at CA3-CA1 and mossy fiber-CA3 synapses (Bolshakov and Siegelbaum, 1994; Kobayashi et al., 1996) as well as at dentate gyrus (O'Mara et al., 1995). Through the use of mGluR2 knockout mice, LTD at the mossy fiber-CA3 synapse has been shown to require group II mGluR activation (Yokoi et al., 1996).

4. *IN VIVO* STUDIES ON THE CLINICAL ROLE OF mGLURS

The results of behavioral investigations on learning and memory are consistent with the findings of the LTP studies which implicate group I mGluRs as having a critical role. Knockout mice lacking mGluR1 have spatial learning deficits as well as impaired hippocampal LTP and cerebellar LTD (Conquet et al., 1994). Additionally, L-AP4-sensitive mGluRs (most likely belonging to group III) seem to play a selective role in learning and memory formation of the rat (Holscher et al., 1996).

Recent studies have also begun to examine the potentially critical role of these receptors in other clinical conditions. Those investigating mechanisms for stroke prevention have elucidated a protective role for group II mGluR activation against neuronal injury via oxygen and glucose deprivation (Small et al., 1996). Other groups interested in clinical sequelae of head trauma have uncovered a significantly enhanced responsiveness to ACPD in both hippocampi as late as 15 days after percussion injury, suggesting a possible mechanism for the cognitive impairment or seizures that may follow head trauma (Delahunty et al., 1995).

Finally, a number of laboratories (ours among them) have been investigating the role of mGluRs in the production of epileptiform activity and seizures. The remainder of this chapter will therefore focus on this particular field, and the critical role the various mGluR subgroups play.

5. EPILEPSY: CLINICAL AND ELECTROPHYSIOLOGICAL OVERVIEW

Epilepsy is the clinical condition in which the affected individual suffers from recurrent, unprovoked seizures. Seizures are the clinical manifestation of a abnormal, hypersynchronous discharge of cerebrocortical neurons. These seizure discharges, also known as ictal discharges, can be recorded on an electroencephalogram (EEG), and are usually seconds to minutes in duration.

Between seizures, the patient with epilepsy often has no clinical manifestations, but may have asymptomatic interictal discharges on his EEG recording during this time.

These discharges, identified as interictal spikes or sharp waves, are usually under 200 ms in duration. Animal studies have revealed the intracellular correlate of the interictal spike to be the paroxysmal depolarization shift, or PDS.

Numerous methods have been developed to elicit interictal and ictal activity in animal models *in vitro* to allow the examination of the mechanisms underlying the generation of such events. These include electrical high-frequency stimulation (kindling), the introduction of agents that suppress $GABA_A$ inhibition (picrotoxin, penicillin), the enhancement of NMDA receptor-mediated excitation (low extracellular magnesium), and the induction of generalized depolarization (elevated extracellular potassium). Over the past two decades, these models have provided information regarding the roles of ionotropic (NMDA and AMPA) glutamate receptor activation as well as $GABA_A$ and $GABA_B$ receptor-mediated inhibition in the production of the interictal discharge (see Traub et al., 1987, for more detailed review). The discovery of the metabotropic glutamate receptors opened a new avenue of exploration which, over recent years, has begun to provide insights into the mechanisms for the interictal-to-ictal transition as well as epileptogenesis, the process in which previously normal brain tissue develops a tendency to produce recurrent seizure discharges.

6. EPILEPTOGENESIS IN THE HIPPOCAMPUS: ROLE OF mGLURS

In vitro studies in rodent hippocampal and neocortical slices have been performed using a variety of methods to elicit baseline epileptiform activity in order to examine the modulatory role of mGluR activation on such activities. Interictal activity in CA3 pyramidal cells induced by elevated extracellular K^+ concentration was attenuated by MCPG, suggesting a role for mGluRs in the maintenance of interictal activity (McBain, 1994). Burke and Hablitz revealed that low concentrations of ACPD had a facilitatory effect on bicuculline-induced epileptiform activity in neocortex, increasing burst frequency (Burke and Hablitz, 1994; 1995); we found the same effect on picrotoxin-induced interictal bursting in hippocampus, and further demonstrated that the effect is mediated by group II activation (Fig. 1; Merlin et al., 1995). In spite of this action on interictal activity, *in vivo* studies reveal that group II agonists have a predominantly anticonvulsant effect (Dalby and Thomsen, 1996; Attwell et al., 1995).

In contrast to the effect of group II activation, group I mGluR agonists are convulsant agents. Early *in vivo* studies in rats revealed that the development of epilepsy by high concentrations of ACPD could be suppressed with dantrolene, an inhibitor of intracellular calcium mobilization, suggesting that the induction of epilepsy required this second-messenger process linked to group I mGluR activation (McDonald et al., 1993). Furthermore, the delay in seizure onset following intrahippocampal ACPD injection suggested that long-term neuronal modification had occurred to induce this effect (Sacaan and Schoepp, 1992).

Consistent with these findings, we have found that ACPD (in elevated concentrations) or selective group I mGluR agonists will elicit prolonged epileptiform discharges in hippocampal slices (Taylor et al., 1995; Merlin and Wong, 1997). Furthermore, when group I mGluR agonist is applied in the presence of ongoing picrotoxin-induced epileptiform activity, the typically brief (<500 ms) picrotoxin-induced synchronized discharges are converted into prolonged ictaform events (Fig. 2; Merlin and Wong, 1997).

These data suggested to us a possible role for group I mGluRs in supporting the interictal-to-ictal transition. Arvanov et al. came to the same conclusion based on their findings in amygdala studies using 4-aminopyridine (4-AP)-induced epileptiform activity. In

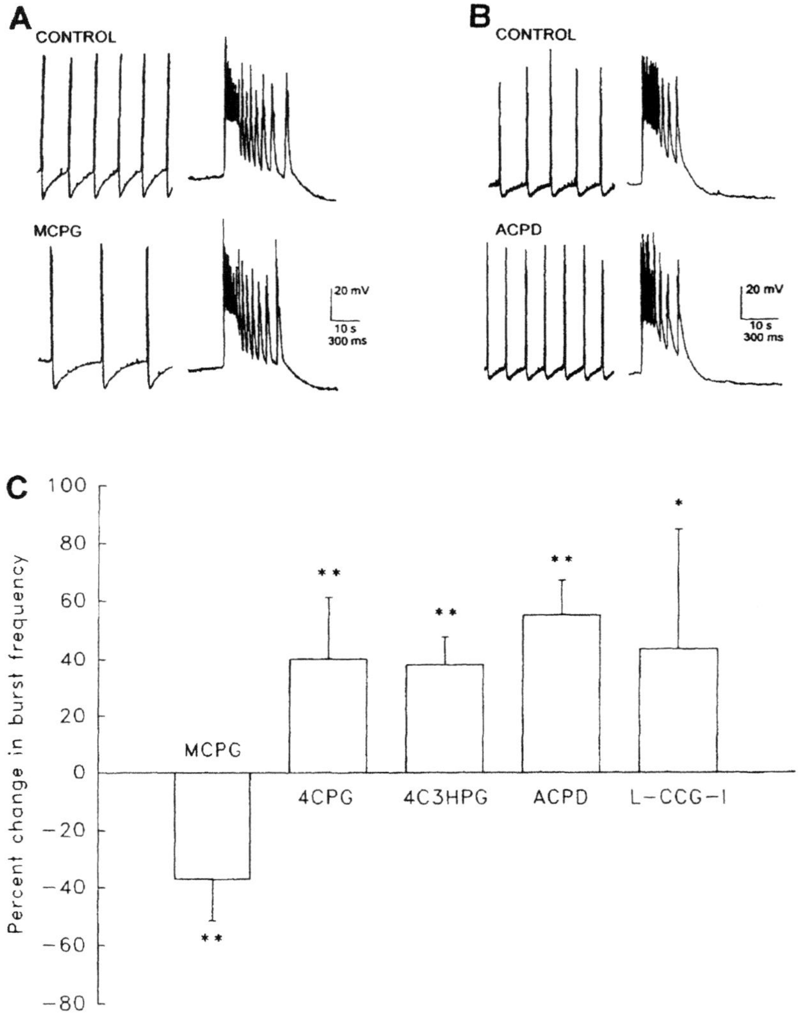

Figure 1. Group II mGluR activation increases interictal burst frequency. A & B. Intracellular recordings from CA3 pyramidal cells in guinea pig hippocampal slices. Control epileptiform activity elicited with bath application of 50 μM picrotoxin. The addition of MCPG (antagonist of group I and group II mGluRs; 750 μM) or ACPD (mGluR agonist; 5 μM) resulted in modulation of the interictal burst frequency without affecting burst duration, membrane potential, or burst afterhyperpolarization. C. MCPG, 700–1000 μM; 4CPG, 750–1000 μM (group II agonist and group I antagonist); 4C3HPG, 200–250 μM (group II agonist and group I antagonist); ACPD, 5 μM; L-CCG-1, 10 μM (group II agonist). Error bars = standard deviations. One asterisk (*) indicates significance at the $P < 0.05$ level. Two asterisks indicate significance at the $P < 0.005$ level n=3–6 slices for each agent tested. (Modified with permission from Merlin et al., 1995.)

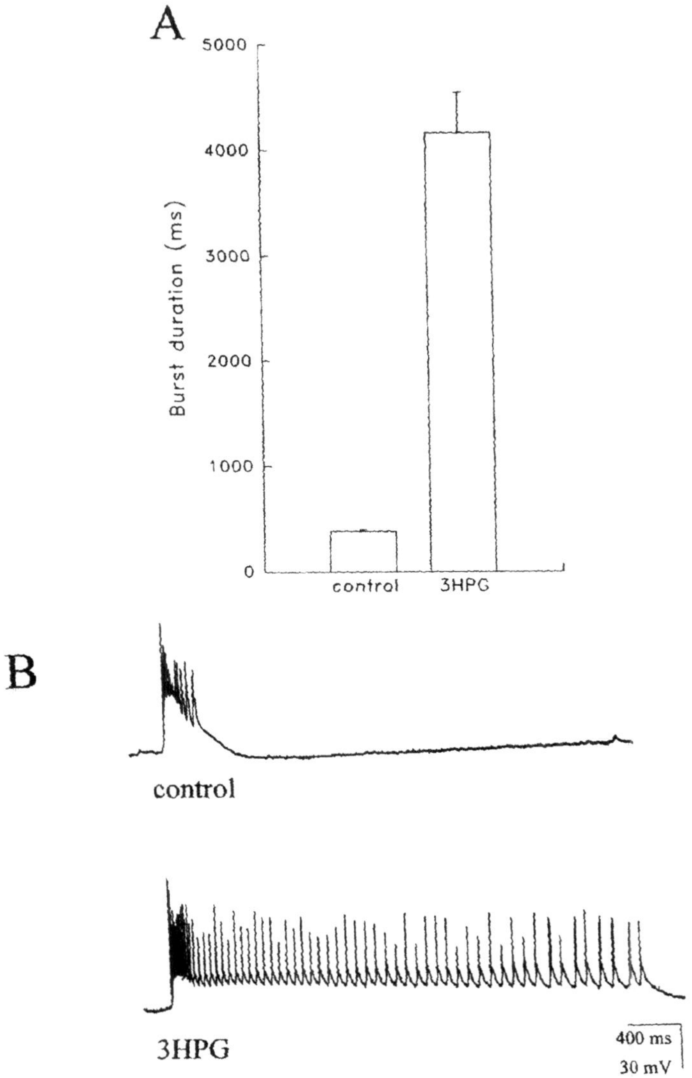

Figure 2. Group I mGluR activation increases epileptiform burst duration, converting interictal bursts into ictal-length discharges. A. Summary graph demonstrating mean duration of synchronized bursts elicited by picrotoxin (control) and maximum duration of prolonged bursts during application of the selective group I mGluR agonist (S)3HPG (250–500 μM, n = 7 slices). Error bars: mean ± SE. B. Intracellular recording from a CA3 pyramidal cell before and during exposure to 250 μM 3HPG. Similar activity could be elicited with either (S)DHPG (25–50 μM) or ACPD (50–100 μM). (Reproduced with permission from Merlin and Wong, 1997)

this *in vitro* model, which normally produces a mixture of interictal and ictal discharges, the mGluR antagonist MCPG failed to suppress either type of activity once the activity was fully induced. However, if MCPG was preapplied, subsequent application of 4-AP produced interictal activity only, suggesting that mGluR activation plays a highly significant role in the induction of ictal events.

Thus, our data reveal that while group II mGluR activation accelerates interictal burst frequency, group I activation prolongs burst duration (Merlin et al., 1995; Merlin and Wong, 1997; Taylor et al., 1995). But more noteworthy is the fact that, unlike the group II effects which are readily reversible, the group I effect (when elicited with selective group I agonists) appears to be long-lasting (Fig. 3, Merlin and Wong, 1997). We have already indicated that group I mGluR activation results in IP3 generation and increases intracellular free calcium. We've also discussed the association between this mGluR subgroup and LTP. One would then be led to presume that the underlying mechanism allowing for the persistence of the group I effect (i.e. epileptiform burst prolongation) is the induction of LTP or perhaps the persistent enhancement of NMDA or AMPA responses (see above). However, what we discovered was that the persistent prolonged discharges maintained in the absence of exogenous agonist could be suppressed with group I mGluR antagonist (Fig. 4; Merlin and Wong, 1997).

Based on the data shown in Fig 4, it was evident that endogenous transmitter (i.e. glutamate) was now able to activate group I mGluRs to maintain the production of the prolonged discharges once such discharges had been induced via transient exogenous agonist application. Thus, group I mGluR activation appears capable of autopotentiation, resulting in enhanced group I mGluR responsiveness and providing a mechanism for epileptogenesis. We have yet to demonstrate whether this effect occurs presynaptically or postsynaptically. Studies reveal that mGluR1 is frequently located perisynaptically (Baude et al., 1993), allowing one to hypothesize that glutamate will be able to recruit group I mGluR activation only under conditions of exceptionally strong presynaptic activity, allowing sufficient glutamate release to occur to spillover onto the perisynaptic areas. If presynaptic group I activation increased glutamate release, it could have the above described result. However, few laboratories have found evidence for presynaptically-localized group I mGluRs, and these studies suggest a suppressive effect of these receptors on neurotransmission (Gereau and Conn, 1995; Harvey et al., 1996). It is therefore more likely that the site of modification responsible for the autopotentiation we've observed is postsynaptic, increasing cellular responsiveness to group I mGluR activation. This could occur via an upregulation of activatable group I receptors, or increased receptor sensitivity to glutamate, or a receptor modulation (e.g. phosphorylation) which may allow enhanced action. Many of these changes could be linked to the group I mGluR-induced IP3 generation, rise in intracellular calcium, and protein kinase activation, inducing long-lasting changes in the pyramidal cells. It has also been shown that ACPD induces immediate early gene expression in lateral septal neurons, possibly mediated by phospholipase A2 activation (Kaatz and Albin, 1996).

Other *in vivo* studies support our hypothesis that epileptogenesis may be associated with an autopotentiation of group I mGluR responsiveness. Following either kainate- or kindling-induced seizures, there has been an observed enhancement of phosphoinositol production in response to group I mGluR agonists (Mayat et al., 1994; Akiyama et al., 1992), and an upregulation of the group I mGluR-mediated response has been observed in the neurons of the basolateral amygdala of kindled rats (Holmes et al., 1996).

The neuronal hyperexcitability and epileptogenesis mediated by group I mGluR activation may involve a protein synthesis-dependent step. Our preliminary results support the

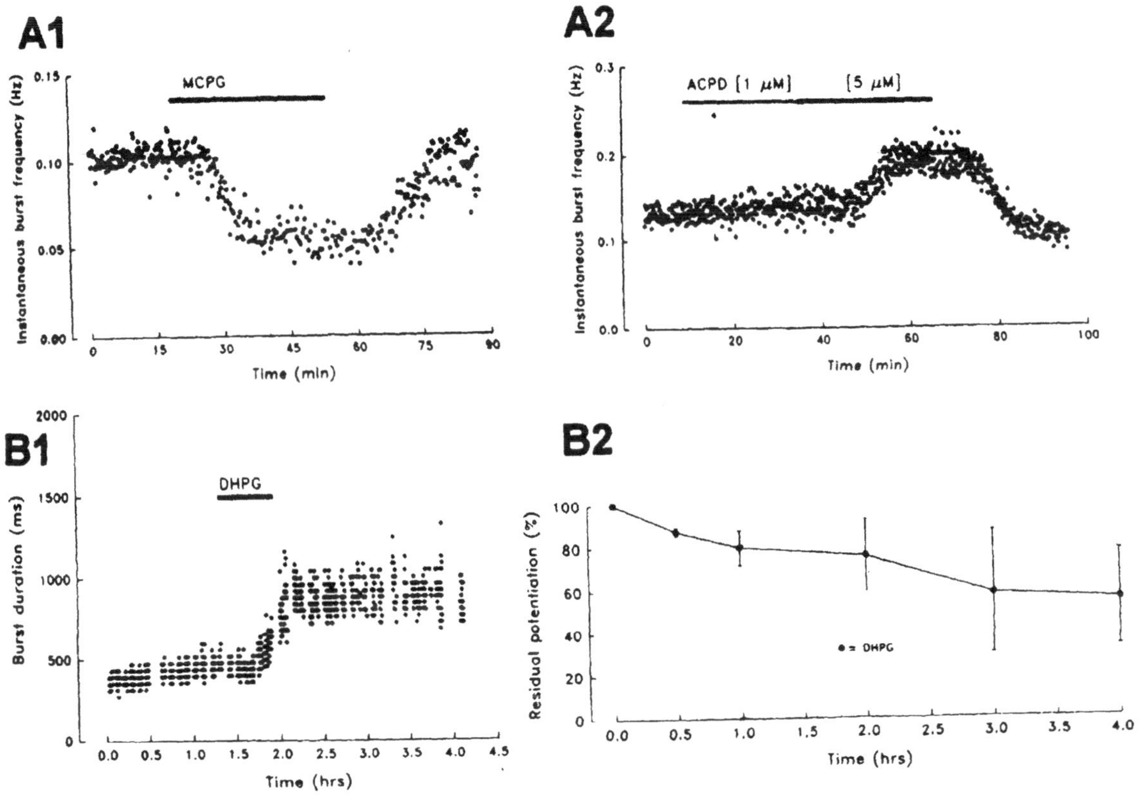

Figure 3. Group II mGluR-mediated effects are reversible, but group I effects are persistent. A1 & A2. Time course of frequency modification induced during transient application of MCPG and ACPD, respectively. MCPG, 750 μM; ACPD, 5 μM. Each filled circle represents an epileptiform burst. (Modified with permission from Merlin et al., 1995).B1 & B2. Time course of modification of burst duration (BD) induced by transient exposure to DHPG (50 μM). Burst prolongation persists for hours following removal of agonist. Summary data in B2 for post-exposure persistence of potentiation (i.e. prolonged bursts) is normalized relative to peak potentiated component [(BD - control BD) x 100 / (peak - control BD)]. N = 6 at 0.5–2 hours; n = 3 at 3 and 4 hours. Studies using 3HPG examined up to 3 hours following removal of agonist showed a similar persistence of effect. (Modified with permission from Merlin and Wong, 1997).

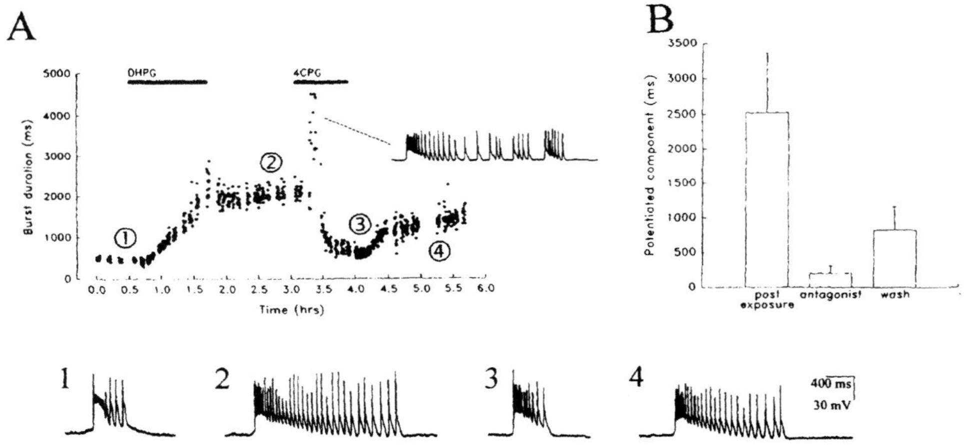

Figure 4. Persistent effects of transient group I mGluR activation are mediated by autopotentiation of group I mGluR responsiveness. A. Example of DHPG-induced postexposure potentiation as revealed via continuous recording from a CA3 pyramidal cell. DHPG, 100 μM (R,S mixture); 4CPG, 400 μM. 4CPG initially fractionated (see inset) then markedly suppressed potentiated bursts. Washout of antagonist resulted in a prompt restoration of prolonged bursts, indicating that the bursts are generated by synaptically-released glutamate acting at group I mGluRs. Traces 1–4 correspond to circled numbers on graph. B. Summary data, n = 3 slices. Percent change in burst duration was measured relative to mean burst duration preceding agonist application. "Post exposure" burst duration was measured 1 h after washout of agonist. "Wash" was measured at 1 h after washout of antagonist (3–4 hours after agonist removal). Application of antagonist significantly attenuated prolonged bursts ($P < 0.05$). After antagonist washout, burst prolongation reappeared to a level that was no longer significantly different from postexposure level ($P > 0.05$); it was, however, significantly different from the abbreviated bursts during antagonist application ($P < 0.05$). (Reproduced with permission from Merlin and Wong 1997).

role of protein synthesis in the induction of long-term group I mGluR-induced changes, since the protein synthesis inhibitor anisomycin disrupts the induction but has little impact on the maintenance of the group I mGluR agonist-induced prolonged epileptiform discharges (Merlin *et al.*, 1998).

Finally, it is important to note that the mRNAs for mGluR1 and mGluR5, the two receptor subtypes belonging to group I, are differentially regulated by kindling (mGluR1 mRNA is upregulated, while mGluR5 mRNA is downregulated following kindling epileptogenesis), indicating that the expression of each of these receptors is under independent regulatory control (Akbar et al., 1996). Thus, experiments need to be designed to examine these two receptors separately, and previous conflicting results using group I agonists in different brain regions may be the product of the preferential activation of mGluR1 or mGluR5.

CONCLUSION

The metabotropic glutamate receptors, although eight in number, can be conveniently classified and studied by subgroup. Group II and group III mGluRs, negatively coupled to adenylate cyclase, are largely presynaptic, and function to suppress neuronal excitability through the inhibition of transmitter release. LTD in some systems requires the activation of these receptors. Furthermore, agonists at group II or group III mGluRs demonstrate a 30-fold increase in potency in kindled neurons, indicating that these receptors may also develop enhanced responsiveness accompanying epileptogenesis (Neugebauer et al., 1997). Thus, activation of these receptors can be neuroprotective and may suppress seizure activity; long-term enhanced responsiveness of these mGluR subgroups may serve as an endogenous compensatory anticonvulsive mechanism.

In contrast, group I mGluRs, linked to IP3 and PKC generation, serve to increase neuronal excitability. This occurs primarily via a postsynaptic action which lead to potentiation of iGluR and mGluR-mediated responses. A critical role in LTP and learning and memory has recently been established, and we've begun to see a highly significant role in the production of seizures as well. Our work suggests that excessive activation may lead to a permanently epileptic state (epileptogenesis), and the underlying mechanism for this modification is protein synthesis-dependent, which provides a focus for further study.

The development of clinically-useful pharmaceutical agents directed at these receptor subtypes may allow for better seizure control with fewer side effects, as well as provide a means for prophylactic therapy in high risk patients such as those sustaining significant head trauma.

REFERENCES

Abe T, Sugihara H, Nawa H, Shigemoto R, Mizuno N, Nakanishi S (1992) Molecular characterization of a novel metabotropic glutamate receptor mGluR5 coupled to inositol phosphate/Ca^{2+} signal transduction. J.Biol.Chem. 267:13361–13368.

Akbar MT, Rattray M, Powell JF, Meldrum BS (1996) Altered expression of group I metabotropic glutamate receptors in the hippocampus of amygdala-kindled rats. Molec.Brain Res. 43:105–116.

Akiyama K, Daigen A, Yamada N, Itoh T, Kohira I, Ujike H, Otsuki S (1992) Long-lasting enhancement of metabotropic excitatory amino acid receptor-mediated polyphosphoinositide hydrolysis in the amygdala/pyriform cortex of deep prepyriform cortical kindled rats. Brain Res. 569:71–77.

Aniksztejn L, Bregestovski P, Ben-Ari Y (1991) Selective activation of quisqualate metabotropic receptor potentiates NMDA but not AMPA responses. Eur.J.Pharmacol. 205:327–328.

Aniksztejn L, Otani S, Ben-Ari Y (1992) Quisqualate metabotropic receptors modulate NMDA currents and facilitate induction of long-term potentiation through protein kinase C. Eur.J.Neurosci. 4:500–505.

Attwell PJE, Kaura S, Sigala G, Bradford HF, Croucher MJ, Jane DE, Watkins JC (1995) Blockade of both epileptogenesis and glutamate release by (1S,3S)-ACPD, a presynaptic glutamate receptor agonist. Brain Res. 698:155–162.

Bashir ZI, Bortolotto ZA, Davies CH, Berretta N, Irving AJ, Seal AJ, Henley JM, Jane DE, Watkins JC, Collingridge GL (1993) Induction of LTP in the hippocampus needs synaptic activation of glutamate metabotropic receptors. Nature 363:347–350.

Baskys A, Malenka RC (1991) Agonists at metabotropic glutamate receptors presynaptically inhibit EPSCs in neonatal rat hippocampus. J.Physiol. 444:687–701.

Baude A, Nusser Z, Roberts JDB, Mulvihill E, McIlhinney RAJ, Somogyi P (1993) The metabotropic glutamate receptor (mGluR1alpha) is concentrated at perisynaptic membrane of neuronal subpopulations as detected by immunogold reaction. Neuron 11:771–787.

Bolshakov VY, Siegelbaum SA (1994) Postsynaptic induction and presynaptic expression of hippocampal long-term depression. Science 264:1152

Bordi F (1996) Reduced long-term potentiation in the dentate gyrus of mGlu1 receptor-mutant mice in vivo. Eur.J.Pharmacol. 301:R15-R16

Bortolotto ZA, Bashir ZI, Davies CH, Collingridge GL (1994) A molecular switch activated by metabotropic glutamate receptors regulates induction of long-term potentiation. Nature 368:740–743.

Bortolotto ZA, Collingridge GL (1993) Characterisation of LTP induced by the activation of glutamate metabotropic receptors in area CA1 of the hippocampus. Neuropharmacology 32:1–9.

Bortolotto ZA, Collingridge GL (1995) On the mechanism of long-term potentiation induced by (1*S*,3*R*)-1-aminocyclopentane-1,3-dicarboxylic acid (ACPD) in rat hippocampal slices. Neuropharmacology 34:1003–1014.

Breakwell NA, Rowan MJ, Anwyl R (1996) Metabotropic glutamate receptor dependent EPSP and EPSP-spike potentiation in area CA1 of the submerged rat hippocampal slice. J.Neurophysiol. 76:3126–3135.

Burke JP, Hablitz JJ (1994) Metabotropic glutamate receptor activation decreases epileptiform activity in rat neocortex. Neurosci.Lett. 174:29–33.

Burke JP, Hablitz JJ (1995) Modulation of epileptiform activity by metabotropic glutamate receptors in immature rat neocortex. J.Neurophysiol. 73:205–217.

Cartmell J, Curtis AR, Kemp JA, Kendall DA, Alexander SPH (1993) Subtypes of metabotropic excitatory amino acid receptor distinguished by stereoisomers of the rigid glutamate analogue, 1-aminocyclopentane-1,3-dicarboxylate. Neurosci.Lett. 153:107–110.

Cohen AS, Abraham WC (1996) Facilitation of long-term potentiation by prior activation of metabotropic glutamate receptors. J.Neurophysiol. 76:953–962.

Conquet F, Bashir ZI, Davies CH, Daniel H, Ferraguti F, Bordi F, Franz-Bacon K, Reggiani A, Matarese V, Condé F, Collingridge GL, Crépel F (1994) Motor deficit and impairment of synaptic plasticity in mice lacking mGluR1. Nature 372:237–243.

Dalby NO, Thomsen C (1996) Modulation of seizure activity in mice by metabotropic glutamate receptor ligands. J.Pharmacol.Exp.Ther. 276:516–522.

Davis S, Laroche S (1996) Activation of metabotropic glutamate receptors induce differential effects on synaptic transmission in the dentate gyrus and CA1 of the hippocampus in the anaesthetized rat. Neuropharmacology 35:337–346.

Delahunty TM, Jiang JY, Gong QZ, Black RT, Lyeth BG (1995) Differential consequences of lateral and central fluid percussion brain injury on receptor coupling in rat hippocampus. J.Neurotrauma 12:1045–1057.

Desai MA, McBain CJ, Kauer JA, Conn PJ (1994) Metabotropic glutamate receptor-induced disinhibition is mediated by reduced transmission at excitatory synapses onto interneurons and inhibitory synapses onto pyramidal cells. Neurosci.Lett. 181:78–82.

Duvoisin RM, Zhang C, Ramonell K (1995) A novel metabotropic glutamate receptor expressed in the retina and olfactory bulb. J.Neurosci. 15:3075–3083.

Eaton SA, Jane DE, Jones PLStJ, Porter RHP, Pook PCK, Sunter DC, Udvarhelyi PM, Roberts PJ, Salt TE, Watkins JC (1993) Competitive antagonism at metabotropic glutamate receptors by (S)-4-carboxyphenylglycine and (RS)-alpha-methyl-4-carboxyphenylglycine. Eur.J.Pharmacol. 244:195–197.

Fitzjohn SM, Irving AJ, Palmer MJ, Harvey J, Lodge D, Collingridge GL (1996) Activation of Group I mGluRs potentiates NMDA responses in rat hippocampal slices. Neurosci.Lett. 203:211–213.

Forsythe ID, Clements JD (1990) Presynaptic glutamate receptors depress excitatory monosynaptic transmission between mouse hippocampal neurones. J.Physiol. 429:1–16.

Garciamunoz M, Patino P, Masliah E, Young SJ, Groves PM (1996) Glutamate-dependent long-term presynaptic changes in corticostriatal excitability. Neuroscience 73:109–119.

Gereau RWI, Conn PJ (1995) Multiple presynaptic metabotropic glutamate receptors modulate axcitatory and inhibitory synaptic transmission in hippocampal area CA1. J.Neurosci. 15:6879–6889.

Harvey J, Palmer MJ, Irving AJ, Clarke VRJ, Collingridge GL (1996) NMDA receptor dependence of mGlu-mediated depression of synaptic transmission in the CA1 region of the rat hippocampus. Br.J.Pharmacol. 119:1239–1247.

Hayashi Y, Sekiyama N, Nakanishi S, Jane DE, Sunter DC, Birse EF, Udvarhelyi PM, Watkins JC (1994) Analysis of agonist and antagonist activities of phenylglycine derivatives for different cloned metabotropic glutamate receptor subtypes. J.Neurosci. 14:3370–3377.

Holmes KH, Keele NB, Shinnick-Gallagher P (1996) Loss of mGluR-mediated hyperpolarizations and increase in mGluR depolarizations in basolateral amygdala neurons in kindling-induced epilepsy. J.Neurophysiol. 76:2808–2812.

Holscher C, McGlinchey L, Rowan MJ (1996) L-AP4 (L-(+)-2-amino-4-phosphonobutyric acid) induced impairment of spatial learning in the rat is antagonized by MAP4 ((S)-2-amino-2-methyl-4-phosphonobutanoic acid). Behav.Brain Res. 81:69–79.

Houamed KM, Kuijper JL, Gilbert TL, Haldeman BA, O'Hara PJ, Mulvihill ER, Almers W, Hagen FS (1991) Cloning, expression, and gene structure of a G protein-coupled glutamate receptor from rat brain. Science 252:1318–1321.

Jane DE, Jones PLStJ, Pook PCK, Salt TE, Sunter DC, Watkins JC (1993) Stereospecific antagonism by (+)-alpha-methyl-4-carboxyphenylglycine (MCPG) of (1S,3R)-ACPD-induced effects in neonatal rat motoneurones and rat thalamic neurones. Neuropharmacology 32:725–727.

Kaatz KW, Albin RL (1996) Intraseptal administration of (1S,3R)-1-aminocyclopentane-1,3-dicarboxylic acid induces immediate early gene expression in lateral septal neurons. Brain Res. 709:205–214.

Kamiya H, Shinozaki H, Yamamoto C (1996) Activation of metabotropic glutamate receptor type 2/3 suppresses transmission at rat hippocampal mossy fibre synapses. J.Physiol. 493:447–455.

Kobayashi K, Manabe T, Takahashi T (1996) Presynaptic long-term depression at the hippocampal mossy fiber-CA3 synapse. Science 273:648–650.

Liu Y-B, Disterhoft JF, Slater NT (1993) Activation of metabotropic glutamate receptors induces long-term depression of GABAergic inhibition in hippocampus. J.Neurophysiol. 69:1000–1004.

Lombardi G, Leonardi P, Moroni F (1996) Metabotropic glutamate receptors, transmitter output and fatty acids - studies in rat brain slices. Br.J.Pharmacol. 117:189–195.

Manahan-Vaughan D, Reiser M, Pin J-P, Wilsch V, Bockaert J, Reymann KG, Riedel G (1996) Physiological and pharmacological profile of trans-azetidine-2,4-dicarboxylic acid - metabotropic glutamate receptor agonism and effects on long-term potentiation. Neuroscience 72:999–1008.

Masu M, Tanabe Y, Tsuchida K, Shigemoto R, Nakanishi S (1991) Sequence and expression of a metabotropic glutamate receptor. Nature 349:760–765.

Mayat E, Lerner-Natoli M, Rondouin G, Lebrun F, Sassetti I, Reasens M (1994) Kainate-induced status epilepticus leads to a delayed increase in various specific metabotropic glutamate receptor responses in the hippocampus. Brain Res. 645:186–200.

McBain CJ (1994) Hippocampal inhibitory neuron activity in the elevated potassium model of epilepsy. J.Neurophysiol. 72:2853–2863.

McDonald JW, Fix AS, Tizzano JP, Schoepp DD (1993) Seizures and brain injury in neonatal rats induced by 1*S*,3*R*-ACPD, a metabotropic glutamate receptor agonist. J.Neurosci. 13:4445–4455.

Merlin LR, Taylor GW, Wong RKS (1995) Role of metabotropic glutamate receptor subtypes in the patterning of epileptiform activities in vitro. J.Neurophysiol. 74:896–900.

Merlin LR, Bergold PJ, Wong RKS (1998) Requirement of protein synthesis for group I mGlur-mediated induction of epileptiform discharges. J.Neurophysiol. 80: 989-993

Merlin LR, Wong RKS (1997) Role of Group I metabotropic glutamate receptors in the patterning of epileptiform activities in vitro. J.Neurophysiol. 78:539–544.

Neugebauer V, Keele NB, Shinnick-Gallagher P (1997) Epileptogenesis *in vivo* enhances the sensitivity of inhibitory presynaptic metabotropic glutamate receptors in basolateral amygdala neurons *in vitro*. J.Neurosci. 17:983–995.

O'Connor JJ, Rowan MJ, Anwyl R (1994) Long-lasting enhancement of NMDA receptor-mediated synaptic transmission by metabotropic glutamate receptor activation. Nature 367:557–559.

O'Leary DM, O'Connor JJ (1997) Potentiation of synaptic transmission in the rat dentate gyrus in vitro by (S)-3,5-dihydroxyphenylglycine ((S)-DHPG). Neurosci.Lett. 229:29–32.

O'Mara SM, Rowan MJ, Anwyl R (1995) Metabotropic glutamate receptor-induced homosynaptic long-term depression and depotentiation in the dentate gyrus of the rat hippocampus *in vitro*. Neuropharmacology 34:983–989.

Petrozzino JJ, Connor JA (1994) Dendritic Ca^{2+} accumulations and metabotropic glutamate receptor activation associated with an N-methyl-D-aspartate receptor-independent long-term potentiation in hippocampal CA1 neurons. Hippocampus 4:546–558.

Platt B, Behnisch T, Reymann KG (1995) Metabotropic glutamate receptors are involved in TEA-induced long-term potentiation in area CA1 of the hippocampus. Neuropharmacology 34:1339–1341.

Riedel G, Manahan-Vaughan D, Kozikowski AP, Reymann KG (1995) Metabotropic glutamate receptor agonist *trans*-azetidine-2,4-dicarboxylic acid facilitates maintenance of LTP in the dentate gyrus *in vivo*. Neuropharmacology 34:1107–1109.

Riedel G, Reymann KG (1996) Metabotropic glutamate receptors in hippocampal long-term potentiation and learning and memory. Acta Physiol.Scand. 157:1–19.

Sacaan AI, Schoepp DD (1992) Activation of hippocampal metabotropic excitatory amino acid receptors leads to seizures and neuronal damage. Neurosci.Lett. 139:77–82.

Sagratella S, Decarolis AS, Domenici MR, Lorenzini P, Fortuna S, Michalek H (1996) Glutamate-dependent mechanisms in the induction of a calcium long-term potentiation-like phenomenon. Brain Res. Bull. 41:193–200.

Schoepp DD, Johnson BG, True RA, Monn JA (1991) Comparison of (1S,3R)-1-aminocyclopentane-1,3-dicarboxylic acid (1S,3R-ACPD)- and 1R,3S-ACPD-stimulated brain phosphoinositide hydrolysis. Eur.J.Pharmacol. 207:351–353.

Small DL, Monette R, Chakravarthy B, Durkin J, Barbe G, Mealing G, Morley P, Buchan AM (1996) Mechanisms of 1S,3R-ACPD-induced neuroprotection in rat hippocampal slices subjected to oxygen and glucose deprivation. Neuropharmacology 35:1037–1048.

Tanabe Y, Masu M, Ishii T, Shigemoto R, Nakanishi S (1992) A family of metabotropic glutamate receptors. Neuron 8:169–179.

Taylor GW, Merlin LR, Wong RKS (1995) Synchronized oscillations in hippocampal CA3 neurons induced by metabotropic glutamate receptor activation. J.Neurosci. 15:8039–8052.

Traub RD, Wong RKS, Miles R. (1987) In vitro models of epilepsy. In: Jobe PC, Laird HE II, eds. *Neurotransmitters and epilepsy*. Clifton: Humana Press, 161–190.

Vignes M, Clarke VRJ, Davies CH, Chambers A, Jane DE, Watkins JC, Collingridge GL (1995) Pharmacological evidence for an involvement of group II and group III mGluRs in the presynaptic regulation of excitatory synaptic responses in the CA1 region of rat hippocampal slices. Neuropharmacology 34:973–982.

Watkins JC, Collingridge GL (1994) Phenylglycine derivatives as antagonists of metabotropic glutamate receptors. Trends Pharmacol.Sci. 15:333–342.

Wong RKS, Bianchi R, Taylor GW, Merlin LR (in press) Role of metabotropic glutamate receptors in epilepsy. In: Delgado-Escueta AV, Wilson WA, Olsen RW, Porter RJ, eds. *Jasper's basic mechanisms of the epilepsies, 3rd edition.* Philadelphia: Lippincott, Williams & Wilkins.

Yokoi M, Kobayashi K, Manabe T, Takahashi T, Sakaguchi I, Katsuura G, Shigemoto R, Ohishi H, Nomura S, Nakamura K, Nakao K, Katsuki M, Nakanishi S (1996) Impairment of hippocampal mossy fiber LTD in mice lacking mGluR2. Science 273:645–647.

Yoshino M, Sawada S, Yamamoto C, Kamiya H (1996) A metabotropic glutamate receptor agonist DCG-IV suppresses synaptic transmission at mossy fiber pathway of the guinea pig hippocampus. J.Physiol. 493:447–455.

Zheng F, Gallagher JP, Connor JA (1996) Activation of a metabotropic excitatory amino acid receptor potentiates spike-driven calcium increases in neurons of the dorsolateral septum. J.Neurosci. 16:6079–6088.

9

FAMILIAL AMYOTROPHIC LATERAL SCLEROSIS AND ALZHEIMER'S DISEASE

Transgenic Models

Philip C. Wong,[1,2] David R. Borchelt,[1,2] Michael K. Lee,[1,2]
Carlos A. Pardo,[1,2] Gopal Thinakaran,[1,2] Lee J. Martin,[1,2,3]
Sangram S. Sisodia,[1,2,3] and Donald L. Price[1,2,3,4*]

[1]Department of Pathology
[2]Neuropathology Laboratory
[3]Department of Neuroscience
[4]Department of Neurology
The Johns Hopkins University School of Medicine
Baltimore, Maryland

INTRODUCTION

Amyotrophic lateral sclerosis (ALS) and Alzheimer's disease (AD) are two classical neurodegenerative disorders. These age-related, chronic, progressive diseases are accompanied by clinical signs that reflect the vulnerability and death of specific populations of neurons. Unfortunately, there are no satisfactory treatments for these diseases. Both ALS and AD may be inherited in an autosomal dominant fashion: some cases of familial ALS (FALS) are linked to mutations in superoxide dismutase 1 (SOD1); and some individuals with familial AD (FAD) have mutations in genes encoding the amyloid precursor protein (APP) or presenilins (PS1 and PS2). Products of these mutant genes, thought to be associated with the formation of improperly folded or processed proteins, impact upon specific subsets of neural cells and cause characteristic clinical manifestations. For example, in ALS, damage to upper and lower motor neurons results in spasticity and weakness/muscle atrophy, respectively; in AD, the involvement of a variety of brain regions/neuronal popu-

* Donald L. Price, M.D. Neuropathology Laboratory The Johns Hopkins University School of Medicine 558 Ross Research Building 720 Rutland Avenue Baltimore, Maryland 21205-2196 (410) 955-5632 FAX (410) 955-9777 Email: ADRC@welchlink.welch.jhu.edu

Molecular and Cellular Mechanisms of Neuronal Plasticity, edited by Ehrlich. Plenum Press, New York, 1998.

lations is reflected in loss of memory, cognitive/behavioral impairments, and, eventually, profound dementia.

For neurological disorders, it is particularly important to establish model systems that faithfully reproduce disease. Until recently, with some exceptions, only surrogate animal models were available for studies of the pathogenic mechanisms that cause degeneration of neurons in these disorders. With the recent identification of gene mutations linked to several of these diseases, transgenic (Tg) strategies have allowed investigators to reproduce features of genetic human neurodegenerative disorders in mice. In turn, models are being used to clarify the reasons for the selective vulnerability of specific cell types, to define the character and spatial/temporal evolution of the cellular pathology, to delineate the mechanisms that cause abnormalities in neural cells, and to test new therapies.

AMYOTROPHIC LATERAL SCLEROSIS AND TRANSGENIC MICE WITH SOD1 MUTATIONS

Human Disease

ALS and FALS are characterized by paralysis, muscular atrophy, spasticity, and a variety of other motor signs. Electrodiagnostic studies disclose fibrillations, fasciculations, and giant polyphasic potentials; muscle biopsies demonstrate denervation atrophy with many small angulated fibers and fiber-type grouping. Spasticity, hyperreflexia, and extensor plantar signs are associated with lesions of upper motor neurons, whereas weakness and atrophy are attributed to abnormalities of large α-motor neurons of the brainstem and spinal cord.

Lower motor neurons show a variety of abnormalities including ubiquitin and phosphorylated neurofilament (NF) immunoreactivity in cell bodies, swollen axons (spheroids) with maloriented arrays of NF, and fragmentation of the Golgi apparatus (Carpenter, 1968; Hirano et al. 1969; Chou, 1992; Gonatas et al. 1992; Hirano and Kato, 1992; Matsumoto et al. 1992; Rouleau et al. 1996). In some cases of SOD1-linked FALS, motor neurons may contain SOD1-immunoreactive intracytoplasmic inclusions, which reflect presumably improperly folded or aggregated proteins (Shibata et al. 1993; Rouleau et al. 1996; Shibata et al. 1996). In both ALS and FALS, axons undergo Wallerian degeneration (Bradley et al. 1983), and denervated muscles show grouped atrophy. In end-stage disease, the numbers of motor neurons are reduced in the spinal cord and brainstem nuclei as are numbers of large pyramidal cells in deeper layers of the motor cortex. There is descending degeneration in corticospinal pathways. Neurons that control eye movements and sphincters (i.e., Onuf's nucleus) are usually relatively spared.

Genetics

It has been estimated that ~10% of adult-onset cases of ALS are familial with autosomal dominant disorder inheritance and age-dependent penetrance (Siddique et al. 1991). Approximately ~20% of cases of FALS are linked to mutations in SOD1 (Rosen et al. 1993; Bading et al. 1995), a member of a family of metalloenzymes that act as free radical scavengers by catalyzing the formation of H_2O_2 through the dismutation of $\bullet O_2$ (Fridovich, 1986). SOD1 is abundant in nervous tissue (0.7% of soluble protein in spinal cord) (Pardo et al. 1995). Initially, the finding of SOD1 mutations in some cases of FALS was interpreted to indicate that these mutations reduced enzyme activity leading to $\bullet O_2$-

mediated damage of motor neurons (Deng et al. 1993). However, recent studies suggest that cell injury is associated with the gain of a toxic property by mutant SOD1.

The concept that SOD1-linked FALS is caused by the gain of a toxic property by the mutant enzyme is supported by several lines of evidence: some mutant SOD1 possess near-normal levels of enzyme activity/stability *in vitro* (Borchelt et al. 1994) and/or restore SOD1 null yeast to the wild-type (wt) phenotype (Rabizadeh et al. 1995); SOD1 mutations do not have a dominant negative effect on wt SOD1 (Borchelt et al. 1995); and SOD1 null mice develop normally but may show a motor neuropathy (R Scott, Cephalon, West Chester, Pennsylvania) and enhanced susceptibility to retrograde degeneration after facial axotomy (Reaume et al. 1996). Some of these mutant SOD1, including those with normal SOD1 activity, accelerate the death of nerve cells *in vitro* (Rabizadeh et al. 1995); wt human (Hu)SOD1 Tg mice do not develop overt motor neuron disease (Wong et al. 1995). Finally, as outlined below, Tg mice with the H93A, G93A, or G85R SOD1 mutations develop motor neuron disease in spite of the presence of normal SOD1 activity (Gurney et al. 1994; Ripps et al. 1995; Wong et al. 1995).

Tg Models of FALS

Over the past decade, studies of a variety of animal models of ALS have provided important information concerning the biology of dysfunction and death of motor neurons, but only recently have several groups of investigators produced Tg mice with FALS-linked SOD1 mutations (Gurney et al. 1994; Ripps et al. 1995; Wong et al. 1995). For example, lines of Tg mice that express the G37R HuSOD1 mutation at 3–12x levels of endogenous SOD1 in the spinal cord invariably develop motor neuron disease (Wong et al. 1995). At four months of age, these Tg mice show reduced spontaneous movements, difficulty moving their hindlimbs, and muscle wasting as well as electromyographic patterns and muscle biopsies consistent with motor neuron disease. Eventually, the forelimbs become weak, and hindlimbs are completely paralyzed. These mice have significantly elevated levels of SOD1, and activity gels show increases in SOD1 activity, confirming that the G37R mutation synthesized *in vivo* retains full specific activity (Wong et al. 1995). In mice expressing high levels of mutant transgene product, motor neurons accumulate SOD1 immunoreactivity in axons (Borchelt et al. 1996b). Vacuoles, usually associated with swollen mitochondria in various stages of degeneration (Wong et al. 1995), are conspicuous in dendrites and axons greater than in cell bodies. Ubiquitin and phosphorylated NF immunoreactivities accumulate in cell bodies and dendrites as occurs in ALS. Motor axons, some of which also show abnormalities of cytoskeletal elements, undergo Wallerian degeneration; and muscle fibers exhibit evidence of denervation.

Gurney and colleagues (Gurney et al. 1994), who were the first to publish Tg mice with a SOD1-linked mutation, have documented a very similar motor neuron disease in the G1H line of G93A HuSOD1 (Dal Canto and Gurney, 1994a; Dal Canto and Gurney, 1994b; Tu et al. 1996). These mice express the transgene at relatively high levels (~5x endogenous), and the neuropathology reported in these animals resembles that occurring in the G37R SOD1 Tg mice that express mutant SOD1 at high levels. Significantly, these motor neurons also show abnormalities in the Golgi apparatus and neurofibrillary axonal abnormalities similar to those occurring in cases of ALS (Mourelatos et al. 1996; Tu et al. 1996). Moreover, the administration of vitamin E and selenium to G93A SOD1 mice was associated with a very modest delay in both the onset and progression of the disease without affecting survival; in contrast, the putative antiexcitotoxic agents riluzole and gabapentin did not influence the onset or progression of the disease but did increase survival slightly (Gurney et al. 1996). Finally, Tg mice with the missense mutation (G86R) in the mouse (Mo)SOD1 gene that corresponds to G85R HuSOD1 develop a rapidly progressive paralysis at 92–118 days that rapidly leads to death (Ripps et al. 1995; Morrison et al. 1996); the transgene was expressed in all tissues, but significant changes were not reported in SOD1 activity relative to controls. Autopsy findings consistent with motor neuron disease included phosphorylated filaments in cell bodies and loss of motor neurons. Some neurons in the anterior horn, especially cholinergic nerve cells expressing phosphorylated NF, appear to be particularly vulnerable, whereas other nerve cells, particularly those expressing calbindin, appear to be preserved (Morrison et al. 1996).

Pathogenic Mechanisms of SOD-1 Linked FALS

Because multiple lines of Tg mice with mutant SOD1 develop motor neuron disease and because several of our lines of wt HuSOD1 mice with similar or higher levels of enzyme in identical tissue distributions do not develop clinical signs or pathology (Wong et al. 1995), the disease occurring with some of these mutations must be caused by a dominant gain of an adverse property by mutant SOD1 (Wong et al. 1995). The different mutations, widespread throughout the enzyme, could induce a conformational change to allow access of an inappropriate substrate to the active site resulting in the generation of radicals that could then target other molecules in the cell. One pathogenic hypothesis suggests that mutant SOD1 acts as a peroxidase generating hydroxyl radical capable of oxidizing a variety of targets (Wiedau-Pazos et al. 1996). *In vitro*, A4V and G93A HuSOD1, but not free Cu^{++} or the SOD1 apoenzyme, catalyze the oxidation of a spin-trap compound (5,5-dimethyl-1-pyroline *N*-oxide) at rates higher than those occurring with wt SOD1 and induce cell death (Wiedau-Pazos et al. 1996); these effects are attenuated by treatment with Cu^{++} chelators (Wiedau-Pazos et al. 1996). A second hypothesis suggests that mutant SOD1 has the ability to utilize peroxynitrite to form nitronium ions that can nitrate tyrosine residues (Beckman et al. 1993). The oxidation or nitration of proteins important in the biology of motor neurons (NF, glutamate transporters, trkB, etc.) could have significant impacts on the structure/function of these cells.

High levels of SOD1 in neurons suggest that SOD1 may have other functions (i.e., Cu^{++} buffering/shielding) in addition to its enzymatic activities (Liu and Cizewski Culotta, 1994; Wong et al. 1995). Significantly, many FALS-linked SOD1 mutations do not impact on the metal-binding domains of SOD1 (i.e., Cu^{++} remains bound as a cofactor, and the enzyme retains significant activity) (Borchelt et al. 1994). Nevertheless, the mutant proteins may be prone to misfolding and could aggregate and sequester normal proteins or, by improper shielding of copper, participate in inappropriate and toxic copper chemistries. Both of the above-outlined hypotheses are consistent with the premise that FALS mutations alter protein structure in such a way that the mutant SOD1 has a toxic effect on substrates critical for the survival of motor neurons.

ALZHEIMER'S DISEASE AND TRANSGENIC MICE WITH APP OR PS1 MUTATIONS

Human Disease

AD, the most common cause of dementia occurring in mid-to-late life, affects 5–10% of individuals >65 years of age and perhaps 40% of persons >80 years of age

(Evans et al. 1989). Clinical signs include memory loss, impairment of higher cortical functions, behavioral/psychiatric abnormalities, and, eventually, profound dementia. Increases in life expectancy will inevitably lead to an increased prevalence of AD, and the number of people with AD will triple over the next 25 years as will the cost (Ernst and Hay, 1994).

AD selectively affects populations of neurons in certain brain regions, including those in the hippocampus, neocortex, and basal forebrain cholinergic system (Arnold et al. 1991; Braak and Braak, 1994; Price et al. 1996).

In the cortex and hippocampus, there are deposits of Aß amyloid derived from APP, a single transmembrane protein encoded by a gene on chromosome 21 (see below) (Glenner and Wong, 1984a; Masters et al. 1985a; Goldgaber et al. 1987; Kang et al. 1987; Robakis et al. 1987; Kitaguchi et al. 1988; Ponte et al. 1988; Tanzi et al. 1988). This type-I integral membrane glycoprotein contains the Aß region comprised of 28 amino acids of the ectodomain and 11–14 amino acids of the adjacent transmembrane domain of APP (Glenner and Wong, 1984a; Glenner and Wong, 1984b; Masters et al. 1985a; Masters et al. 1985b). Aß is produced by two endorproteolytic cleavage activities termed ß- and γ- secretase which cleave at the N- and C-terminal portions of Aß respectively. The most common of peptide species are Aß1–40, AßX-40, Aß1–42(43), and AßX-42(43) [termed Aß40 and Aß42(43), hereafter]. In the cortices of individuals with AD or Down's syndrome, that amyloid Aß deposition begins with Aß42(43) and not Aß40 (Iwatsubo et al. 1994; Iwatsubo et al. 1995; Shinkai et al. 1995; Lemere et al. 1996a). Physicochemical studies argue that Aß1–42 and/or Aß1–43, rather than Aß1–40, may be more fibrillogenic (Burdick et al. 1992) and provide a substrate for amyloid deposits (Jarrett and Lansbury, Jr. 1993; Jarrett et al. 1993).

Many vulnerable nerve cells show abnormalities of the cytoskeleton manifest as accumulations of straight and paired helical filaments in cell bodies and proximal dendrites (neurofibrillary tangles), neurites (axons/terminals), and neuropil threads (distal dendrites) (Lee, 1995; Thinakaran et al. 1996b). These filaments are comprised principally of phosphorylated tau, a low molecular weight microtubule-associated protein. However, tau hyperphosphorylation, which can interfere with the binding of microtubules, appears to be neither necessary or sufficient for the formation of paired helical filaments. *In vitro* studies (Goedert et al. 1996) have shown that exposure of nonphosphylated three and four repeat recombinant tau to sulphated glycosaminoglycans can lead to the formation of paired and straight helical filaments, respectively. These findings suggest that interactions of tau and glycosaminoglycans may play an important role in the formation of neurofibrillary tangles *in vivo*. Eventually, neurofibrillary tangle-bearing cells die, possibly by mechanisms that involve apoptotic pathways (Su et al. 1994; Troncoso et al. 1996).

Genetics

Approximately 10% of cases of AD are familial and show autosomal dominant inheritance (Bird, 1994; Schellenberg, 1995). Subsets of these cases are linked to mutations of specific genes (Bondi et al. 1994; Schellenberg, 1995), including the APP and PS1 and PS2 genes; some cases of late-onset disease are associated with the presence of copies of the apoE4 allele (Roses, 1995).

APP. In a small fraction of cases of FAD, mutations have been documented in the APP gene (located on chromosome 21) (Chartier-Harlin et al. 1991; Goate et al. 1991; Naruse et al. 1991; Hendricks et al. 1992; Mullan et al. 1992). In some families, the valine residue at position 717 (four amino acids downstream of the C-terminus of Aß1–42) is substituted with either isoleucine, glycine, or phenylalanine (Chartier-Harlin et al. 1991; Goate et al. 1991; Naruse et al. 1991). Cells that express APP harboring 717 mutations do not secrete higher levels of Aß but do secrete a higher fraction of longer Aß peptides (i.e., extending to Aß residue 42) relative to cells that express wt APP (Suzuki et al. 1994). In two large, related, early-onset AD families from Sweden, Lys-Met codons immediately N-terminal to Asp 1 of Aß are substituted within Asn-Leu (Mullan et al. 1992). Cells that express APP harboring the "Swedish" substitutions (APPswe) secrete higher levels of Aß species as compared to cells expressing wt constructs (Citron et al. 1992; Cai et al. 1993; Thinakaran et al. 1996c).

PS1 and 2

Encoded by genes on chromosomes 14 and 1, PS1 and PS2, respectively, are highly homologous proteins (Sherrington et al. 1995) predicted to contain eight transmembrane helices (Doan et al. 1996a). PS1 MRA protein is expressed in the brain, particularly in neurons (Kovacs et al. 1996; Lee et al. 1996). In cultured cells and brains (human, monkey, and mouse), PS1 is endoproteolytically processed as an N-terminal ~28-kD fragment and a C-terminal ~18-kD fragment (Thinakaran et al. 1996c). The topology of PS1 suggests that a specific cytoplasmic loop domain interacts with cytosolic constituents. Because of the paucity of accumulated full-length PS and the generality of PS processing across tissues, we have suggested that PS fragments are functional units. Although the biological functions of PS are not yet well understood, the finding of significant homology between PS and sel-12, a gene product that plays a role in the determination of cell fates during development in *C. elegans* (Levitan and Greenwald, 1995), suggests that PS may be important in developmental processes in mammalian tissues.

The majority (~70%) of cases of early-onset FAD are linked to mutations in the *PS1* gene located on chromosome 14 (St George-Hyslop et al. 1992; Alzheimer's Disease Collaborative Group, 1995; Sherrington et al. 1995). Approximately 50% of mutations occur within or immediately adjacent to the predicted loop domain (Sherrington et al. 1995; Alzheimer's Disease Collaborative Group, 1995; Doan et al., 1996b). Cases with the Glu 280 Ala mutation show massive deposits of Aß42 in many brain regions (Lemere et al. 1996b). The PS2 gene, localized on chromosome 1, has substantial homology with PS1, and PS2 mutations have been reported to cause autosomal dominant AD in Volga German kindreds and in an Italian pedigree (Levy-Lahad et al. 1995a; Levy-Lahad et al. 1995b; Li et al. 1995; Rogaev et al. 1995).

Tg Models of FAD

Tg Mice with APP Mutations. Many groups have attempted to create APP Tg mice (Higgins et al. 1994; Price and Sisodia, 1994; Howland et al. 1995; Hsiao et al. 1995; LaFerla et al. 1995; Moran et al. 1995; Oster-Granite et al. 1996). Several of the most interesting lines of Tg mice are discussed below.

In one line of mice, the PDGF ß-promoter was used to drive the expression of a HuAPP minigene encoding the FAD-linked APP (717V→F) mutation; the construct contained portions of APP introns 6–8 that allow alternative splicing of exons 7 and 8. Levels of HuAPP mRNA and protein were significantly greater than endogenous levels. Transcripts encoding the three major splicing variants, particularly the KPI-coding form (Rockenstein et al. 1995), and levels of the transgene product were ~4–5x higher than endogenous MoAPP (Games et al. 1995). Aß, in diffuse deposits and in plaques, was present in hippocampus, corpus callosum, and cerebral cortex. Extracellular Aß fibrils were abundant in these mice. Dystrophic neurites were often present in proximity to plaques, but there were no tau-positive neurites or neurofibrillary tangles.

In collaboration with Dr. K Hsiao, we produced multiple lines of Tg FVB mice that express a variety of Hu- and MoAPP transgenes (wt and mutant) placed under the transcriptional control of the hamster prion gene promoter (Hsiao et al. 1995). FVB/N mice were chosen for these studies because they are an inbred strain, have an easily injectable and prominent pronucleus in single-cell embryos, are highly fertile, and have been used successfully in producing Tg models of prion diseases (Hsiao et al. 1994). As observed in previous studies with the SHaPrP.neo vector (Scott et al. 1989; Scott et al. 1992), the expression of transgene-encoded APP was largely copy-number dependent (Hsiao et al. 1995). Mice with >30 copies produced 3–5x endogenous levels of APP. Tg mice expressing these high levels of APP showed behavioral abnormalities including: inactivity, agitation, neophobia, and seizures; impaired spatial alterations in a Y maze; and diminished glucose utilization in cortical-limbic areas (in these brain regions, there was significant gliosis). These abnormalities appeared earlier with increasing levels of APP in brain. No extracellular amyloid was detected. A similar neurological disorder develops naturally in older non-Tg FVB mice, and our work indicates that this age- and strain-related phenotype is exacerbated by the presence of the APP transgene. FVB mice appear to be sensitive to transgene-encoded APP, the presence of which exacerbate the strain-related phenotype and is associated with premature death. Thus, despite achieving high levels of the transgene product (up to 5x endogenous), the FVB strain is not ideal for attempts to produce mutant APP Tg mice with AD-type pathology.

Success in achieving high levels of APP expression but failure to observe an AD pathology motivated us to modify our approach in two ways: use of a newly modified vector; and introduction of the transgene into another strain of mice. The SHaPrP.neo vector, which we used initially (Hsiao et al. 1995), is a large 45-kb cosmid; hence, genetic manipulation of this vector is difficult. To overcome this difficulty, we obtained a much smaller fragment of the Mo prion protein (PrP) genome (phgPrP) (Dr. Charles Weissmann, University of Zurich) in which the second 10-kb intron and 15 kb of 3' untranslated sequence were removed; this vector expresses PrP in a virtually identical pattern as a full-length genomic clone (cos6.11) (Fischer et al. 1996). The open reading frame of the coding exon in the phgPrP plasmid was replaced with a unique Xho I restriction endonuclease site, allowing for insertion of a cDNA that encodes a murine 695-amino acid isoform of the APP with a "humanized" Aß domain and mutations (K595N, M596L) linked to Swedish pedigrees of FAD (Mullan et al. 1992). Using this strategy, we have recently produced mice with Aß deposits (Drs. DR Borchelt, SS Sisodia, and DL Price, personal observations)

Dr. Hsiao and colleagues (Hsiao et al. 1996) used a similar approach to introduce a construct containing the prion promotor driving the overexpression of APP-695 containing the Swedish double mutation (i.e., $Lys^{670} \rightarrow Asn, Met^{671} \rightarrow Leu$) in Tg mice. At three months of age, these mice showed normal learning and memory in spatial reference and alternation tasks, but, by 9–10 months of age, animals were impaired on these tasks. Levels of Aß40 and Aß42 were increased fivefold and 14-fold, respectively, and plaques with Aß deposits and abundant vacuoles were present in hippocampus and cortex. The features that distinguish these several lines APP Tg animals from other Tg mice are not fully understood, but achieving high levels of expression of the mutant APP transgene in nervous tissue appears to be very important.

Tg Mice with PS1 and APP Mutations. The brains of Tg mice expressing high levels of wt or mutant HuPS1 transgene products show the two PS1 fragments (Thinakaran et al. 1996a). As of this writing, mice expressing the A246E PS1 variant are only eight months old and have not developed an FAD phenotype. To determine whether PS1 mutations influence APP/Aß processing, particularly levels of the toxic Aß42, we mated mutant PS1 Tg mice with mutant APP Tg animals. The brains of young Tg animals that coexpress mutant PS1 APPswe showed elevated ratios of Aß42:Aß40 as compared to Tg mice that express APPswe alone or that coexpress wt PS1 and APPswe (Borchelt et al. 1996a). Similarly, Tg mice expressing mutant PS1 alone (but not mice expressing wt PS1) have an increase in Aß42 in brain (Duff et al. 1996). Several lines of evidence indicate that one mechanism by which mutant PS1 can cause AD is the acquisition (or enhancement) of properties that influence APP processing in a manner that leads to increased extracellular concentrations of Aß42. The APP and PS Tg models will undoubtedly prove to be of extraordinary value for studies designed to clarify the cellular and biochemical mechanisms of disease and for testing novel therapies.

Pathogenesis of APP and PS-Linked FAD

Varying levels of newly synthesized APP molecules appear at the cell surface (Weidemann et al. 1989; Haass et al. 1992; Sisodia, 1992); some of these molecules are cleaved endoproteolytically between positions 16 and 17 of Aß by APP "α-secretase" (Esch et al. 1990; Sisodia et al. 1990; Anderson et al. 1991; Wang et al. 1991) to release the ectodomain of APP, including residues 1–16 of Aß, into the medium. This α-secretase cleavage within the Aß domain precludes the formation of Aß. The APswe and 717 mutations lead to increased levels of Aß or increased ratios of Aß42:40, respectively. Thus, these mutations enhance amyloidogenesis, presumably by inflencing β and γ-secretase activity respectively. The reinternalization of APP from the cell surface appears to be obligatory for the generation of a significant fraction of Aß (Koo and Squazzo, 1994). Aggregated forms of the synthetic Aß peptide are neurotoxic (Pike et al. 1993; Cotman and Pike, 1994; Simmons et al. 1994; Busciglio et al. 1995), and in vivo Aß42 is believed to be particularly toxic to nerve cells (Yankner, 1996). The presence of Aß deposits is associated with neurites and neurofibrillary tangles, degeneration of nerve cells, and alterations in neural circuits in brain regions critical for cognitive, memory, and behavioral functions.

The mechanisms by which mutations in PS1 and 2 predispose individuals to FAD are not clear. However, the absence of deletions or truncation mutations in PS1 and PS2 leads to the suggestion that early-onset FAD in these affected pedigrees is the result of a toxic effect promoted by the mutant PS protein. Recent studies indicate that the mutation in PS1 influences levels of Aß42; plasma and conditioned media from fibroblasts obtained from carriers of PS1 and PS2 mutations have elevated levels of Aß42 species as compared to samples from unaffected family members (Scheuner et al. 1996). Moreover, *in vitro* studies of transfected cell lines expressing PS1 mutations (A246E, M146L, or ΔE9) have increased ratios of Aß42:Aß40 as compared to ratios in the media of cells expressing wt PS1 (Borchelt et al. 1996a). Moreover, the brains of young Tg mice coexpressing a chimeric APP and an FAD-linked PS1 variant show elevated Aß42/Aß40 ratios as compared to ratios in the brains of Tg mice expressing APP alone or Tg mice coexpressing wt human PS1 and APP (Borchelt et al. 1996a). The results of these studies provide support for the view that one mechanism by which these mutant PS1 cause AD is by influencing amyloidogenic cleavage increasing the extracellular concentration of Aß42 peptides that are prone to deposition/aggregation and these peptides are thought to be the toxic Aß species.

CONCLUSIONS

AD and ALS are prototypes of the human neurodegenerative diseases. In ALS, weakness and atrophy reflect dysfunction/death of motor neurons; in AD, memory loss and dementia are the result of neurofibrillary tangles, Aß42 amyloid deposits, and death of neurons in cortex and hippocampus. Subsets of cases of FALS and FAD often show dominant inheritance: some cases of FALS are linked to mutations in the SOD1 gene; and some pedigrees with FAD exhibit mutations in genes encoding either APP or PS1 and PS2. Current evidence suggests that mutant proteins may be misfolded and influence the aggregation properties of these proteins or the processing events that are critical for the normal biology of these cells. Much has been learned about the biology of these mutant transgene products by recent *in vitro* and *in vivo* studies. For example, *in vitro*, some mutant SOD1 show normal enzyme activity but kill cells, presumably by the acquisition of neurotoxic properties. Similarly, *in vitro* studies have provided new information concerning APP and PS topology, processing, and the effects of mutations on Aß. Recently, exciting advances have been made using Tg strategies that allow investigators to reproduce features of these human disorders in mice. Tg mice with SOD1 mutations develop weakness and muscle atrophy associated with degenerative changes in motor neurons that result from the acquisition of toxic properties by mutant SOD1. Similarly, Tg mice that express mutant human FAD-linked genes show behavioral impairments and plaque (i.e., Aß42 deposits associated with dystrophic neurites). The formation of Aß42, derived by the aberrant processing of APP, is influenced by APP mutations that increase the amount, length, and fibrillogenic properties of Aß. Mutations of PS1 increase levels of Aß42. These Tg models are of great value for investigations of mechanisms of disease and will be critical for testing novel therapies that, if efficacious in model systems, can be introduced rapidly into clinical trials.

ACKNOWLEDGMENTS

The authors thank Drs. Don Cleveland, Lucie Bruijn, Jeffrey Rothstein for their discussions on ALS and Steve Younkin and C. Eckman for their help with analysis of Aß in our APP and PS1 Tg mice. Drs. Nancy Jenkins and Neil Copeland have been invaluable collaborators in our efforts to create transgenic models of neurodegenerative disease.

This work was supported by grants from the U.S. Public Health Service (NIH NS 20471, AG 05146) as well as the Metropolitan Life Foundation, the Adler Foundation, the Alzheimer's Association, the Develbiss fund, the American Health Assistance Foundation, the Amyotrophic Lateral Sclerosis Association, and the Claster family fund. Dr. Price is the recipient of a Javits Neuroscience Investigator Award (NIH NS 10580) and a Leadership & Excellence in Alzheimer's Disease (LEAD) Award (NIH AG 07914). Dr. Sisodia is the recipient of a Zenith Award from the Alzheimer's Association.

REFERENCES

Alzheimer's Disease Collaborative Group (1995) The structure of the presenilin 1 (*S182*) gene and identification of six novel mutations in early onset AD families, *Nature Genetics* 11, 219–222.

Anderson, J.P., F.S. Esch, P.S. Keim, K. Sambamurti, I. Lieberburg and N.K. Robakis (1991) Exact cleavage site of Alzheimer amyloid precursor in neuronal PC-12 cells, *Neurosci. Lett.* 128, 126–128.

Arnold, S.E., B.T. Hyman, J. Flory, A.R. Damasio and G.W. Van Hoesen (1991) The topographical and neuroanatomical distribution of neurofibrillary tangles and neuritic plaques in the cerebral cortex of patients with Alzheimer's disease, *Cereb. Cortex* 1, 103–116.

Bading, H., M.M. Segal, N.J. Sucher, H. Dudek, S.A. Lipton and M.E. Greenberg (1995) *N*-methyl-D-aspartate receptors are critical for mediating the effects of glutamate on intracellular calcium concentration and immediate early gene expression in cultured hippocampal neurons, *Neuroscience* 64, 653–664.

Beckman, J.S., M. Carson, C.D. Smith and W.H. Koppenol (1993) ALS, SOD and peroxynitrite, *Nature* 364, 584

Bird, T.D. (1994) Clinical genetics of familial Alzheimer disease, in: *in Alzheimer Disease*, (R.D. Terry, R. Katzman and K.L. Bick, eds), pp. 65–74, Raven Press, New York.

Bondi, M.W., D.P. Salmon and N.M. Butters (1994) Neuropsychological features of memory disorders in Alzheimer disease, in: *in Alzheimer Disease*, (R.D. Terry, R. Katzman and K.L. Bick, eds), pp. 41–63, Raven Press, New York.

Borchelt, D.R., M.K. Lee, H.H. Slunt, M. Guarnieri, Z.-S. Xu, P.C. Wong, R.H. Brown,Jr., D.L. Price, S.S. Sisodia and D.W. Cleveland (1994) Superoxide dismutase 1 with mutations linked to familial amyotrophic lateral sclerosis possesses significant activity, *Proc. Natl. Acad. Sci. USA* 91, 8292–8296.

Borchelt, D.R., M. Guarnieri, P.C. Wong, M.K. Lee, H.S. Slunt, Z. Xu, S.S. Sisodia, D.L. Price and D.W. Cleveland (1995) Superoxide dismutase 1 subunits with mutations linked to familial amyotrophic lateral sclerosis do not affect wild-type subunit function, *J. Biol. Chem.* 270, 3234–3238.

Borchelt, D.R., G. Thinakaran, C.B. Eckman, M.K. Lee, F. Davenport, T. Ratovitsky, C.-M. Prada, G. Kim, S. Seekins, D. Yager, H.H. Slunt, R. Wang, M. Seeger, A.I. Levey, S.E. Gandy, N.G. Copeland, N.A. Jenkins, D.L. Price, S.G. Younkin and S.S. Sisodia (1996a) Familial Alzheimer's disease-linked presenilin 1 variants elevate Aβ1–42/1–40 ratio in vitro and in vivo, *Neuron* 17, 1005–1013.

Borchelt, D.R., P.C. Wong, M.W. Becher, C.A. Pardo, M.K. Lee, Z.-S. Xu, G. Thinakaran, N.A. Jenkins, N.G. Copeland, S.S. Sisodia, D.W. Cleveland, D.L. Price and P.N. Hoffman (1996b) Early axonal abnormalities and axonal transport of mutant superoxide dismutase 1 in a transgenic model of familial amyotrophic lateral sclerosis, *J. Neurosci.*

Braak, H. and E. Braak (1994) Pathology of Alzheimer's disease, in: *in Neurodegenerative Diseases*, (D.B. Calne, ed), pp. 585–613, W.B. Saunders, Philadelphia.

Bradley, W.G., P. Good, C.G. Rasool and L.S. Adelman (1983) Morphometric and biochemical studies of peripheral nerves in amyotrophic lateral sclerosis, *Ann. Neurol.* 14, 267–277.

Burdick, D., B. Soreghan, M. Kwon, J. Kosmoski, M. Knauer, A. Henschen, J. Yates, C. Cotman and C. Glabe (1992) Assembly and aggregation properties of synthetic Alzheimer's A4/β amyloid peptide analogs, *J. Biol. Chem.* 267, 546–554.

Busciglio, J., A. Lorenzo, J. Yeh and B.A. Yankner (1995) β-amyloid fibrils induce tau phosphorylation and loss of microtubule binding, *Neuron* 14, 879–888.

Cai, X.-D., T.E. Golde and S.G. Younkin (1993) Release of excess amyloid β protein from a mutant amyloid β protein precursor, *Science* 259, 514–516.

Carpenter, S. (1968) Proximal axonal enlargement in motor neuron disease, *Neurology* 18, 841–851.

Chartier-Harlin, M.-C., F. Crawford, H. Houlden, A. Warren, D. Hughes, L. Fidani, A. Goate, M. Rossor, P. Roques, J. Hardy and M. Mullan (1991) Early-onset Alzheimer's disease caused by mutations at codon 717 of the β-amyloid precursor protein gene, *Nature* 353, 844–846.

Chou, S.M. (1992) Pathology-light microscopy of amyotrophic lateral sclerosis, in: *in Handbook of Amyotrophic Lateral Sclerosis*, (R.A. Smith, ed), pp. 133–181, Marcel Dekker, New York.

Citron, M., T. Oltersdorf, C. Haass, L. McConlogue, A.Y. Hung, P. Seubert, C. Vigo-Pelfrey, I. Lieberburg and D.J. Selkoe (1992) Mutation of the β-amyloid precursor protein in familial Alzheimer's disease increases β-protein production, *Nature* 360, 672–674.

Cotman, C.W. and C.J. Pike (1994) β-amyloid and its contributions to neurodegeneration in Alzheimer disease, in: *in Alzheimer Disease*, (R.D. Terry, R. Katzman and K.L. Bick, eds), pp. 305–315, Raven Press, New York.

Dal Canto, M.C. and M.E. Gurney (1994a) The development of central nervous system pathology in a murine transgenic model of human amyotrophic lateral sclerosis, *Am. J. Pathol.* 145, 1–9.

Dal Canto, M.C. and M.E. Gurney (1994b) Development of central nervous system pathology in a murine transgenic model of human amyotrophic lateral sclerosis, *Am. J. Pathol.* 145, 1271–1280.

Deng, H.-X., A. Hentati, J.A. Tainer, Z. Iqbal, A. Cayabyab, W.-Y. Hung, E.D. Getzoff, P. Hu, B. Herzfeldt, R.P. Roos, C. Warner, G. Deng, E. Soriano, C. Smyth, H.E. Parge, A. Ahmed, A.D. Roses, R.A. Hallewell, M.A. Pericak-Vance and T. Siddique (1993) Amyotrophic lateral sclerosis and structural defects in Cu,Zn superoxide dismutase, *Science* 261, 1047–1051.

Doan, A., G. Thinakaran, D.R. Borchelt, H.H. Slunt, T. Ratovitsky, M. Podlisny, D.J. Selkoe, M. Seeger, S.E. Gandy, D.L. Price and S.S. Sisodia (1996a) Protein topology of presenilin 1, *Neuron* 17, 1023–1030.

Doan, A., G. Thinakaran, A. Lanahan, D.L. Price, S.S. Sisodia and P.E. Worley (1996b) Identification and characterization of a presenilin 1 interacting protein, *Soc. Neurosci. Abstr.* 22, 728

Duff, K., C. Eckman, C. Zehr, X. Yu, C.-M. Prada, J. Perez-Tur, M. Hutton, L. Buee, Y. Harigaya, D. Yager, D. Morgan, M.N. Gordon, L. Holcomb, L. Refolo, B. Zenk, J. Hardy and S. Younkin (1996) Increased amyloid-β42(43) in brains of mice expressing mutant presenilin 1, *Nature* 383, 710–713.

Ernst, R.L. and J.W. Hay (1994) The US economic and social costs of Alzheimer's disease revisited, *Am. J. Public Health* 84, 1261–1264.

Esch, F.S., P.S. Keim, E.C. Beattie, R.W. Blacher, A.R. Culwell, T. Oltersdorf, D. McClure and P.J. Ward (1990) Cleavage of amyloid β peptide during constitutive processing of its precursor, *Science* 248, 1122–1124.

Evans, D.A., H.H. Funkenstein, M.S. Albert, P.A. Scherr, N.R. Cook, M.J. Chown, L.E. Hebert, C.H. Hennekens and J.O. Taylor (1989) Prevalence of Alzheimer's disease in a community population of older persons. Higher than previously reported, *JAMA* 262, 2551–2556.

Fischer, M., T. Rülicke, A. Raeber, A. Sailer, M. Moser, B. Oesch, S. Brandner, A. Aguzzi and C. Weissmann (1996) Prion protein (PrP) with amino-proximal deletions restoring susceptibility of PrP knockout mice to scrapie, *EMBO J.* 15, 1255–1264.

Fridovich, I. (1986) Superoxide dismutases, *Adv. Enzymol. Relat. Areas Mol. Biol.* 58, 61–97.

Games, D., D. Adams, R. Alessandrini, R. Barbour, P. Berthelette, C. Blackwell, T. Carr, J. Clemens, T. Donaldson, F. Gillespie, T. Guido, S. Hagopian, K. Johnson-Wood, K. Khan, M. Lee, P. Leibowitz, I. Lieberburg, S. Little, E. Masliah, L. McConlogue, M. Montoya-Zavala, L. Mucke, L. Paganini, E. Penniman, M. Power, D. Schenk, P. Seubert, B. Snyder, F. Soriano, H. Tan, J. Vitale, S. Wadsworth, B. Wolozin and J. Zhao (1995) Alzheimer-type neuropathology in transgenic mice overexpressing V717F β-amyloid precursor protein, *Nature* 373, 523–527.

Glenner, G.G. and C.W. Wong (1984a) Alzheimer's disease: initial report of the purification and characterization of a novel cerebrovascular amyloid protein, *Biochem. Biophys. Res. Commun.* 120, 885–890.

Glenner, G.G. and C.W. Wong (1984b) Alzheimer's disease and Down's syndrome: sharing of a unique cerebrovascular amyloid fibril protein, *Biochem. Biophys. Res. Commun.* 122, 1131–1135.

Goate, A., M.-C. Chartier-Harlin, M. Mullan, J. Brown, F. Crawford, L. Fidani, L. Giuffra, A. Haynes, N. Irving, L. James, R. Mant, P. Newton, K. Rooke, P. Roques, C. Talbot, M. Pericak-Vance, A. Roses, R. Williamson, M. Rossor, M. Owen and J. Hardy (1991) Segregation of a missense mutation in the amyloid precursor protein gene with familial Alzheimer's disease, *Nature* 349, 704–706.

Goedert, M., R. Jakes, M.G. Spillantini, M. Hasegawa, M.J. Smith and R.A. Crowther (1996) Assembly of microtubule- associated protein tau into Alzheimer-like filaments induced by sulphated glycosaminoglycans, *Nature* 383, 550–553.

Goldgaber, D., M.I. Lerman, O.W. McBride, U. Saffiotti and D.C. Gajdusek (1987) Characterization and chromosomal localization of a cDNA encoding brain amyloid of Alzheimer's disease, *Science* 235, 877–880.

Gonatas, N.K., A. Stieber, Z. Mourelatos, Y. Chen, J.O. Gonatas, S.H. Appel, A.P. Hays, W.F. Hickey and J.-J. Hauw (1992) Fragmentation of the Golgi apparatus of motor neurons in amyotrophic lateral sclerosis, *Am. J. Pathol.* 140, 731–737.

Gurney, M.E., H. Pu, A.Y. Chiu, M.C. Dal Canto, C.Y. Polchow, D.D. Alexander, J. Caliendo, A. Hentati, Y.W. Kwon, H.-X. Deng, W. Chen, P. Zhai, R.L. Sufit and T. Siddique (1994) Motor neuron degeneration in mice that express a human Cu,Zn superoxide dismutase mutation, *Science* 264, 1772–1775.

Gurney, M.E., F.B. Cuttings, P. Zhai, A. Doble, C.P. Taylor, P.K. Andrus and E.D. Hall (1996) Benefit of vitamin E, riluzole, and gabapentin in a transgenic model of familial amyotrophic lateral sclerosis, *Ann. Neurol.* 39, 147–157.

Haass, C., E.H. Koo, A. Mellon, A.Y. Hung and D.J. Selkoe (1992) Targeting of cell-surface β-amyloid precursor protein to lysosomes: alternative processing into amyloid-bearing fragments, *Nature* 357, 500–503.

Hendricks, L., C.M. van Duijn, P. Cras, M. Cruts, W. Van Hul, F. van Harskamp, A. Warren, M.G. McInnis, S.E. Antonarakis, J.-J. Martin, A. Hofman and C. Van Broeckhoven (1992) Presenile dementia and cerebal haemorrhage linked to a mutation at codon 692 of the β-amyloid precursor protein gene, *Nature Genetics* 1, 218–221.

Higgins, L.S., D.M. Holtzman, J. Rabin, W.C. Mobley and B. Cordell (1994) Transgenic mouse brain histopathology resembles early Alzheimer's disease, *Ann. Neurol.* 35, 598–607.

Hirano, A., N. Malamud, L.T. Kurland and H.M. Zimmerman (1969) A review of the pathologic findings in amyotrophic lateral sclerosis, in: *in Motor Neuron Diseases. Research on Amyotrophic Lateral Sclerosis and Other Disorders*, (F.H. Norris and L.T. Kurland, eds), pp. 51–60, Grune & Stratton, New York.

Hirano, A. and S. Kato (1992) Fine structural study of sporadic and familial amyotrophic lateral sclerosis, in: *in Handbook of Amyotrophic Lateral Sclerosis*, (R.A. Smith, ed), pp. 183–192, Marcel Dekker, New York.

Howland, D.S., M.J. Savage, F.A. Huntress, R.E. Wallace, D.A. Schwartz, T. Loh, R.H. Melloni,Jr., L.J. DeGennaro, B.D. Greenberg, R. Siman, M.E. Swanson and R.W. Scott (1995) Mutant and native human β-amyloid precursor proteins in transgenic mouse brain, *Neurobiol. Aging* 16, 685–699.

Hsiao, K., P. Chapman, S. Nilsen, C. Eckman, Y. Harigaya, S. Younkin, F. Yang and G. Cole (1996) Correlative memory deficts, Aβ elevation and amyloid plaques in transgenic mice, *Science* 274, 99–102.

Hsiao, K.K., D. Groth, M. Scott, S.-L. Yang, H. Serban, D. Rapp, D. Foster, M. Torchia, S.J. DeArmond and S.B. Prusiner (1994) Serial transmission in rodents in neurodegeneration from transgenic mice expressing mutant prion proteins, *Proc. Natl. Acad. Sci. USA* 91, 9126–9130.

Hsiao, K.K., D.R. Borchelt, K. Olson, R. Johannsdottir, C. Kitt, W. Yunis, S. Xu, C. Eckman, S. Younkin, D. Price, C. Iadecola, H.B. Clark and G. Carlson (1995) Age-related CNS disorder and early death in transgenic FVB/N mice overexpressing Alzheimer amyloid precursor proteins, *Neuron* 15, 1203–1218.

Iwatsubo, T., A. Odaka, N. Suzuki, H. Mizusawa, N. Nukina and Y. Ihara (1994) Visualization of Aβ42(43)-positive and Aβ40-positive senile plaques with end-specific Aβ-monoclonal antibodies: evidence that an initially deposited Aβ species is Aβ1–42(43), *Neuron* 13, 45–53.

Iwatsubo, T., D.M.A. Mann, A. Odaka, N. Suzuki and Y. Ihara (1995) Amyloid β protein (Aβ) deposition: Aβ42(43) preceds Aβ40 in Down syndrome. *Ann. Neurol.* 37, 294–299.

Jarrett, J.T., E.P. Berger and P.T. Lansbury,Jr. (1993) The carboxy terminus of the β amyloid protein is critical for the seeding of amyloid formation: implications for the pathogenesis of Alzheimer's disease, *Biochemistry* 32, 4693–4697.

Jarrett, J.T. and P.T. Lansbury,Jr. (1993) Seeding "one-dimensional crystallization" of amyloid: a pathogenic mechanism in Alzheimer's disease and scrapie? *Cell* 73, 1055–1058.

Kang, J., H.-G. Lemaire, A. Unterbeck, J.M. Salbaum, C.L. Masters, K.-H. Grzeschik, G. Multhaup, K. Beyreuther and B. Müller-Hill (1987) The precursor of Alzheimer's disease amyloid A4 protein resembles a cell-surface receptor, *Nature* 325, 733–736.

Kitaguchi, N., Y. Takahashi, Y. Tokushima, S. Shiojiri and H. Ito (1988) Novel precursor of Alzheimer's disease amyloid protein shows protease inhibitory activity, *Nature* 331, 530–532.

Koo, E.H. and S.L. Squazzo (1994) Evidence that production and release of amyloid β-protein involves the endocytic pathway, *J. Biol. Chem.* 269, 17386–17389.

Kovacs, D.M., H.J. Fausett, K.J. Page, T.-W. Kim, R.D. Moir, D.E. Merriam, R.D. Hollister, O.G. Hallmark, R. Mancini, K.M. Felsenstein, B.T. Hyman, R.E. Tanzi and W. Wasco (1996) Alzheimer-associated presenilins 1 and 2: neuronal expression in brain and localization to intracellular membranes in mammalian cells, *Nature Med.* 2, 224–229.

LaFerla, F.M., B.T. Tinkle, C.J. Bieberich, C.C. Haudenschild and G. Jay (1995) The Alzheimer's Aβ peptide induces neurodegeneration and apoptotic cell death in transgenic mice, *Nature Genetics* 9, 21–30.

Lee, M.K., H.H. Slunt, L.J. Martin, G. Thinakaran, G. Kim, S.E. Gandy, M. Seeger, E. Koo, D.L. Price and S.S. Sisodia (1996) Expression of presenilin 1 and 2 (PS1 and PS2) in human and murine tissues, *J. Neurosci.* 16, 7513–7525.

Lee, V.M.-Y. (1995) Disruption of the cytoskeleton in Alzheimer's disease, *Curr. Opin. Neurobiol.* 5, 663–668.

Lemere, C.A., J.K. Blusztajn, H. Yamaguchi, T. Wisniewski, T.C. Saido and D.J. Selkoe (1996a) Sequence of deposition of heterogeneous amyloid β-peptides and APO E in Down syndrome: implications for initial events in amyloid plaque formation, *Neurobiol. Dis.* 3, 16–32.

Lemere, C.A., F. Lopera, K.S. Kosik, C.L. Lendon, J. Ossa, T.C. Saido, H. Yamaguchi, A. Ruiz, A. Martinez, L. Madrigal, L. Hincapie, J.C.L. Arango, D.C. Anthony, E.H. Koo, A.M. Goate, D.J. Selkoe and J.C.V. Arango (1996b) The E280A presenilin 1 Alzheimer mutation produces increased Aβ42 deposition and severe cerebellar pathology, *Nature Med.* 2, 1146–1150.

Levitan, D. and I. Greenwald (1995) Facilitation of *lin-12*-mediated signalling by *sel-12*, a *Caenorhabditis elegans* S182 Alzheimer's disease gene, *Nature* 377, 351–354.

Levy-Lahad, E., W. Wasco, P. Poorkaj, D.M. Romano, J. Oshima, W.H. Pettingell, C.-E. Yu, P.D. Jondro, S.D. Schmidt, K. Wang, A.C. Crowley, Y.-H. Fu, S.Y. Guenette, D. Galas, E. Nemens, E.M. Wijsman, T.D. Bird, G.D. Schellenberg and R.E. Tanzi (1995a) Candidate gene for the chromosome 1 familial Alzheimer's disease locus, *Science* 269, 973–977.

Levy-Lahad, E., E.M. Wijsman, E. Nemens, L. Anderson, K.A.B. Goddard, J.L. Weber, T.D. Bird and G.D. Schellenberg (1995b) A familial Alzheimer's disease locus on chromosome 1, *Science* 269, 970–973.

Li, J., J. Ma and H. Potter (1995) Identification and expression analysis of a potential familial Alzheimer disease gene on chromosome 1 related to *AD3*, *Proc. Natl. Acad. Sci. USA* 92, 12180–12184.

Liu, X.F. and V. Cizewski Culotta (1994) The requirement for yeast superoxide dismutase is bypassed through mutations in *BSD2*, a novel metal homeostasis gene, *Mol. Cell. Biol.* 14, 7037–7045.

Masters, C.L., G. Multhaup, G. Simms, J. Pottgiesser, R.N. Martins and K. Beyreuther (1985a) Neuronal origin of a cerebral amyloid: neurofibrillary tangles of Alzheimer's disease contain the same protein as the amyloid of plaque cores and blood vessels, *EMBO J.* 4, 2757–2763.

Masters, C.L., G. Simms, N.A. Weinman, G. Multhaup, B.L. McDonald and K. Beyreuther (1985b) Amyloid plaque core protein in Alzheimer disease and Down syndrome, *Proc. Natl. Acad. Sci. USA* 82, 4245–4249.

Matsumoto, S., H. Kusaka, N. Murakami, Y. Hashizume, H. Okazaki and A. Hirano (1992) Basophilic inclusions in sporadic juvenile amyotrophic lateral sclerosis: an immunocytochemical and ultrastructural study, *Acta Neuropathol.* 83, 579–583.

Moran, P.M., L.S. Higgins, B. Cordell and P.C. Moser (1995) Age-related learning deficits in transgenic mice expressing the 751-amino acid isoform of human β-amyloid precursor protein, *Proc. Natl. Acad. Sci. USA* 92, 5341–5345.

Morrison, B.M., J.W. Gordon, M.E. Ripps and J.H. Morrison (1996) Quantitative immunocytochemical analysis of the spinal cord in G86R superoxide dismutase transgenic mice: neurochemical correlates of selective vulnerability, *J. Comp. Neurol.* 373, 619–631.

Mourelatos, Z., N.K. Gonatas, A. Stieber, M.E. Gurney and M.C. Dal Canto (1996) The Golgi apparatus of spinal cord motor neurons in transgenic mice expressing mutant Cu,Zn superoxide dismutase becomes fragmented in early, preclinical stages of the disease, *Proc. Natl. Acad. Sci. USA* 93, 5472–5477.

Mullan, M., F. Crawford, K. Axelman, H. Houlden, L. Lillius, B. Winblad and L. Lannfelt (1992) A pathogenic mutation for probable Alzheimer's disease in the APP gene at the N-terminus of β-amyloid, *Nature Genetics* 1, 345–347.

Naruse, S., S. Igarashi, H. Kobayashi, K. Aoki, T. Inuzuka, K. Kaneko, T. Shimizu, K. Iihara, T. Kojima, T. Miyatake and S. Tsuji (1991) Mis-sense mutation Val->Ile in exon 17 of amyloid precursor protein gene in Japanese familial Alzheimer's disease, *Lancet* 337, 978–979.

Oster-Granite, M.L., D.L. McPhie, J. Greenan and R.L. Neve (1996) Age-dependent neuronal and synaptic degeneration in mice transgenic for the C terminus of the amyloid precursor protein, *J. Neurosci.* 16, 6732–6741.

Pardo, C.A., Z. Xu, D.R. Borchelt, D.L. Price, S.S. Sisodia and D.W. Cleveland (1995) Superoxide dismutase is an abundant component in cell bodies, dendrites, and axons of motor neurons and in a subset of other neurons, *Proc. Natl. Acad. Sci. USA* 92, 954–958.

Pike, C.J., D. Burdick, A.J. Walencewicz, C.G. Glabe and C.W. Cotman (1993) Neurodegeneration induced by β-amyloid peptides *in vitro*: the role of peptide assembly state, *J. Neurosci.* 13, 1676–1687.

Ponte, P., P. Gonzalez-DeWhitt, J. Schilling, J. Miller, D. Hsu, B. Greenberg, K. Davis, W. Wallace, I. Lieberburg, F. Fuller and B. Cordell (1988) A new A4 amyloid mRNA contains a domain homologous to serine proteinase inhibitors, *Nature* 331, 525–532.

Price, D.L., C.H. Kawas and S.S. Sisodia (1996) Aging of the brain and dementia of the Alzheimer's type, in: *in Principles of Neural Science*, (E.R. Kandel, J.H. Schwartz and T.M. Jessell, eds), Elsevier, New York.

Price, D.L. and S.S. Sisodia (1994) Cellular and molecular biology of Alzheimer's disease and animal models, *Annu. Rev. Med.* 45, 435–446.

Rabizadeh, S., E. Butler Gralla, D.R. Borchelt, R. Gwinn, J. Selverstone Valentine, S. Sisodia, P. Wong, M. Lee, H. Hahn and D.E. Bredesen (1995) Mutations associated with amyotrophic lateral sclerosis convert superoxide dismutase from an antiapoptotic gene to a proapoptotic gene: studies in yeast and neural cells, *Proc. Natl. Acad. Sci. USA* 92, 3024–3028.

Reaume, A.G., J.L. Elliott, E.K. Hoffman, N.W. Kowall, R.J. Ferrante, D.F. Siwek, H.M. Wilcox, D.G. Flood, M.F. Beal, R.H. Brown,Jr., R.W. Scott and W.D. Snider (1996) Motor neurons in Cu/Zn superoxide dismutase-deficient mice develop normally but exhibit enhanced cell death after axonal injury, *Nature Genetics* 13, 43–47.

Ripps, M.E., G.W. Huntley, P.R. Hof, J.H. Morrison and J.W. Gordon (1995) Transgenic mice expressing an altered murine superoxide dismutase gene provide an animal model of amyotrophic lateral sclerosis, *Proc. Natl. Acad. Sci. USA* 92, 689–693.

Robakis, N.K., N. Ramakrishna, G. Wolfe and H.M. Wisniewski (1987) Molecular cloning and characterization of a cDNA encoding the cerebrovascular and the neuritic plaque amyloid peptides, *Proc. Natl. Acad. Sci. USA* 84, 4190–4194.

Rockenstein, E.M., L. McConlogue, H. Tan, M. Power, E. Masliah and L. Mucke (1995) Levels and alternative splicing of amyloid β protein precursor (APP) transcripts in brains of APP transgenic mice and humans with Alzheimer's disease, *J. Biol. Chem.* 270, 28257–28267.

Rogaev, E.I., R. Sherrington, E.A. Rogaeva, G. Levesque, M. Ikeda, Y. Liang, H. Chi, C. Lin, K. Holman, T. Tsuda, L. Mar, S. Sorbi, B. Nacmias, S. Piacentini, L. Amaducci, I. Chumakov, D. Cohen, L. Lannfelt, P.E. Fraser, J.M. Rommens and P.H. St George-Hyslop (1995) Familial Alzheimer's disease in kindreds with missense mutations in a gene on chromosome 1 related to the Alzheimer's disease type 3 gene, *Nature* 376, 775–778.

Rosen, D.R., T. Siddique, D. Patterson, D.A. Figlewicz, P. Sapp, A. Hentati, D. Donaldson, J. Goto, J.P. O'Regan, H.-X. Deng, Z. Rahmani, A. Krizus, D. McKenna-Yasek, A. Cayabyab, S.M. Gaston, R. Berger, R.E. Tanzi, J.J. Halperin, B. Herzfeldt, R. Van den Bergh, W.-Y. Hung, T. Bird, G. Deng, D.W. Mulder, C. Smyth, N.G. Laing, E. Soriano, M.A. Pericak-Vance, J. Haines, G.A. Rouleau, J.S. Gusella, H.R. Horvitz and R.H. Brown,Jr. (1993) Mutations in Cu/Zn superoxide dismutase gene are associated with familial amyotrophic lateral sclerosis, *Nature* 362, 59–62.

Roses, A.D. (1995) Apolipoprotein E genotyping in the differential diagnosis, not prediction, of Alzheimer's disease, *Ann. Neurol.* 38, 6–14.

Rouleau, G.A., A.W. Clark, K. Rooke, A. Pramatarova, A. Krizus, O. Suchowersky, J.-P. Julien and D. Figlewicz (1996) SOD1 mutation is associated with accumulation of neurofilaments in amyotrophic lateral sclerosis, *Ann. Neurol.* 39, 128–131.

Schellenberg, G.D. (1995) Progress in Alzheimer's disease genetics, *Curr. Opin. Neurol.* 8, 262–267.

Scheuner, D., C. Eckman, M. Jensen, X. Song, M. Citron, N. Suzuki, T.D. Bird, J. Hardy, M. Hutton, W. Kukull, E. Larson, E. Levy-Lahad, M. Viitanen, E. Peskind, P. Poorkaj, G. Schellenberg, R. Tanzi, W. Wasco, L. Lannfelt, D. Selkoe and S. Younkin (1996) Secreted amyloid β-protein similar to that in the senile plaques of Alzheimer's disease is increased *in vivo* by the presenilin 1 and 2 and *APP* mutations linked to familial Alzheimer's disease, *Nature Med.* 2, 864–852.

Scott, M., D. Foster, C. Mirenda, D. Serban, F. Coufal, M. Wälchli, M. Torchia, D. Groth, G. Carlson, S.J. DeArmond, D. Westaway and S.B. Prusiner (1989) Transgenic mice expressing hamster prion protein produce species-specific scrapie infectivity and amyloid plaques, *Cell* 59, 847–857.

Scott, M.R., R. Köhler, D. Foster and S.B. Prusiner (1992) Chimeric prion protein expression in cultured cells and transgenic mice, *Protein Sci.* 1, 986–997.

Sherrington, R., E.I. Rogaev, Y. Liang, E.A. Rogaeva, G. Levesque, M. Ikeda, H. Chi, C. Lin, G. Li, K. Holman, T. Tsuda, L. Mar, J.-F. Foncin, A.C. Bruni, M.P. Montesi, S. Sorbi, I. Rainero, L. Pinessi, L. Nee, I. Chumakov, D. Pollen, A. Brookes, P. Sanseau, R.J. Polinsky, W. Wasco, H.A.R. Da Silva, J.L. Haines, M.A. Pericak-Vance, R.E. Tanzi, A.D. Roses, P.E. Fraser, J.M. Rommens and P.H. St George-Hyslop (1995) Cloning of a gene bearing missense mutations in early-onset familial Alzheimer's disease, *Nature* 375, 754–760.

Shibata, N., M. Hirano and K. Kobayashi (1993) Immunohistochemical demonstration of Cu/Zn superoxide dismutase in the spinal cord of patients with familial amyotrophic lateral sclerosis, *Acta Histochem. Cytochem.* 26, 619–622.

Shibata, N., A. Hirano, M. Kobayashi, T. Siddique, H.-X. Deng, W.-Y. Hung, T. Kato and K. Asayama (1996) Intense superoxide dismutase-1 immunoreactivity in intracytoplasmic hyaline inclusions of familial amyotrophic lateral sclerosis with posterior column involvement, *J. Neuropathol. Exp. Neurol.* 55, 481–490.

Shinkai, Y., M. Yoshimura, Y. Ito, A. Odaka, N. Suzuki, K. Yanagisawa and Y. Ihara (1995) Amyloid β-proteins 1–40 and 1–42(43) in the soluble fraction of extra- and intracranial blood vessels, *Ann. Neurol.* 38, 421–428.

Siddique, T., D.A. Figlewicz, M.A. Pericak-Vance, J.L. Haines, G. Rouleau, A.J. Jeffers, P. Sapp, W.-Y. Hung, J. Bebout, D. McKenna-Yasek, G. Deng, H.B. Horvitz, J.F. Gusella, R.H. Brown,Jr., A.D. Roses, R.P. Roos, D.B. Williams, D.W. Mulder, P.C. Watkins, R. Noore, G. Nicholson, R. Reed, B.R. Brooks, B. Festoff, J.P. Antel, R. Tandan, T.L. Munsat, N.G. Laing, J.J. Halperin, F.H. Norris, R. Van den Bergh, L. Swerts, R.E. Tanzi, B. Jubelt, K.D. Mathews and E.P. Bosch (1991) Linkage of a gene causing familial amyotrophic lateral sclerosis to chromosome 21 and evidence of genetic-locus heterogeneity, *N. Engl. J. Med.* 324, 1381–1384.

Simmons, L.K., P.C. May, K.J. Tomaselli, R.E. Rydel, K.S. Fuson, E.F. Brigham, S. Wright, I. Lieberburg, G.W. Becker, D.N. Brems and et al (1994) Secondary structure of amyloid beta peptide correlates with neurotoxic activity in vitro, *Mol. Pharmacol.* 45, 373–379.

Sisodia, S.S., E.H. Koo, K. Beyreuther, A. Unterbeck and D.L. Price (1990) Evidence that β-amyloid protein in Alzheimer's disease is not derived by normal processing, *Science* 248, 492–495.

Sisodia, S.S. (1992) β-amyloid precursor protein cleavage by a membrane-bound protease, *Proc. Natl. Acad. Sci. USA* 89, 6075–6079.

St George-Hyslop, P.H., P. Haines, E. Rogaev, M. Mortilla, G. Vaula, M. Pericak-Vance, J.-F. Foncin, M. Montesi, A. Bruni, S. Sorbi, I. Rainero, L. Pinessi, D. Pollen, R. Polinsky, L. Nee, J. Kennedy, F. Macciardi, E. Rogaeva, Y. Liang, N. Alexandrova, W. Lukiw, K. Schlumpf, R. Tanzi, T. Tsuda, L. Farrer, J.-M. Cantu, R.

Duara, L. Amaducci, L. Bergamini, J. Gusella, A. Roses and D. Crapper McLachlan (1992) Genetic evidence for a novel familial Alzheimer's disease locus on chromosome 14. *Nature Genetics* 2, 330–334.

Su, J.H., A.J. Anderson, B.J. Cummings and C.W. Cotman (1994) Immunohistochemical evidence for apoptosis in Alzheimer's disease, *Neuroreport* 5, 2529–2533.

Suzuki, N., T.T. Cheung, X.-D. Cai, A. Odaka, L. Otvos,Jr., C. Eckman, T.E. Golde and S.G. Younkin (1994) An increased percentage of long amyloid β protein secreted by familial amyloid β protein precursor (βAPP_{717}) mutants, *Science* 264, 1336–1340.

Tanzi, R.E., A.I. McClatchey, E.D. Lampert, L. Villa-Komaroff, J.F. Gusella and R.L. Neve (1988) Protease inhibitor domain encoded by an amyloid protein precursor mRNA associated with Alzheimer's disease, *Nature* 331, 528–530.

Thinakaran, G., D.R. Borchelt, M.K. Lee, H.H. Slunt, L. Spitzer, G. Kim, T. Ratovitski, F. Davenport, C. Nordstedt, M. Seeger, J. Hardy, A.I. Levey, S.E. Gandy, N. Jenkins, N. Copeland, D.L. Price and S.S. Sisodia (1996a) Endoproteolysis of presenilin 1 and accumulation of processed derivatives *in vivo*, *Neuron* 17, 181–190.

Thinakaran, G., L.J. Martin, D.R. Borchelt, S.E. Gandy, S.S. Sisodia and D.L. Price (1996b) Studies of Aβ amyloidogenesis in model systems of Alzheimer's disease, in: *in The Dementias, Association for Research in Nervous and Mental Disease, American Psychiatric Press, Washington, D.C.*

Thinakaran, G., D.B. Teplow, R. Siman, B. Greenberg and S.S. Sisodia (1996c) Metabolism of the APP variant in Neuro2A (N2a) cells: evidence that cleavage at the "β-secretase" site occurs in the Golgi apparatus, *J. Biol. Chem.* 271, 9390–9397.

Troncoso, J.C., R.R. Sukhov, C.H. Kawas and V.E. Koliatsos (1996) *In situ* labeling of dying cortical neurons in normal aging and in Alzheimer's disease: correlations with senile plaques and disease progression, *J. Neuropathol. Exp. Neurol.* 55, 1134–1142.

Tu, P.-H., P. Raju, K.A. Robinson, M.E. Gurney, J.Q. Trojanowski and V.M.-Y. Lee (1996) Transgenic mice carrying a human mutant superoxide dismutase transgene develop neuronal cytoskeletal pathology resembling human amyotrophic lateral sclerosis lesions, *Proc. Natl. Acad. Sci. USA* 93, 3155–3160.

Wang, R., J.F. Meschia, R.J. Cotter and S.S. Sisodia (1991) Secretion of the β/A4 amyloid precursor protein. Identification of a cleavage site in cultured mammalian cells, *J. Biol. Chem.* 266, 16960–16964.

Weidemann, A., G. König, D. Bunke, P. Fischer, J.M. Salbaum, C.L. Masters and K. Beyreuther (1989) Identification, biogenesis, and localization of precursors of Alzheimer's disease A4 amyloid protein, *Cell* 57, 115–126.

Wiedau-Pazos, M., J.J. Goto, S. Rabizadeh, E.B. Gralla, J.A. Roe, M.K. Lee, J.S. Valentine and D.E. Bredesen (1996) Altered reactivity of superoxide dismutase in familial amyotrophic lateral sclerosis, *Science* 271, 515–518.

Wong, P.C., C.A. Pardo, D.R. Borchelt, M.K. Lee, N.G. Copeland, N.A. Jenkins, S.S. Sisodia, D.W. Cleveland and D.L. Price (1995) An adverse property of a familial ALS-linked SOD1 mutation causes motor neuron disease characterized by vacuolar degeneration of mitochondria, *Neuron* 14, 1105–1116.

Yankner, B.A. (1996) Mechanisms of neuronal degeneration in Alzheimer's disease, *Neuron* 16, 921–932.

10

PROTEINS IMPLICATED IN ALZHEIMER DISEASE

The Role of FE65, a New Adapter which Binds to β-Amyloid Precursor Protein

Kira S. Ermekova,[1] Alex Chang,[1] Nicola Zambrano,[2] Paola de Candia,[2] Tommaso Russo[2*] and Marius Sudol[1*]

[1]Mount Sinai School of Medicine
Department of Biochemistry and Molecular Biology
One Gustave Levy Place
New York, New York 10029-6574,
[2]Università di Napoli Federico II
Dipartimento di Biochimica e Biotecnologie Mediche
Via S. Pansini 5
I-80131 Napoli, Italy

1. ALZHEIMER DISEASE: A NEURODEGENERATIVE DISORDER WHOSE ETIOLOGY INVOLVES A NUMBER OF DIFFERENT PROTEINS

Alzheimer disease (AD) is a devastating neurodegenerative disorder which affects several million people in the world. It is characterized by progressive memory loss and cognitive deficits. In the U.S. alone, the direct and indirect costs of AD are exorbitant, at an estimated 90 billion dollars per year. The etiology and the molecular basis of AD are unknown. The apparent heterogeneity of the molecules implicated in the pathogenesis of AD is consistent with the hypothesis that the disease could be due to independent molecular defects. However, these apparently unrelated molecular lesions seem to exert similar effects because there are three pathological features common to all cases of AD: i) The presence of extracellular senile plaques in the brain, ii) The appearance of neuro-fibrillary tangles in neurons, and ultimately, iii) Massive neuronal loss. To date, the majority of AD

* Università di Napoli Federico II, Dipartimento di Biochimica e Biotecnologie Mediche, Via S. Pansini 5, I-80131 Napoli, Italy Phone: +39817463131,

Molecular and Cellular Mechanisms of Neuronal Plasticity, edited by Ehrlich. Plenum Press, New York, 1998.

research has focused on senile plaques, neurofibrillary tangles, and their principal components: the β-amyloid peptide (Aβ), its precursor protein (β-APP), and the tau-protein[1,2,3,4,5]. The repertoire of proteins which play an etiological role in AD increased when linkage studies led to the identification of two genes encoding membrane proteins named presenilin 1 (PS1) and presenilin 2 (PS2). Mutations of PS1 and PS2 were demonstrated to be responsible for familial forms of AD, which account for at least 10% of all AD cases. In addition, clear genetic evidence indicates that a naturally occurring allele of the apolipoprotein E gene (ε4) is associated with a high risk of AD.

With the exception of subfamily members, the proteins implicated in AD are apparently unrelated to each other, and little is known about their normal functions. The aim of this review is to discuss several of these proteins and their putative roles in AD etiology. Specifically, we will concentrate on FE65, a new adapter protein which interacts with the intracellular domain of β-APP and may provide clues to its metabolism in normal and AD-affected neural tissues. One of the messages which we would like to convey here is that adapter type signal transduction molecules may play an important role as key regulators of the molecular events underlying AD.

2. PROTEINS AND GENES IMPLICATED IN THE PATHOGENESIS OF AD

2.1. β-App

Extensive literature exists on β-APP. Its possible functions and its processing, which result in the generation of the Aβ, have been exhaustively described in several review articles[1,3,6]. As summarized in Figure 1, the β-APP is a large N-glycosylated transmembrane protein of about 120 kDa with the general structure of a surface molecule[7,8,9,10,11]. It possesses a short cytoplasmic tail (47 residues long) which contains the NPTY sequence motif, which was proven to be an internalization signal for various transmembrane proteins, and the recognition core for one of the protein modules (PID/PTB, see below)[12]. At least four β-APP isoforms exist (563, 695, 751, and 770 amino acids long), derived by alternative splicing of the primary transcript of the β-APP gene, located on chromosome 21. Two other proteins which are closely related to β-APP and which are its possible paralogs were recently identified and named: APLP1, and APLP2[13,14,15]. Like β-APP, APLP1 and APLP2 are membrane-associated glycoproteins with a similar pattern of expression in the brain and throughout the body[13,16]. Little is known about the putative extracellular ligands of the β-APP protein family or about the molecular function of these membrane molecules. The fact that they share some characteristics with integrin receptors invites various speculations and specific experimental approaches to study their roles in cell signaling. The secreted form of β-APP seems to function as a neurotrophic factor, by inducing cell proliferation and/or cell adhesion (for a review see ref. 3). Aβ is derived from β-APP by complex proteolytic processing[7,9,17]. It represents a 40–42 amino acid long, internal region of β-APP, and it is located in the extracellular juxtamembrane region and the N-terminal half of the transmembrane region (Figure 1). β-APP can undergo proteolytic processing by two major pathways: (i) After normal constitutive cleavage by α-secretase, the large secreted amino-terminal portion of the protein (s-APP) is generated along with a small 10 kDa membrane-associated carboxyterminal fragment. This posttranslational modification occurs within the Aβ sequence, thus preventing the formation of the peptide. (ii) During both normal and abnormal processing, formation of the 4 kDa amyloid β-peptide occurs as a result of the two proteolytic cleavages at the N- terminus by β-secretase, and the other at the C-terminus by γ-secretase[17,18,19,20,21,22]. In pathological conditions, the level of Aβ production

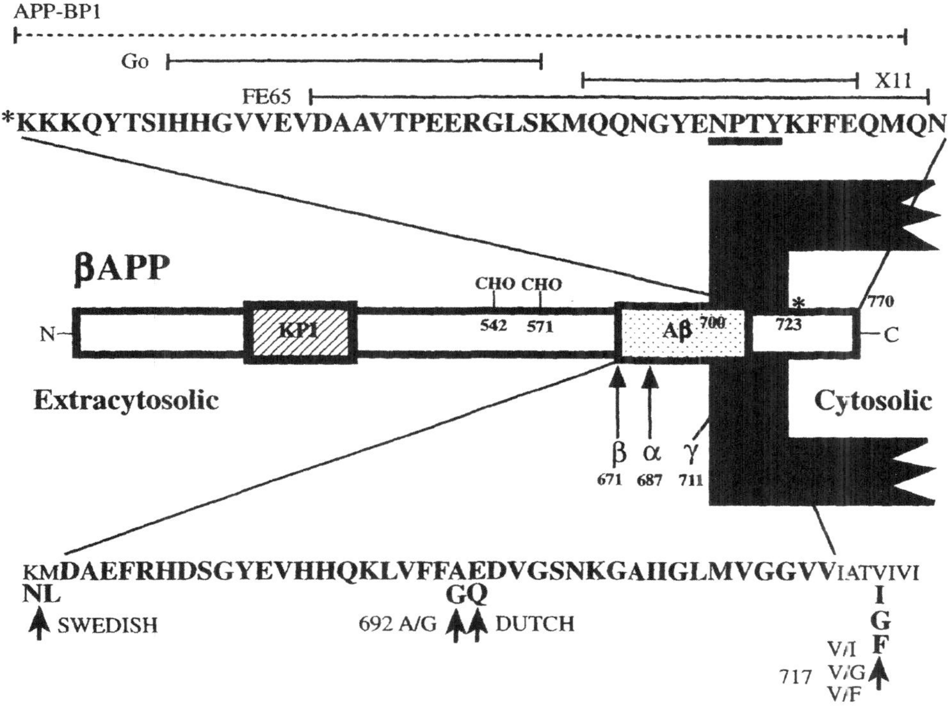

Figure 1. Schematic representation of β-APP. Numbers 1 and 770 denote the N- and the C-terminal ends (APP770 isoform). The precursor molecule is inserted in a phospholipid bilayer, the N-terminal region being extracytosolic and the short C-terminal portion being cytosolic. The Kunitz Protease Inhibitor (KPI) region, encoded by an alternatively spliced exon, is indicated, followed by a sequence (patterned box) encoded by a second alternatively spliced exon. Two sites for the glycosylation are also indicated (CHO) on the extracytosolic domain. The β-amyloid peptide (Aβ) is shown and the cleavage positions by the β-, α- and γ-secretase are indicated by arrows. Its sequence is magnified, in the lower part of the figure: the Aβ sequence is in bold and the altered residues of the mutant forms of the precursor associated with the Familial Alzheimer's Disease (FAD) are indicated by arrows. Note that the Swedish and the codon 717 mutations lie outside the Aβ peptide region. The C-terminal, intracellular domain of the β-APP molecule is magnified, in the upper part of the figure. The NPTY sequence is underlined. The solid lines in the upper side of the figure indicate the minimal peptide regions for the binding of β-APP with FE65, X11 and Go; for APP-BP1 the whole C-terminal tail is indicated by a dashed line, since the minimal peptide of β-APP required for the interaction with APP-BP1 is not known to date. This and remaining figures of this review are available at the Web site : *http://www.mssm.edu/biochem/ms.html*

and accumulation is dramatically increased[4,23,24,25,26,27]. α-, β-, and γ-secretases are proteases or proteolytic complexes, in some cases shown to be cell-specific, which have been implicated in β-APP processing, but which have not been purified as yet. Aβ is known to be the principal constituent of the depositions which form extracellular senile lesions and cerebrovascular amyloid deposits in the brain of AD and Down Syndrome (DS) patients. By cell culture experiments and transgenic mice approaches, the aggregated Aβ peptide was shown to be neurotoxic, causing neuronal death by apoptosis[28,29,30]. In a recent study, transgenic mice overexpressing mutant form of β-APP were generated[31]. These mice had an approximate 10-fold increase in Aβ levels. As expected, when subjected to behavioral testing, they demonstrated abnormalities in learning and memory by age of 9 to 10 months relative to the control mice. In another report, Aβ formation in cell-free preparations has been described[32]. This system provides an attractive bioassay for studying gene products that participate in and/or control the formation of Aβ.

The genetic linkage between AD and mutations in β-APP close to the Aβ segment was found[1,33]. Characterization of specific mutations in a small number of cases of familial AD suggests that some mutations of β-APP may cause the AD pathology. In spite of the efforts of many laboratories, our understanding of the molecular mechanisms of Aβ formation and the physiological role of β-APP and its posttranslational processing are still quite limited.

2.2. Presenilins

Presenilin 1 (PS1) and presenilin 2 (PS2) are integral membrane proteins with unknown physiological functions, which are implicated in the early onset familial AD (for reviews see refs. 34, 35). Presenilins are proteins of 45–65 kDa which contain multiple trans-membrane domains spaced by hydrophilic loops and with the N-terminus and the C-terminus oriented toward the cytoplasm[34,36,37]. PS1 and PS2 are expressed ubiquitously; in the brain they accumulate in cortical and hippocampal neurons, and to a lesser extent in the subcortical areas. Incidentally, this pattern of expression overlaps almost completely with that of β-APP and of FE65[38,39]. The most homologous protein to the presenilins is the *C. elegans* protein SEL-12, a member of the Notch receptor family of transmembrane proteins which transduce various signals from cell surface to the nucleus[34]. Considering the power of genetics in the nematode system, interesting clues to signaling pathways as initiated by presenilins at the cellular membranes, may come from the study of SEL-12 gene.

Presenilin genes were identified by linkage analysis of several AD families. Two types of mutations occurring in presenilin genes lead to early onset AD: missense mutations and mutations affecting splicing[40,41,42]. The molecular mechanism which relates presenilins and familial AD is not yet known, but the relationship between PSs and Aβ production is now emerging. In fact, it was shown that mutated forms of PS1 and PS2 highly increase the production of Aβ in transfected cells[43]. Moreover, in PS1 transgenic mice, a significant overexpression of Aβ was observed in the brain[43,44,45,46]. In contrast, the overexpression of wild type PS genes in transfected cells or in transgenic mice did not alter the rate of Aβ production. It seems that genetic lesions in PSs result in a gain of function mutations that increase the generation of Aβ in the brain, probably altering the proteolytic processing of β-APP[46]. Furthermore, PSs could contribute to neuronal loss in AD by increasing the apoptosis induced by Aβ or by neurotrophic factor withdrawal[26]. This is consistent with the recent observation which classifies normal PS2 as a functional component of programmed cell death machinery[48].

2.3. Tau

Neurofibrillary tangles result from an accumulation of paired helical filaments (PHF) whose main subunit constituents are the Tau proteins. These are microtubule-associated phosphoproteins (MAP) whose function in the brain is to modulate neuronal microtubule assembly from tubulin. Tau in solution does not seem to have a defined shape or a regular secondary structure, but rather resembles a denatured protein. Based on its amino acid sequence, the following domains/motifs of Tau may be defined: acidic (A), basic (B), proline rich (P1 and P2), tubulin-binding repeats (R1–4, R'), and tail (C)[49,50,51]. The R repeats of Tau are primarily responsible for binding to tubulin; the flanking regions have been shown to increase the binding affinity probably by maintaining optimal conformation of the assayed fusion proteins[50]. Hyperphosphorylated Tau (21 phosphorylation sites) is the principal component of the AD-PHF (PHF-Tau) present in neurofibrillary tangles of AD-affected neurons. However, it is not clear whether the hyperphosphorylated state of Tau is attributed to reduced phosphatase levels or increased kinase activities in AD neurons. Two types of phosphorylation sites are present on the PHF-Tau: Ser/Thr-Pro and Ser/Thr-Xaa where Xaa represents Ser, Arg, Thr, Gly, or Asn. *In vitro* studies indicate that phosphorylation by proline-dependent protein kinases (PDPKs), such as MAP kinase, GSK-3, cdk2, and cdk5, as well as non-PDPKs, such as PKA, PKC, and casein kinase II, are involved[50,51,52,53]. Although most of the additional phosphorylation sites of PHF-Tau are of Ser/Thr-Pro type, *in vitro* experiments demonstrate that this hyperphosphorylation only affects the binding of Tau to microtubules. In contrast, the phosphorylation at the unique PHF-Tau site Ser262 (R1) disrupts Tau-microtubule interactions and confers microtubule instability. This phosphorylation event which changes properties of the Tau protein is probably carried out by MAP/microtubule affinity regulating kinase (MARK)[50]. Phosphorylation/dephosphorylation is known to be a key regulatory event of many molecular processes in a cell, including development and neurodegeneration. The elucidation of the Tau phosphorylation mechanism and the full spectrum of molecular consequences of this reaction will lead to an understanding of the role of Tau in neurodegenerative pathology. It is important to understand the relationship among multiple components and processes such as Tau, PHF generation, and the roles of other molecules implicated in AD, including β-APP and PSs. For example, recent reports indicate that the Aβ peptide was able to activate Tau protein kinase, I/GSK-3, which in turn led to hyperphosphorylation of Tau and consequently resulted in neuronal cell death[53,54].

2.4. ApoE

ApoE is a plasma protein of 34 kDa molecular mass, whose function in the brain is to supply lipids to neuronal membranes during growth and regeneration. Based on genetic evidence, the apolipoprotein E gene, located on chromosome 19, is associated with AD[55,56,57]. Specifically, the allele ε4 has been implicated, based on the observation that individuals who have inherited ε4 have higher levels of accumulated Aβ deposits as well as an earlier age of onset of both sporadic and familial AD when compared to ε3 homozygotes[56]. The three major isoforms of apoE (E2, E3, and E4) are encoded by different alleles, and their corresponding protein products differ by a single amino acid[58,59,60,61]. Cell-surface receptors mediate internalization of apoE-containing complexes via clathrin-coated pits; these are low density lipoprotein receptors (LDL-r) and very low density lipoprotein receptors (VLDL-r). Of interest is that VLDL-r itself has been reported to have genetic link with AD[62,63]. Moreover, apoE was shown to be present in senile plaques and has been suggested to interact with Aβ and/or Tau protein[56].

3. FE65, A NEURAL PROTEIN WITH FEATURES OF A DOCKING PROTEIN AND TRANSCRIPTIONAL REGULATOR

FE65 protein was identified as a product of a brain-specific transcript from a differential screening of a rat brain cDNA library [64]. FE65 is named for the initials of the researcher who isolated the cDNA clone[64] and for the specific number of the isolate from among the first selection of potentially brain-specific clones, 65. The FE65 gene is ex-

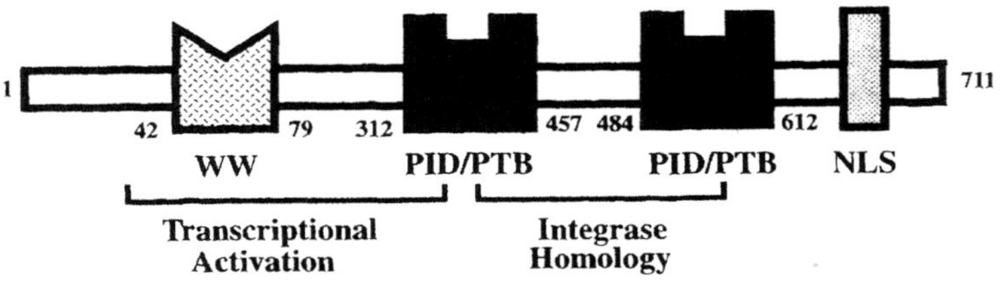

WW Domain: 42 SD**LP**A**GW**MRVQDTS**G**TY**Y**WHIP**T**G**TT**QWEP**P**GRASPS 79

WW Domain: Mutant 42 ...........................***F***..***A***...... 79

Figure 2. Schematic diagram of the rat FE65 protein. Numbers 1 and 711 indicate the N- and the C-terminal positions of the protein, respectively. The WW, PID/PTB1, and C-terminal PID/PTB2 domains are indicated. A possible nuclear localization signal (NLS) is present in the C-terminal end of the protein. The lines below the protein diagram indicate the transcriptional activation and the integrase-homology regions of the protein. Sequences of the WW domain and its mutant (rendered inactive in binding to WW ligands) are shown.

pressed in distinct regions of the brain including the cortex and the hippocampus, and in all the ganglia of the nervous system and sense organs[65]. Transcriptional studies demonstrated that a TATA-less promoter of the FE65 gene drives an efficient transcription of FE65 in neural cells[66]. At least two other proteins, named FE65-L1 and FE65-L2, have been identified which share with FE65 both structural and functional characteristics (67, Russo T. et al., unpublished results).

The FE65 protein was studied by using three different antibodies which always recognize three closely-spaced protein bands of about 90 kDa on Western blots. This heterogeneity is probably due to post-translational modifications[68]. A computer-aided analysis of the FE65 ORF indicated a sequence region which showed high similarity to a segment of all the known retroviral integrases and is thought to be involved in the interactions with the host DNA (Figure 2). Since the aminoterminal domain of FE65 flanking the integrase homology was shown to activate the transcription of a reporter gene when fused to a heterologous DNA-binding domain, it was hypothesized that FE65 could be a transcriptional factor[69]. However, additional possibilities of FE65 function were introduced when two protein-protein interaction domains (WW and PID/PTB), as depicted in Figure 2, were found within its sequence[70,71,72].

3.1. The Protein-Protein Interaction Domains of FE65

Several protein domains, such as SH2 (Src homology 2), SH3 (Src homology 3), PH (pleckstrin homology), and PID/PTB (phosphotyrosine interaction/phosphotyrosine binding), are implicated in protein-protein interactions and in signal transduction pathways[73]. These amino acid sequences fold into globular structures and have prominent binding

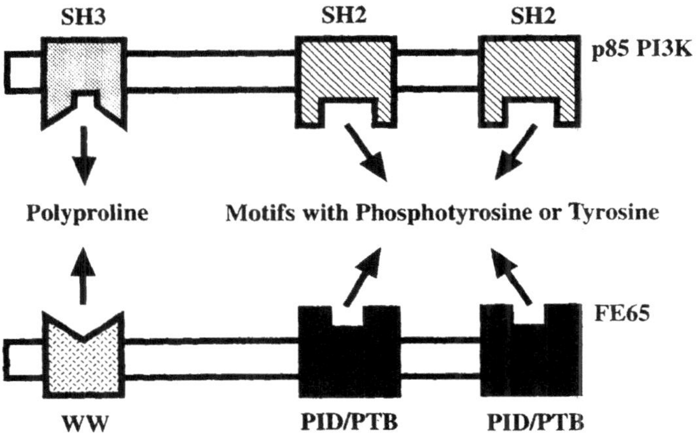

Figure 3. Schematic representation of the two molecules, PI3 kinase-associated protein (p85 PI3K) and FE65, containing different protein-protein interaction domains. We would like to demonstrate an analogy between SH3 and WW domains and between SH2 and PID/PTB domains. The first pair binds to polyprolines, the latter both bind to phosphotyrosine or tyrosine-containing motifs. The interaction of PID/PTB domains with its ligands depends on aminoterminal residues of the motif, NPXY, whereas SH2 binding specificity is determined by the residues which lie carboxyterminally to phosphotyrosine. SH3 domains bind to proline-rich ligands with the consensus PxxP. WW domains of YAP and Nedd-4 interact with peptides containing an xPPxY motif (where P signifies proline, Y signifies tyrosine, and x represents any amino acid). WW domains of other proteins might have slightly different binding preferences, for example, PPLP.

properties usually well conserved through evolution. Various combinations of these domains are frequently present on the same protein (adapters, docking proteins) which enable it to assemble multicomponent complexes that participate in signaling pathways. Some proteins contain regions with catalytic activities along with the modular protein-binding domains. Others, like FE65, consist mostly of interactive domains and act as molecular adapters (Figure 3).

3.1.1. The WW Domain. The region of FE65 spanning from residue 254 to residue 290 contains the WW domain. This structural domain was termed the WW domain (Figure 2) for the two conserved tryptophans within its sequence. It was originally identified in YAP (Yes-kinase Associated Protein), a binder to the SH3 domain of the Yes oncoprotein[71,74]. The WW domain is a universal module that mediates protein-protein interaction by binding to molecules containing xPPxY or PPLP consensus sequences. Similar to other intracellular modules, one of the important features of the WW domain is its widespread occurrence. The WW domain is shared by proteins of diverse functions including structural, regulatory, and signaling proteins in yeast, nematode, and mammals. Among more than 60 examples of proteins which contain the WW domain are dystrophin (implicated in Duchenne and Becker dystrophies), RSP5 (a transcriptional regulator), and ESS1 (a modulator of the cell cycle).

The WW sequence occurs frequently in multiple repeats within the same molecule and is composed of 38 amino acids, four of which are conserved aromatic residues surrounded by beta-strands. Analysis of the primary sequence of different WW domains shows that two or more subvariants of the domain exist[74,75,76]. For instance, the WW domains of YAP and Nedd-4, a protein characterized by a specific pattern of expression at early stages of central nervous system development, demonstrate high similarity to each other, whereas the WW domains of FE65 and ESS1 are distinct from the first group but sequence-wise are similar to each other. It is likely that domains belonging to the same group also share certain functional features, such as interaction with similar ligands. By screening a cDNA expression library, two putative ligands of the WW domain of YAP, WBP-1 and WBP-2, have been identified[77]. Peptide sequence comparison showed homologous proline-rich motifs in the sequence of WBP-1 and WBP-2. The sequence xPPxY, termed for a short PY motif, appears to be a preliminary consensus identified from binding assays and site-specific mutagenesis for binding to the WW domains[77,78]. Interestingly, the PY motif binds to the WW domains of YAP and Nedd-4 but does not interact with the WW domains of FE65 or ESS1 (Ermekova and Sudol, unpublished data).

Recently Leder and his colleagues hypothesized that the SH3 and WW domains can modulate each others' functions by competing for the same proline-rich target motifs[79,80]. They demonstrated that polyproline regions of formin can bind both WW and SH3 domains. Their experimental data showed that the WW modules of the formin-binding proteins compete with the SH3 domain of Abl tyrosine kinase in binding to proline-rich sequences of formin.

Although the WW domain has been identified relatively recently, it has already been implicated in several genetic diseases and biological processes, including a hereditary form of hypertension (Liddle's syndrome), Muscular Dystrophy, and retroviral budding[81,82,83].

3.1.2. The PID/PTB Domains. The carboxyterminal region of FE65 contains two PID/PTB domains (Figure 2). Originally discovered in the oncoprotein Shc, PID/PTB domains were also shown to be present in Shc-related proteins, in IRS (insulin receptor substrate), in X11 neuronal factor, and in several other proteins[72,84,85,86,87]. Upon stimulation of various receptors, Shc binds to phosphorylated tyrosines on these receptors (e.g. EGFR) and subsequently becomes tyrosine-phosphorylated, in which form it interacts with SH2 domain of Grb2. Shc has been viewed as an adapter protein, containing a tyrosine phosphorylation site as well as two modules: SH2 in the carboxy terminus and PID/PTB in the amino terminus[88]. The PID/PTB domain is a globular structure with an average size of 160 amino acids composed of antiparallel beta-sheets[73]. Both SH2 and PID/PTB domains bind to tyrosine-phosphorylated sites with high affinity. However, it was shown by mutagenetic analysis that the specificity of interaction of the PID/PTB domain with phosphotyrosine containing ligands depends on aminoterminal residues of the motif, whereas SH2 domains recognize 3–5 amino acids following the phosphotyrosine[89,90]. Thus, PID/PTB domains bind to NPxY(P)-containing motifs found in a variety of proteins.

4. THE CYTOPLASMIC DOMAIN OF APP IS THE LIGAND OF THE PID/PTB DOMAIN OF FE65

To understand the molecular function of the FE65 protein, identification of the proteins which bind to its PID/PTB domains was attempted. Using the two-hybrid system in yeast, it was shown that both PID/PTB domains of FE65 interact with the intracellular domain of the β-APP and APP-like proteins[70]. Shortly thereafter, these results were confirmed by other laboratories. Using the carboxyterminal cytoplasmic domain of β-APP as bait in the "interaction trap", the human FE65 protein was identified as a cognate ligand[67,91,92,93]. Immunoprecipitation experiments performed in CHO and PC12 cells, expressing the full length APP751 and FE65 transcripts, demonstrated that the interaction, previously shown *in vitro* and in the yeast two-hybrid system, also takes place *in vivo* [68]. In contrast to interactions mediated by other PID/PTB domains[89], the FE65-APP interaction does not depend upon the phosphorylation of the tyrosine present in the NPXY motif of β-APP, and the minimal sequence of β-APP required for the binding is longer than the NPXY core element (see Figure 1 and ref. 68). The differences between FE65 PID/PTB domains and that of Shc were further supported by the observation that despite the high degree of homology between their PID/PTB domains, it was impossible to create chimeric proteins able to interact with both EGF-R and β-APP[68]. Another difference between FE65 and Shc is the presence of two PID/PTB domains in FE65 versus one in Shc. More detailed analysis demonstrated that only the most C-terminal PID/PTB domain of FE65 is necessary for the binding to APP[68], and, recently, a potential ligand for the other PID/PTB domain of FE65 was isolated and characterized. Its sequence and functional features are distinct from APP.

5. POSSIBLE LIGANDS OF THE FE65 WW DOMAIN

To identify binders to the WW domain of FE65, we performed functional screening of mouse embryo expression libraries. Mutant GST-FE65 WW fusion protein was generated (as shown in Figure 2B) in which tryptophan (W) at position 69 was substituted with phenylalanine (F), and proline (P) at position 72 was substituted with alanine (A). These two amino acids are among the most conserved within the domain sequence and are known to be part of the ligand binding interface. Binding of the YAP WW domain to its ligands is abolished if residues at these positions are mutated[78]. Nuclear Magnetic Resonance data show that within the structure of the WW domain these amino acids are critical for appropriate folding of the domain and are directly involved in ligand binding[78,94]. Four partial cDNA clones were isolated using the GST-FE65 WW fusion protein as a probe (Table 1). Three cDNA clones, 13, 48, 9–1/7, have been characterized. They represent sequences that do not share homology with known sequences in the databases. To identify an amino acid motif that interacts specifically with FE65 WW, we have used three complementary approaches: a binding assay with GST-fusion proteins and deletion constructs (Western ligand blotting), a phage-displayed peptide library screening, and the 'SPOTs' technique of generating repertoires of peptides. Three of the clones appeared to contain similar polyproline motifs which are responsible for interaction with the FE65 WW domain. A fourth clone encodes a segment of the mbh1 (myc basic motif homolog-1)/gCap39 (macrophage capping protein 39) protein, which is a known cytoskeletal component belonging to the gelsolin/severin family[965,96,97,98,99,100]. It is an actin-binding protein of about 40 kDa, phosphorylated at multiple serines and threonines[100,101]. In an *in*

Table 1. Summary of cDNA clones isolated from mouse embryo expression libraries screened with radiolabeled GST-FE65 WW fusion protein

Clone number	Times isolated	Length of insert	Source
mbh1	2	470 bp	12, 16d, cDNA library
9-1/7	16	700 bp	12d, 14d, 16d, cDNA library
13	1	1300 bp	12d cDNA library
48	1	2000 bp	14d cDNA library

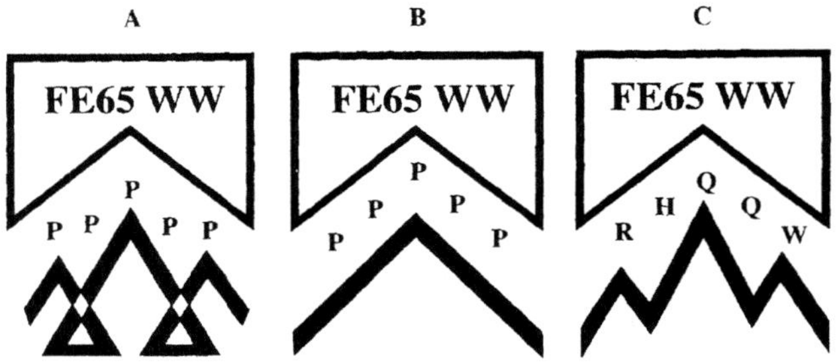

Figure 4. Models of FE65 WW domain interactions with its ligands. A. Non-linear prolines in the polypeptide sequence are converted to a polyproline motif by an ordered structure. Experimental data obtained from binding studies with GST-mbh1 and GST-FE65 WW fusion proteins suggest that this could be the case in vitro and possibly in vivo. B. Proline-rich sequences usually form rigid structures which commonly reside on a protein surface and are able to interact with protein domains, such as SH3 and WW. Clones 9–1/7, 13, and 48, isolated from a functional screen of a phage expression library with GST-FE65 WW fusion protein used as a probe (summarized in Table 1), containing stretches of consequtive prolines. Deletion analysis mapped the binding to the very proline-rich motifs. Our data support our model B in which the FE65 WW domain interacts with a ligand containing linear polyproline motifs. C. Model C represents a putative mimotope of the polyproline ligand, which may bind to the WW domain of FE65 by virtue of "fitting" in the binding pocket of the domain. The primary sequence of the mimotope would not neceserilly contain polyprolines.

vitro binding assay, GST-mbh1 fusion protein showed binding to the GST-FE65 WW protein but not to its mutated versions. However, we were not able to confirm the existence of an *in vivo* binding between FE65 and mbh1/gCap39 by co-immunoprecipitation experiments using extracts of PC12 cells or mouse brain. Furthermore, this protein does not contain polyproline regions; however, it can be hypothesized that interaction occurring *in vitro* is due to the presence of a helix-loop-helix motif within mbh1/gCap39. The models proposed in Figure 4 describe one possible interaction (Figure 4A) in which distant amino acids, including prolines, could be brought together in space by an ordered structure and form a motif composed of an array of prolines sufficient for binding to the WW domain of FE65 under favorable, *in vitro*, conditions.

6. THE CYTODOMAIN OF β-APP ALSO INTERACTS WITH OTHER PROTEINS

Three other proteins were recently shown to interact with the intracellular domain of β-APP. These include the heterotrimeric G protein Go[102], a 59 kDa ubiquitously expressed protein named APP-BP1[103] and the neuron-specific X11 protein[91]. It was shown that intact β-APP binds to oligomeric Go protein and that the intracellular region of β-APP spanning residues 657 to 676 activates Go[102,104]. Furthermore, the interaction of APP with a monoclonal antibody directed against its extracellular domain mimics a ligand-receptor binding which triggers Go activation[104]. Also, the mutant β-APP forms of FAD constitutively activate Go[105], thus inducing various effects, including DNA fragmentation[106]. The characterization of the regions of APP interacting with Go and with FE65 demonstrated that they overlap, and this suggests that the two binding events cannot occur simultaneously (see Figure 1). APP-BP1 interacts both *in vitro* and *in vivo* with the carboxyterminal region of β-APP. Very little is known about this peculiar protein, which is homologous to the product of the *Arabidopsis* auxin resistance gene, AXR1, and to a *C. elegans* protein of unknown function[103].

The third protein interacting with the cytodomain of β-APP, X11, deserves attention. The X11 is a protein of approximately 110–130 kDa, which is expressed primarily in neurons of the cerebellum, the hippocampus, and the spinal cord. Based on amino acid sequence analysis, X11, similar to FE65, could function as an adapter molecule as it has one PID/PTB domain, two PDZ domains, proline-rich segments, and a carboxyterminal putative transmembrane segment. PDZ domains (also known as Discs-Large homology repeats (DHR)), which have been implicated in protein-protein interactions, are usually identified in cell junction-associated proteins, including synapses of the central nervous system. By coimmunoprecipitation in living cells, and in binding assays using purified proteins expressed in bacteria, the PID/PTB domain of X11 was demonstrated to bind to β-APP, specifically to the YENPTY motif[91]. Similar to FE65, this interaction does not depend on phosphorylation of the carboxyterminal tyrosine. The PID/PTB domain of Shc does not bind to β-APP[68,91]. The PID/PTB domains of FE65 and X11 PID/PTB differ in their binding specificities as shown by mutagenesis within the YENPTY motif[91]. Also, site-directed mutagenesis of PID/PTB domain demonstrated that substitution of the phenylalanine (conserved among all PID/PTB domains but that of FE65) at position 479 of X11 with valine eliminates binding to β-APP. FE65 contains cysteine at this position (Cys 443).

7. POSSIBLE FUNCTIONAL RELATIONSHIPS BETWEEN FE65 AND β-APP

The modular protein-binding domains of FE65 and its adapter-like topology clearly suggest possible mechanisms and functional consequences of FE65 interaction with the β-APP cytodomain. As described above, the β-APP molecule is a substrate of several post-translational activities. The β-APP is modified by glycosylation, phosphorylation and sulfation during its movements through the secretory pathway[7,9,17,19,26]. Some of the β-APP molecules reach the cell surface, where they can act as substrates for α-secretase activity. As a result of the proteolytic cleavage of the precursor, a soluble ectodomain fragment is generated (α-APPs), ending at the lysine 687, and corresponding to residue 16 of the Aβ peptide region. A C-terminal, membrane-anchored 10 kDa peptide is the second product of the cleavage, from which a different secretase(s), the γ-secretase, releases a 3 kDa peptide containing the C-terminal tail of the Aβ peptide which is unable to aggregate. Some membrane-anchored β-APP molecules are internalized in clathrin-coated endocytic vesicles[107]; the internalization of the precursor seems to be mediated by the NPTY motif on the β-APP cytodomain[108]. Experimental evidence implicates the involvement of the endocytic pathway of APP in the generation of amyloidogenic C-terminal fragments by β-secretase. In fact, following internalization, the membrane-bound β-APP, is rapidly recycled to the cell surface. While this process takes place, by cleaving the peptide bond between methionine 671 and aspartic 672, the β-secretase generates the soluble ectodomain fragment known as β-APPs. A 12 kDa, C-terminal fragment containing at its N-terminal end the Aβ peptide sequence, and 4 kDa Aβ peptides, identical to the peptides found in the senile plaques of AD-affected patients, are also finally released by the action of the γ-secretase. It is important to note that this phenomenon also takes place in normal cells[18]. The deletion of the β-APP cytoplasmic domain, as well as a shorter deletion involving the NPTY motif[20] leads to a decrease in the production of the Aβ peptide, supporting the role of the endocytic pathway in the generation of the Aβ peptide. However, recent studies indicate that the

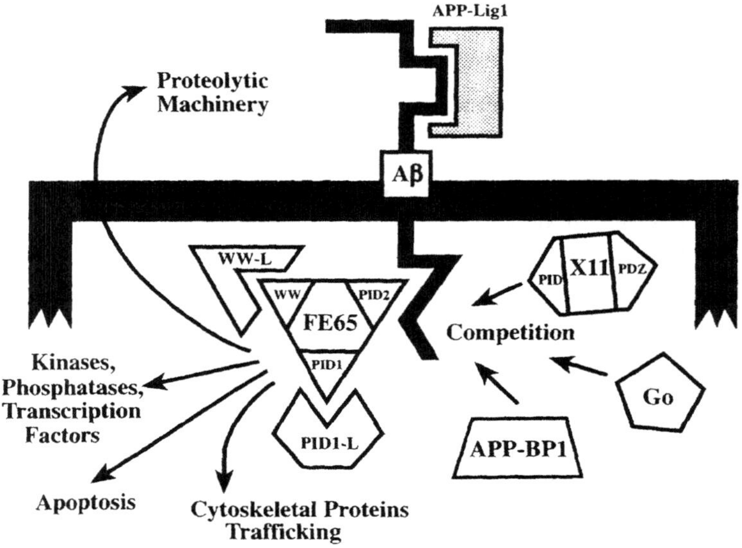

Figure 5. Cellular proteins which interact with the β-APP cellular and extracellular domains. FE65, X11, Go and APP-BP1 are the intracytosolic ligands of β-APP. Due to the limited size of the β-APP cytosolic domain, the binding sites for the different proteins overlap, generating a system in which the competition for the interaction sites may evoke different biochemical and cellular effects. The interaction of the FE65 WW and PID/PTB1/2 domains with other proteins could result in the regulation of the β-APP processing.

generation of the Aβ peptide by the β-secretase pathway also involves the secretory pathway[109].

The competition among protein ligands of the APP cytodomain could provide a sophisticated mechanism which controls the sorting of the membrane-associated APP towards the cellular compartments where different secretase activities are located (Figure 5). As described above, at least four different proteins interact with APP through its cytodomain. Among these factors, two PID/PTB-containing proteins, FE65 and X11, interact with β-APP through an amino acid sequence including the internalization motif NPTY. Both factors could play an important role in the regulation of events leading to the internalization of APP and in the generation of the Aβ peptide through the endocytic pathway. Since the NPTY sequence is not sufficient for the interaction between FE65 and APP, but residues that are N-terminal to the motif are also required, it will be important to verify whether the binding of one factor is dependent on the interaction of other proteins with the APP cytodomain, or if single interactions are mutually exclusive. The region of the β-APP cytodomain interacting with APP-BP1 is not yet known; in any case, given the limited length of this cytodomain, it can be speculated that sterical constraints may exist which prevent a simultaneous binding of FE65, X11, Go, and APP-BP1 to β-APP (Figure 1).

The binding of different factors to the β-APP cytodomain could also be responsible for conformational changes in the β-APP extracellular domain which would result in a different sensitivity to proteolytic enzymes acting on its juxta- or trans-membrane domains (Figure 6A). The roles of the individual ligands of β-APP, or of their combinations, on the regulation of the β-secretase pathway should provide important clues to an understanding of the molecular mechanism(s) underlying the AD etiology.

The emerging scenario of the possible extracellular signals which can bind to β-APP invites many speculations. For example, different extracellular ligands of β-APP could induce conformational changes in the cytodomain of β-APP, thus modifying its affinity to various intracellular ligands (Figure 6B). This is exemplified by the interaction of the cytodomain of β-APP with heterotrimeric Go proteins. It was suggested that β-APP is a potential Go-coupled receptor, by demonstrating that the cytodomain of β-APP binds and activates Go *in vitro* and *in vivo*[102,104,110]. Mutant β-APP associated with familial AD (FAD-APP) was also shown to induce apoptosis in the presence of overexpressed Go. This phenomenon was inhibited by dominant negative Go, pertussis toxin, and by the deletion of the cytodomain of β-APP[106]. Furthermore, FAD-APP is able to suppress CRE-regulated transcription, in a pertussis toxin-sensitive manner[111], supporting the hypothesis that FAD-APP is in a conformational status which favors the interaction with Go.

Considering the emerging complexity of the protein network interacting with β-APP at the cytoplasmic side of the membrane, numerous speculations can be made, including the possibility that FE65 (or X11 or APP-BP1) could: i) Directly or indirectly bring the proteolytic machinery in close proximity to β-APP; ii) Participate in, or regulate, the transduction of extracellular signals from the β-APP receptor; iii) Function as an intermediate in the as yet unknown crosstalk between β-APP and the other players of the Alzheimer's puzzle, *i.e.* presenilins, Tau protein and/or apoE (Figure 5).

8. FE65 AND THE ALZHEIMER PHENOTYPE

Co-immunoprecipitation experiments performed on lysates from CHO cells overexpressing mutant forms of APP (FAD type) have shown a poor ability to form complexes with FE65 compared to wild type APP[68]; this phenomenon is particularly evident in the case of the mutation known as Swedish, which, as shown by transfection in mammalian cell lines[112,113], generates 5–10 times more Aβ peptide compared to the levels obtained with wild-type APP constructs. However, these differences could not be visualized in *in vitro* binding assays with bacterially expressed proteins[68]. Different explanations are possible for different behaviors of the mutant APP proteins in *in vivo* versus *in vitro* conditions. For example, the mutants could generate an altered conformation in which a site for the interaction with other components of the complex (X11, APP-BP1, Go) may be exposed. The low affinity interaction with FE65 that occurs *in vivo* in the case of the β-APP mutants could: i) Direct β-APP towards specific cellular compartments; ii) Expose the precursor to the proteolytic activity of the β-secretase machinery; and/or iii) Favor the interaction of β-APP with other intracellular ligands. This last possibility is in agreement with the finding of a constitutive interaction between FAD-APP and Go (see above). An alternative possibility, which is in agreement with the reduced levels of plasma membrane-anchored APP (Swedish), compared to the wild-type APP protein, assumes that the mutant APP molecules are optimal substrates for the β-secretase activity. If this is the case, many substrate molecules will be targeted to the late-occurring endocytic pathway and only a limited number of the mutant APP molecules will be able to reach the cell surface to interact with FE65.

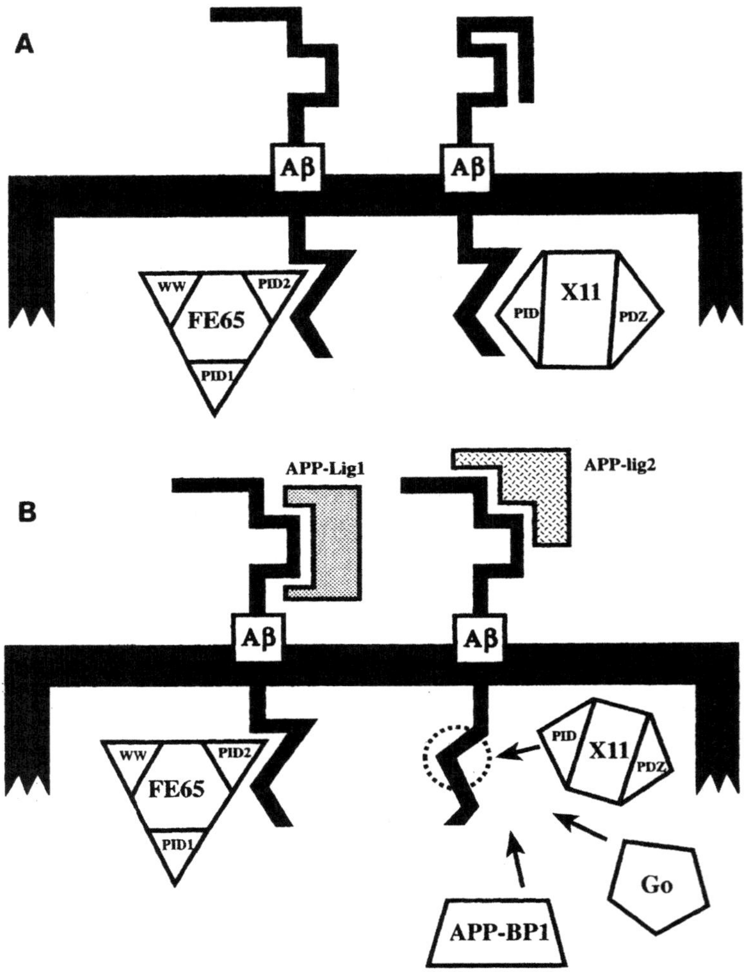

Figure 6. Alternative conformations of the β-APP molecules upon interaction with cytoplasmic (Panel A) or extracytoplasmic (Panel B) ligands. **Panel A**: The interaction of β-APP with the PID2 domain of FE65 (left) or with the X11 PID/PTB domain (right) results in conformational changes in the extracellular domain of β-APP, generating β-APP isoforms differently sensitive to proteolytic cleavage by secretases and/or with different affinity to extracellular ligands. **Panel B**: The interaction of β-APP with different extracytosolic ligands (APP-lig1 and APP-lig2) may generate different conformations of the cytosolic domain with a restricted spectrum of interaction with intracellular ligands. The FAD-APP mutants could assume a specific conformation favoring the interaction with Go, but incompatible for binding to FE65 (see text for discussion).

9. RECENT DEVELOPMENTS AND FUTURE LOOK

Since the completion of this chapter, there have been several important developments in identification of proteins interacting with FE65. We have demonstrated that five proteins of molecular masses 60, 75, 80, 140, and 200kDa could be purified from mouse brain lysates by affinity to the FE65 WW domain[114]. These proteins interact with the WW domain through proline-rich regions with a minimal consensus motif of six consecutive prolines. Furthermore, we have identified two of these five proteins as the 80- and 140-kDa isoforms of Mena, the mammalian homolog of the *Drosophia* Enabled[114, 115], and demonstrated the interaction *in vivo* by coimmunoprecipitation assay.

By interaction trap procedure we isolated a binding partner of the PID/PTB1 domain of FE65, which appeared to be previously identified transcription factor CP2/LSF/LBP1[116]. Coimmunoprecipitation experiments demonstrated that the interaction between FE65 and CP2/LSF/LBP1 takes place *in vivo* between the native molecules.

Recently, two groups have shown that overexpression of X11 protein in HEK293 cells slows cellular β-APP processing and reduces Aβ peptide secretion[117,118].

In addition, an intronic polymorphism within the FE65 gene was shown to be associated with sporadic AD[119].

We expect that the detailed study of proteins implicated in AD, especially those which constitute the β-APP multicomponent complex will help to understand the cellular and molecular mechanisms of β-APP processing which results in the generation of Aβ peptide. The topology, structure, and functional features of several proteins which interact with β-APP point to the characteristics of well defined signaling molecules. For example, FE65 and X11 proteins could be considered as "bona fide" adapters which function as docking proteins to assemble groups of molecules in a sequence-specific manner[120]. The paradigm of adapter molecules is prevalent in signal transduction studies and various approaches used to analyze these molecules are being applied now to study the intricacies of FE65 and X11 proteins. Moreover, the identification of functional domains and motifs that mediate protein-protein interaction among β-APP, FE65 and its ligand(s), and possibly other partner molecules, will point to their specific genomic regions which, if mutated, could contribute to the phenotype of AD. This study may have direct implication for diagnostics and for designing potential therapeutics.

ACKNOWLEDGMENTS

We would like to express special thanks to Greg Elder, Mitch Goldfarb, Hillary Linn, and Stacey Rentschler for valuable comments on the manuscript. Supported by grants: from Telethon (to T. R.) and from National Institutes of Health (CA45757; CA01605), Human Frontier Science Program Organization, and the Klingenstein Award in the Neurosciences (to M. S.).

REFERENCES

1. Selkoe D. J. (1996) Amyloid β-protein and the genetics of Alzheimer's disease. *J. Biol. Chem.* **271**, 18295–18298.
2. Selkoe D. J. (1997) Alzheimer's disease: genotypes, phenotypes and treatments. *Science* **275**, 630–631.
3. Yanker B. A. (1996) Mechanisms of neuronal degeneration in Alzheimer's disease. *Neuron* **16**, 921–932.
4. Lamb B. T. (1995) Making models for Alzheimer's disease. *Nature Genetics* **9**, 4–6.
5. Kelly W. J. (1996) Alternative conformations of amiloidogenic proteins govern their behavior. *Curr. Opin. Struct. Biol.* **6**, 11–17.
6. Selkoe D. J. (1994) Normal and abnormal biology of the β-amyloid precursor protein. *Ann. Rev. Neurosci.* **17**, 489–517.
7. Kang J., Lemaire H., Unterbeck A., Salbaum J. M., Masters C. L., Grzeschik K., Multhaup G., Beyreuther K., and Muller-Hill B. (1987) The precursor of Alzheimer's disease amyloid A4 protein resembles a cell-surface receptor. *Nature* **325**, 733–736.

8. Selkoe D. J., Podlisny M. B., Joachim C. L., Vickers E. A., Lee G., Fritz L. C., and Oltersdorf T. (1988) β-amyloid precursor protein of Alzheimer disease occurs as 110- to 135-kilodalton membrane-associated proteins in neural and nonneural tissues. *Proc. Natl. Acad. Sci. USA* **85**, 7341–7345.
9. Oltersdorf T., Ward P. J., Henriksson T., Beattie E. C., Neve R., Lieberburg I., and Fritz L. C. (1990) The Alzheimer amyloid precursor protein. *J. Biol. Chem.* **265**, 4492–4497.
10. Weidemann A., Konig G., Bunke D., Fischer P., Salbaum J. M., Masters C. L., and Beyreuther K. (1989) Identification, biogenesis, and localization of precursors of Alzheimer's disease A4 amyloid protein. *Cell* **57**, 115–126.
11. Ponte P., Gonzalez-DeWhitt P., Schilling J., Miller J., Hsu D., Greenberg B., Davis K., Wallace W., Lieberburg I., Fuller F., and Cordell B. (1988) A new A4 amyloid mRNA contains a domain homologous to serine proteinase inhibitors. *Nature* **331**, 525–532.
12. Trowbridge I. S., Collawn J. F., and Hopkins, C. R. (1993) Signal-dependent membrane protein trafficking in the endocytic pathway. *Annu. Rev. Cell Biol.* **9**, 129–161.
13. Bush A. I., Pettingell W. H., Jr., de Paradis M., Tanzi E., and Wasco W. (1994) The amyloid β-protein precursor and its mammalian homologs. *J. Biol. Chem.* **269**, 26618–26621.
14. Wasco W., Gurubhagavatula S., de Paradis M., Romano D. M., Sisodia S. S., Hyman B. T., Neve R. L., and Tanzi R. E. (1993) Isolation and characterization of APLP2 encoding a homologue of the Alzheimer's associated amyloid β protein precursor. *Nature Genetics* **5**, 95–100.
15. Wasco W., Bupp K., Magendantz M., Gusella J. F., Tanzi R. E., and Solomon F. (1992) Identification of a mouse brain cDNA that encodes a protein related to the Alzheimer disease-associated amyloid β protein precursor. *Proc. Natl. Acad. Sci. USA* **89**, 10758–10762.
16. Kim T.-W., Wu K., Xu J.-L., McAuliffe G., Tanzi R. E., Wasco W., and Black I. B. (1995) Selective localization of amyloid precursor-like protein 1 in the cerebral cortex postsynaptic density. *Mol. Brain Res.* **32**, 36–44.
17. Esch F. S., Keim P. S., Beattie E. C., Blacher R. W., Culwell A. R., Oltersdorf T., McClure D., and Ward P. J. (1990) Cleavage of amyloid β peptide during constitutive processing of its precursor. *Science* **248**, 1122–1124.
18. Shoji M., Golde T. E., Ghiso J., Cheung T. T., Estus S., Shaffer L. M., Cai X., McKay D. M., Tintner R., Frangione B., and Younkin S. G. (1992) Production of the Alzheimer amyloid β protein by normal proteolytic processing. *Science* **258**, 126–129.
19. Busciglio J., Gabuzda D. H., Matsudaira P., and Yankner B. A. (1993) Generation of β-amyloid in the secretory pathway in neuronal and nonneuronal cells. *Proc. Natl. Acad. Sci. USA* **90**, 2092–2096.
20. Koo E. H. and Squazzo S. L. (1994) Evidence that production and release of amyloid β-protein involves the endocytic pathway. *J. Biol. Chem.* **269**, 17386–17389.
21. Haass C., Schlossmacher M. G., Hung A. Y., Vigo-Pelfrey C., Mellon A., Ostaszewski B. L., Lieberburg I., Koo E. H., Schenk D., Teplow D. B., and Selkoe D. J. (1992) Amyloid β-peptide is produced by cultured cells during normal metabolism. *Nature* **359**, 322–325.
22. Kuo Y-M., Emmerling M. R., Vigo-Pelfrey C., Kasunic T. C., Kirkpatrick J. B., Murdoch G. H., Ball M. J., and Roher A. E. (1996) Water-soluble Aβ (N-40, N-42) oligomers in normal and Alzheimer disease brains. *J. Biol. Chem.* **271**, 4077–4081.
23. Roher A. E., Lowenson J. D., Clarke S., Woods A. S., Cotter R. J., Gowing E., and Ball M. J. (1993) β-amyloid-(1–42) is a major component of cerebrovascular amyloid deposits: implications for the pathology of Alzheimer disease. *Proc. Natl. Acad. Sci. USA* **90**, 10836–10840.
24. Sisodia S. S., Koo E. H., Beyreuther K., Unterbeck A., and Price D. L. (1990) Evidence that β-amyloid protein in Alzheimer's disease is not derived by normal processing. *Science* **248**, 492–495.
25. Hsiao K., Chapman P., Nilsen S., Eckman C., Harigaya Y., Younkin S., Yang F., and Cole G. (1996) Correlative memory deficits, Aβ elevation, and amyloid plaques in transgenic mice. *Science* **274**, 99–102.
26. Xu H., Sweeney D., Greengard P., and Gandy S. (1996) Metabolism of Alzheimer β-amyloid precursor protein: regulation by protein kinase A in intact cells and in a cell-free system. *Proc. Natl. Acad. Sci. USA* **93**, 4081–4084.
27. Cai X.-D., Golde T. E., and Younkin S. G. (1993) Release of excess amyloid β protein from a mutant amyloid β protein precursor. *Science* **259**, 514–516.
28. LaFerla F. M., Tinkle B. T., Bieberich C. J., Haudenschild C. C., and Jay G. (1995) The Alzheimer's Aβ peptide induces neurodegeneration and apoptotic cell death in transgenic mice. *Nature Genetics* **9**, 21–29.
29. Oster-Granite M. L., McPhie D. L., Greenan J., and Neve R. L. (1996) Age-dependent neuronal and synaptic degeneration in mice transgenic for the C terminus of the amyloid precursor protein. *J. Neurosci.* **16**, 6732–6741.
30. Yamatsuji T., Okamoto T., Takeda S., Murayama Y., Tanaka N., and Nishimoto I. (1996) Expression of V642 APP mutant causes cellular apoptosis as Alzheimer trait-linked phenotype. *EMBO J.* **15**, 498–509.

31. Games D., Adams D., Alessandrini R., Barbour R., Berthelett P., Blackwell C., Carr T., Clemens J., Donaldson T., and Gillespie F. (1995) Alzheimer-type neuropathology in transgenic mice overexpressing V717F β-amyloid precursor protein. *Nature* **373**, 523–527.
32. Desdouits F., Buxbaum J. D., Desdouits-Magnen J., Nairn A. C., and Greengard P. (1996) Amyloid β peptide formation in cell-free preparations. *J Biol. Chem.* **271**, 24670–24674.
33. Citron M., Teplow D. B., and Selkoe D. J. (1995) Generation of amyloid β protein from its precursor is sequence specific. *Neuron* **14**, 661–670.
34. Van Broeckhoven C. (1995) Presenilins and Alzheimer disease. *Nature Genetics* **11**, 230–232.
35. Cruts M., Hendrics L., and Van Broeckhoven C. (1996) The presenilin genes: a new gene family involved in Alzheimer disease pathology. *Human Mol. Gen.* **5**, 1449–1455.
36. Doan A., Thinakaran G., Borchelt D. R., Slunt H. H., Ratovitsky T., Podlisny M., Selkoe D. J., Seeger M., Gandy S. E., Price D. L., and Sisodia S. S. (1996) Protein topology of presenilin 1. *Neuron* **17**, 1023–1030.
37. Li X. and Greenwald I. (1996) Membrane topology of the C. elegans SEL-12 presenilin. *Neuron* **17**, 1015–1021.
38. Page K., Hollister R., Tanzi R. E., and Hyman B. T. (1996) *In situ* hybridization analysis of presenilin 1 mRNA in Alzheimer disease and in lesioned rat brain. *Proc. Natl. Acad. Sci. USA* **93**, 14020–14024.
39. Lee M. K., Slunt H. H., Martin L. J., Thinakaran G., Kim G., Gandy S. E., Seeger M., Koo E., Price D. L., and Sisodia S. S. (1996) Expression of presenilin 1 and 2 (PS1 and PS2) in human and murine tissues. *J. Neurosci.* **16**, 7513–7525.
40. Alzheimer's Disease Collaborative Group (1995) The structure of the presenilin 1 (S182) gene and identification of six novel mutations in early onset AD families. *Nature Genetics* **11**, 219–222.
41. Campion D., Flaman J.-M., Brice A., Hannequin D., Dubois B., Martin C., Moreau V., Charbonnier F., Didierjean O., Tardieu S., Penet C., Puel M., Pasquier F., Le Doze F., Bellis G., Calenda A., Heilig R., Martinez M., Mallet J., Bellis M., Clerget-Darpoux F., Agid Y., and Frebourg T. (1995) Mutations of the presenilin 1 gene in families with early-onset Alzheimer's disease. *Hum. Mol. Genet.* **4**, 2373–2377.
42. Anwar R., Moynihan T. P., Ardley H., Brindle Z., Coletta P. L., Cairns N., Markham A. F., and Robinson P. A. (1996) Molecular analysis of the presenilin 1 (S182) gene in "sporadic" cases of Alzheimer's disease: identification and characterization of unusual splice variants. *J. Neurochem.* **66**, 1774–1777.
43. Citron M., Westaway D., Xia W., Carlson G., Diehl T., Levesque G., Johnson-Wood K., Lee M., Seubert P., Davis A., Kholodenko D., Motter R., Sherrington R., Perry B., Yao H., Strome R., Lieberburg I., Rommens J., Kim S., Schenk D., Fraser P., Hyslop P. S. G., and Selkoe D. J. (1996) Mutant presenilins of Alzheimer's disease increase production of 42-residue amyloid beta-protein in both transfected cells and transgenic mice. *Natl. Med.* **3**, 67–72.
44. Borchelt D. R., Thinakaran G., Eckman C. B., Lee M. K., Davenport F., Ratovitsky T., Prada C. M., Kim G., Seekins S., Yager D., Slunt H. H., Wang R., Seeger M., Levey A. I., Gandy S. E., Copeland N. G., Jenkins N. A., Price D. L., Younkin S. G., and Sisodia S. S. (1996) Familial Alzheimer's disease-linked presenilin 1 variants elevate Abeta1–42/1–40 ratio *in vitro* and *in vivo*. *Neuron* **17**, 1005–1013.
45. Duff K., Eckman C., Zher C., Yu X., Prada C. M., Perez-tur J., Hutton M., Buee L., Harigaya Y., Yager D., Morgan D., Gordon M. N., Holcomb L., Refolo L., Zenk B., Hardy J., and Younkin S. (1996) Increased amyloid-beta 42(43) in brains of mice expressing mutant presenilin 1. *Nature*, **383** 710–713.
46. Lemere C. A., Lopera F., Kosik K. S., Lendon C. L., Ossa J., Saido T. C., Yamaguchi H., Ruiz A., Martinez A., Madrigal L., Hincapie L., Arango J. C., Antony D. C., Koo E. H., Goate A. M., Selkoe D. J., and Arango J. C. (1996) The E280A presenilin 1 Alzheimer mutation produces increased A beta 42 deposition and severe cerebellar pathology. *Natl. Med.* **2**, 1146–1150.
47. Wolozin B., Iwasaki K., Vito P., Ganjei J. K., Lacana E., Sunderland T., Zhao B., Kusiak J. W., Wasco W., and D'Adamio L. (1996) Participation of presenilin 2 in apoptosis: enhanced basal activity conferred by an Alzheimer mutation. *Science* **274**, 1710–1713.
48. Vito P., Wolozin B., Ganjei J. K., Iwasaki K., Lacan E., and D'Adamio L. (1996) Requirement of the familial Alzheimer's disease gene PS2 for apoptosis. Opposing effects of ALG-3. *J. Biol. Chem.* **271**, 31025–31028.
49. Mandelkow E.-M., Schweers O., Drewes G., Biernat J., Gustke N., Trinczek B., and Mandelkow E. (1996) Structure, microtubule interactions, and phosphorylation of Tau protein. *Ann. NY Acad. Sci.* **777**, 96–106.
50. Trinczek B., Biernat J., Baumann K., Mandelkow E.-M., and Mandelkow E. (1995) Domains of Tau protein, differential phosphorylation, and dynamic instability of microtubules. *Mol. Biol. Cell* **6**, 1887–1902.
51. Burack M. A. and Halpain S. (1996) Site-specific regulation of Alzheimer-like tau phosphrylation in living neurons. *Neuroscience* **72**, 167–184.
52. Singh T. J., Zaidi T., Grundke-Iqbal I., and Iqbal K. (1996) Non-proline-dependent protein kinases phosphorylate several sites found in tau from Alzheimer disease brain. *Mol. Cell. Biochem.* **154**, 143–151.

53. Hoshi M., Takashima A., Noguchi K., Murayama M., Sato M., Kondo S., Saitoh Y., Ishiguro K., Hoshino T., and Imahori K. (1996) Regulation of mitochondrial pyruvate dehydrogenase activity by tau protein kinase I/glycogen synthase kinase 3β in brain. *Proc. Natl. Acad. Sci. USA* **93**, 2719–2723.
54. Giaccone G., Pedrotti B., Migheli A., Verga L., Perez J., Racagni G., Smith M. A., Perry G., De Gioia L., Selvaggini C., Salmona M., Ghiso J., Frangione B., Islam K., Bugiani O., and Tagliavini F. (1996) βAPP and Tau interaction. A possible link between amyloid and neurofibrillary tangles in Alzheimer's disease. *Am. J. Pathol.* **148**, 79–85.
55. Isoe K., Urakami K., Sato K., and Takahashi K. (1996) Apolipoprotein E in patients with dementia of the Alzheimer type and vascular dementia. *Acta Neurol. Scand.* **93**, 133–137.
56. Marz W., Scharnagl H., Kirca M., Bohl J., Grob W., and Ohm T. G. (1996) Apolipoprotein E polymorphism is associated with both senile plaque load and Alzheimer-type neurofibrillary tangle formation. *Ann. NY Acad. Sci.* **777**, 276–280.
57. Jarvik G. P., Larson E. B., Goddard K., Kukull W. A., Schellenberg G. D., and Wijsman E. M. (1996) Influence of apolipoprotein E genotype on the transmission of Alzheimer disease in a community-based sample. *Am. J. Human. Genet.* **58**, 191–200.
58. Fung W.-P., Howlett G. J., and Schreider G. (1986) Structure and expression of the rat apolipoprotein E gene. *J. Biol. Chem.* **261**, 13777–13783.
59. Zannis V. I., McPherson J., Goldberger G., Karathanasis S. K., and Breslow J. L. (1984) Synthesis, intracellular processing, and signal peptide of human apolipoprotein E. *J. Biol. Chem.* **259**, 5495–5499.
60. Hixson J. E., Cox L. A., and Borenstein S. (1988) The baboon apolipoprotein E gene: structure, expression, and linkage with the gene for apolipoprotein C-1. *Genomics* **2**, 315–323.
61. Hoffer M. J. V., van Eck M. M., Havekes L. M., Hofker M. H., and Frants R. R. (1993) The mouse apolipoprotein C1 gene: structure and expression. *Genomics* **18**, 37–42.
62. Christie R. H., Chung H., Rebeck G. W., Strickland D., and Hyman B. T. (1996) Expression of the very low-density lipoprotein receptor (VLDL-r), an Apolipoprotein-E receptor, in the central nervous system and in Alzheimer's disease. *J. Neuropathol. and Exp. Neurol.* **55**, 491–498.
63. Okuizumi K., Onodera O., Namba Y., Ikeda K., Yamamoto T., Seki K., Ueki A., Nanko S., Tanaka H., Takahashi H., Oyanagi K., Mizuzawa H., Kanazawa I., and Tsuji S. (1995) Genetic association of the very low density lipoprotein (VLDL) receptor gene with sporadic Alzheimer's disease. *Nature Genetics* **11**, 207–209.
64. Esposito F., Ammendola R., Duilio A., Costanzo F., Giordano M., Zambrano N., D'Agostino P., Russo T., and Cimino F. (1990) Isolation of cDNA fragments hybridizing to rat brain-specific mRNAs. *Dev. Neurosci.* **12**, 373–381.
65. Simeone A., Duilio A., Fiore F., Acampora R., De Felice P., Faraonio R., Paolocci S., Cimino E., and Russo T. (1994) Expression of the neuron-specific FE65 gene marks the development of embryo ganglionic derivatives. *Dev. Neurosci.* **16**, 53–60.
66. Faraonio R., Minopoli G., Porcellini A., Costanzo F., Cimino F., and Russo T. (1994) The DNA sequence encompassing the transcription start site of a TATA-less promoter contains enough information to drive neuron-specific transcription. *Nucleic Acids Res.* **22**, 4876–4883.
67. Guenette S. Y., Chen J., Jondro P. D., and Tanzi R. E. (1996) Association of a novel human FE65-like protein with the cytoplasmic domain of the β-amyloid precursor protein. *Proc. Natl. Acad. Sci. USA* **93**, 10832–10837.
68. Zambrano N., Buxbaum J. D., Minopoli G., Fiore F., de Candia P., de Renzis S., Faraonio R, Sabo S., Cheetham J., Sudol M., and Russo, T. (1997) Interaction of the tandem PID/PTB-related domain of FE65 with wild type and mutant Alzheimer's β-amyloid precursor proteins. *J. Biol. Chem.* **272**, 6399–6405.
69. Duilio A., Zambrano N., Mogavero A. R., Ammendola R., Cimino F., and Russo T. (1991) A rat brain mRNA encoding a transcriptional activator homologous to the DNA binding domain of retroviral integrases. *Nucleic Acids Res.* **19**, 5269–5274.
70. Fiore F., Zambrano N., Minopoli G., Donini V., Duilio A., and Russo T. (1995) The regions of the FE65 protein homologous to the phosphotyrosine interaction/ phosphotyrosine binding domain of Shc bind the intracellular domain of the Alzheimer's amyloid precursor protein. *J. Biol. Chem.* **270**, 1–5.
71. Sudol M., Bork P., Einbond A., Kastury K., Druck T., Negrini M., Huebner K., and Lehman D. (1995) Characterization of the mammalian YAP (Yes-associated protein) gene and its role in defining a novel protein module, the WW domain. *J. Biol. Chem.* **270**, 14733–14741.
72. Bork P. and Margolis B. (1995) A phosphotyrosine interaction domain. *Cell* **80**, 693–694.
73. Pawson T. (1995) Protein modules and signalling networks. *Nature* **373**, 573–580.
74. Sudol M., Chen H. I., Bougeret C., Einbond A., and Bork P. (1995) Characterization of a novel protein-binding module - the WW domain. *FEBS Letters* **369**, 1–5.

75. Einbond A. and Sudol M. (1996) Towards prediction of cognate complexes between the WW domain and proline-rich ligands. *FEBS Letters* **384**, 1–8.
76. Sudol M. (1996) Structure and function of the WW domain. *Prog. Biophys. Molec. Biol.* **65**, 113–132.
77. Chen H. I. and Sudol M. (1995) The WW domain of Yes-associated protein binds a proline-rich ligand that differs from the consensus established for Src homology 3-binding modules. *Proc. Natl. Acad. Sci. USA* **92**, 7819–7823.
78. Chen H. I., Einbond A., Kwak S.-J., Linn H., Koepf E., Peterson S., Kelly J. W., and Sudol M. (1997) Characterization of the WW domain of human Yes-associated protein and its ligands, WBP-1 and WBP-2. *J. Biol. Chem.* **272**, 17070–17077.
79. Sudol M. (1996) The WW module competes with the SH3 domain? *TIBS Letters* **21**, 161–163.
80. Chan D. C., Bedford M. T., and Leder P. (1996) Formin binding proteins bear WWP/WW domains that bind proline-rich peptides and functionally resemble SH3 domains. *EMBO J.* **15**, 1045–1054.
81. Staub O., Dho S., Henry P., Correa J., Ishikawa T., McGlade J., and Rotin D. (1996) WW domain of Nedd4 bind to the proline-rich PY motifs in the epithelial Na channel deleted in Liddle's syndrome. *EMBO J.* **15**, 2371–2380.
82. Bork P. and Sudol M. (1994) The WW domain: a signalling site in dystrophin? *TIBS* **19**, 531–533.
83. Garnnier L., Wills J. W., Verderame M. F., and Sudol M. (1996) WW domains and retrovirus budding. *Nature* **381**, 744–745.
84. Zhou M., Ravichandran K. S., Olejniczak E. T., Petros A. M., Meadows R. P., Sattler M., Harlan J. E., Wade W. S., Burakoff S. J., and Fesik S. W. (1995) Structure and ligand recognition of the phosphotyrosine binding domain of Shc. *Nature* **378**, 584–592.
85. Wolf G., Trub T., Ottinger E., Groninga L., Lynch A., White M. F., Miyazaki M., Lee J., and Shoelson S. E. (1995) PTB domains of IRS-1 and Shc have distinct but overlapping binding specificities. *J. Biol. Chem.* **270**, 27407–2741.
86. Isakoff S. J., Yu Y., Su Y., Blaikie P., Yajnik V., Rose E., Weidner K. M., Sachs M., Margolis B., and Skolnik E. Y. (1996) Interaction between the phosphotyrosine binding domain of Shc and the insulin receptor is required for Shc phosphorylation by insulin in vivo. *J. Biol. Chem.* **271**, 3959–3962.
87. Blaikie P., Immanuel D., Wu J., Li N., Yajnik V., and Margolis B. (1994) A region in Shc distinct from the SH2 domain can bind tyrosine-phosphorylated growth factor receptors. *J. Biol. Chem.* **269**, 32031–32034.
88. Chen Y., Grall D., Salcini A. E., Pelicci P. G., Pouyssegur J., and van Obberghen-Schilling E. (1996) Shc adaptor proteins are key transducers of mitogenic signalling mediated by the G protein-coupled thrombin receptor. *EMBO J.* **15**, 1037–1044.
89. Yajnik V., Blaikie P., Bork P., and Margolis B. (1996) Identification of residues within the Shc phosphotyrosine binding/phosphotyrosine interaction domain crucial for phosphopeptide interaction. *J. Biol. Chem.* **271**, 1813–1816.
90. van der Geer P., Wiley S., Gish G. D., Lai V. K., Stephens R., White M. F., Kaplan D., and Pawson T. (1996) Identification of residues that control specific binding of the Shc phosphotyrosine-binding domain to phosphotyrosine sites. *Proc. Natl. Acad. Sci. USA* **93**, 963–968.
91. Borg J.-P., Ooi J., Levy E., and Margolis B. (1996) The PI domains of X11 and FE65 bind to distinct sites on the YENPTY motif of amyloid precursor protein. *Mol. Cell. Biol.* **16**, 6229–6241.
92. Bressler S. L., Gray M. D., Sopher B. L., Hu Q., et al. (1996) cDNA cloning and chromosome mapping of the human FE65 gene: interaction of the conserved cytoplasmic domain of the human β-amyloid precursor protein and its homologues with the mouse FE65 protein. *Human Mol. Gen.* **5**, 1589–1598.
93. McLoughlin D. M. and Miller C. C. J. (1996) The intracellular cytoplasmic domain of the Alzheimer's disease amyloid precursor protein interacts with phosphotyrosine-binding domain proteins in the yeats two hybrid system. *FEBS Letters* **397**, 197–200.
94. Macias M. J., Hyvonen M., Baraldi E., Schultz J., Sudol M., Saraste M., and Oschkinat H. (1996) Structure of the WW domain of a kinase-associate protein complexed with a proline-rich peptide. *Nature* **382**, 646–649.
95. Prendergast G. C. and Ziff E. B. (1991) Mbh1: a novel gelsolin/severin-related protein which binds actin in vitro and exhibits nuclear localization in vivo. *EMBO J.* **10**, 757–766.
96. Dabiri G. A., Young C. L., Rosenbloom J., and Southwick F. S. (1992) Molecular cloning of human macrophage capping protein cDNA. *J. Biol. Chem.* **267**, 16545–16552.
97. Bearer E. L. (1991) Direct observation of actin filament severing by gelsolin and binding by gCap39 and CapZ. *J. Cell Biol.* **115**, 1629–1638.
98. Johnston P. A., Yu F., Reynolds G. A., Yin H. L., Moomaw C. R., Slaughter C. A., and Sudhof T. C. (1990) Purification and expression of gCap39. *J. Biol. Chem.* **265**, 17946–17952.
99. Yu F., Johnston P. A., Sudhof T. C., and Yin H. L. (1990) gCap39, a calcium ion and polyphosphoinositide-regulated actin capping protein. *Science* **250**, 1413–1415.

100. Onoda K. and Yin H. L. (1993) gCap39 is phosphorylated. *J. Biol. Chem.* **268**, 4106–4112.
101. Onoda K., Yu F., and Yin H. L. (1993) gCap39 is a nuclear and cytoplasmic protein. *Cell Motil. and Cytoskel.* **26**, 227–238.
102. Nishimoto I., Okamoto T., Matsuura Y., Takahashi S., Okamoto T., Murayama Y., and Ogata E. (1993) Alzheimer amyloid protein precursor forms a complex with brain GTP binding protein Go. *Nature* **362**, 75–79.
103. Chow N., Koremberg J. R., Chen X.-N., and Neve R.L. (1996) APP-BP1, a novel protein that binds to the carboxyl-terminal region of the amyloid precursor protein. *J. Biol. Chem.* **271**, 11339–11346.
104. Okamoto T., Takeda S., Murayama Y., Ogata E., and Nishimoto I. (1995) Ligand-dependent G protein coupling function of amyloid transmembrane precursor. *J. Biol. Chem.* **270**, 4205–4208
105. Okamoto T., Takeda S., Giambarella U., Murayama Y., Matsui T., Katada T., Matsuura Y., and Nishimoto I. (1996) Intrinsinc signaling function of APP as a novel target of three V642 mutations linked to familial Alzheimers disease. *EMBO J.* **15**, 3769–3777.
106. Yamatsuji T., Matsui T., Okamoto T., Komatsuzaki K., Takeda S., Fukumoto H., Iwatsubo T., Suzuki N., Asami-Okada A., Ireland S., Kinane B., Giambarella U., and Nishimoto I. (1996) G protein-mediated neuronal DNA fragmentation induced by familial Alzheimer's disease-associated mutants of APP. *Science* **272**, 1349–1352.
107. Nordstedt C., Caporaso G. L., Thyberg J., Gandy S. E., and Greengard P. (1993) Identification of the Alzheimer β/A4 amyloid precursor protein in clathrin-coated vesicles purified from PC12 cells. *J. Biol. Chem.* **268**, 608–612.
108. Lai A., Sisodia S. S., and Trowbridge I. S. (1995) Characterization of sorting signals in the beta-amyloid precursor protein cytoplasmic domain. *J. Biol. Chem.* **270**, 3565–3573.
109. Perez R. J., Squazzo S. L., and Koo E. H. (1996) Enhanced release of amyloid beta protein from codon 670/671 "Swedish" mutant beta amyloid precursor protein occurs in both secretory and endocytic pathways. *J. Biol. Chem.* **271**, 9100–9107.
110. Lang J., Nishimoto I., Okamoto T., Sadoul K., Regazzi R., Kiraly K., Weller U., and Wollheim C. B. (1995) Direct control of exocytosis by receptor activation of the heterotrimeric GTPases Gi and Go or the expression of their active Gα-subunits. *EMBO J.*, **14**, 3635–3644.
111. Izezu T., Okamoto T., Komatsuzaki K., Matsui T., Martyn J. A. J., and Nishimoto I. (1996) Negative transactivation of cAMP response element by familial Alzheimer's mutants of APP. *EMBO J.* **15**, 2468–2475.
112. Cai X. D., Golde T. E., and Younkin S. G. (1993) Release of excess amyloid beta protein from a mutant amyloid beta protein precursor. *Science* **259**, 514–516.
113. Citron M., Oltersdorf T., Haas C., McConlogue K., Hung A. Y., Seubert P., Vigo-Pelfrey C., Lieberburg I., and Selkoe D. J. (1992) Mutation of the beta amyloid precursor protein in familial Alzheimer's disease increases beta protein production. *Nature* **360**, 672–674.
114. Ermekova, K.S., Zambrano, N., Linn, H., Minopoli, G., Gertler, F., Russo, T., and Sudol, M. (1997) The WW domain of neural protein FE65 interacts with proline-rich motifs in Mena, the mammalian homolog of *Drosophila* Enabled. *J. Biol. Chem.* **272**, 32869-32877.
115. Gertler, F.B., Niebuhr, K., Reinhard, M., Wehland, J., and Soriano, P. (1996) Mena, a relative of VASP and Drosophila Enabled, is implicated in the control of Microfilament dynamics. *Cell* **87**, 227-239.
116. Zambrano, N., Minopoli, G., de Candia, P., and Russo, T. (1998) The FE65 adaptor protein interacts through its PID1 domain with the transcriptional factor CP2/LSF/LBP1. *J. Biol. Chem.* **273**, 20128-20133.
117. Borg, J-P., Yang, Y., De Taddeo-Borg, M., Margolis, B., and Turner, R.S. (1998) The X11α protein slows cellular amyloid precursor protein processing and reduces Aβ40 and Aβ42 secretion. *J Biol Chem.* **273**, 14761-14766.
118. Sastre, M., Turner, R.S., and Levy, E. (1998) X11 interaction with β-protein secretion. *J Biol Chem.* **273**, 23351-22357.
119. Hu Q., Kukull W.A., Bressler S.L., Gray M.D., Cam J.A., Larson E.B., Martin G.M., Deeb S.S. (1998) The human FE65 gene: genomic structure and intronic polymorphism associated with sporadic dementia of the Alzheimer type. *Hum. Genet.* **103**, 295-303.
120. Sudol, M. (1998) From Src Homology domains to other signaling modules: proposal of the 'protein recognition code'. *Oncogene* **17**, 1469-1474.

11

INFLUENCE OF PHOSPHOLIPIDS AND SEQUENTIAL KINASE ACTIVITIES ON TAU IN VITRO

Thomas B. Shea and Fatma J. Ekinci

Center for Cellular Neurobiology and Neurodegeneration Research
Department of Biological Sciences
University of Massachusetts at Lowell
Lowell, Massachusetts 01854

1. INTRODUCTION

The *in situ* function(s) of the axonal-specific microtubule-associated protein tau have not been fully disclosed, but are the focus of intense study both in terms of neuronal differentiation and homeostasis, and also since tau is the major structural constituent of neurofibrillary tangles in Alzheimer's disease (for reviews, see Goedert, 1993; Mandelkow and Mandelkow, 1993; Kosik, 1993). It has been known for some time that tau promotes MT assembly and renders MTs resistant to depolymerization under cell-free conditions (Cleveland et al., 1977; Dreschel et al., 1992) as well as within in intact cells (Baas et al., 1994; Drubin and Kirschner, 1986; Shea and Beermann, 1994; Takemura et al., 1992). However, more recent analyses revealing that tau is more concentrated in the distal region of growing axons (Black et al., 1996; Brandt et al., 1995; Kempf et al., 1996; Mandell and Banker, 1996), where MTs are the most labile, have prompted the hypothesis that tau *in situ* must have function(s) other than promotion of MT stability, and that the likely site for such additional putative functions is the distal axon and growth cone (Black et al., 1996; Kempf et al., 1996). In this regard, recent studies have demonstrated the apparent association of tau with the plasma membrane of cultured neuronal cells (Brandt et al., 1995; Kempf et al., 1996), where it may participate in as yet undisclosed aspects of signal transduction (Lee et al., 1996a,b). Consistent with such possibilities are that *(1)* the association of tau with the plasma membrane is mediated via its its N-terminal projection domain (Brandt et al., 1995), while MT-mediated association is via its C-terminus (Aizawa et al., 1988; Lee et al., 1989; Himmler et al., 1989), that *(2)* tau is more weakly associated with MTs than other microtubule-associated proteins (MAPs; Black et al., 1996; Kempf et al., 1996), and that *(3)* tau immunoreactivity is best preserved under conditions that maintain the integrity of the plamsa membrane (Black et al., 1996; Kempf et al., 1996).

Molecular and Cellular Mechanisms of Neuronal Plasticity, edited by Ehrlich. Plenum Press, New York, 1998.

During continued analyses of the role of PKC-mediated signal transduction pathways in tau phosphorylation, we noted that the presence of phosphatidyl serine (PS), a necessary co-factor of PKC, influenced the degree of tau immunoreactivity on immunoblots in control samples lacking kinase. In the first portion of this chapter, we present data herein demonstrating that the interaction of tau with PS and other phospholipids alters tau antigenicity, its ability to undergo phosphorylation by a non-PS dependent kinase (MAP kinase), its ability to undergo calpain-mediated proteolysis, and its ability to co-assemble with MTs. We note that all of these phenomena are considered by one or more studies to be intergral to the altered distribution and/or accumulation of tau in AD.

We then turn to considerations of the combined influence of kinases on tau. A number of kinases, including Ca^{2+}-calmodulin kinase, cyclin-dependent kinases, protein kinase C (PKC), mitogen-activated protein (MAP) kinases, glycogen synthase kinases, and proline-directed kinases (e.g., Baudier and Cole, 1987; Baumann et al., 1993; Cressman et al., 1995a; Cressman and Shea, 1995; Drewes et al., 1992; Goedert et al., 1992; Hanger et al., 1992; Kobayashi et al., 1993; Ledesma et al., 1992; Mandelkow et al., 1992; Mulot et al., 1994; Paudel et al., 1993; Shea et al., 1996; Vulliet et al., 1992; for review, see Pelech, 1995), have been shown to phosphorylate tau in a manner that confers the altered antigenicity and electrophoretic migration characteristically observed for tau in AD brains. However, in addition to the demonstration of profound alteration in tau antigenicity and behavior following incubation with individual kinases, an important consideration in unraveling the series of events leading to PHF formation may be the cumulative influence of sequential phosphorylation events by multiple kinases (Morishima-Kawashima et al., 1995). Several studies in this regard have demonstrated that prior phosphorylation of tau by certain kinases alters the susceptibility of tau to undergo subsequent phosphorylation by other kinases (Arioka et al., 1993; Blanchard et al., 1994; Goedert et al., 1994; Raghunandan and Ingram, 1995; Sengupta et al., 1997; Singh et al., 1994 a,b; Steiner et al., 1990). In the second portion of this chapter, we examine the impact of PKC on tau phosphorylation individually and in combination with other kinases. We demonstrate that prior phosphorylation of tau *in situ* alters the ability of PKC, cAMP-dependent protein kinase (PKA) and MAP kinase to generate site-specific AD-like antigenic changes in subsequent cell-free analyses, and, further, that cell-free incubation of tau with PKC alters the influence of MAP kinase and PKA on tau antigenicity.

2. MATERIALS AND METHODS

2.1. Cell-Free Incubation and Phosphorylation of Tau

Human brain tau (generous gift of Drs. Marc Mercken and Ralph Nixon) was isolated by affinity chromatography as described previously (Cressman et al., 1995a). Consistent with previous studies (e.g., (Garver et al., 1994; Matsuo et al., 1994), tau isolated under these conditions displays moderate reactivity with so-called "AD-like" phosphoepitopes, including PHF-1 and ALZ-50, and a marginal degree of reactivity with AT-8. Tau (5ng/ml or 300ng/ml) was diluted in 20 mM Tris (pH 7.5) containing 5mM EDTA, 2mM PMSF and 50μg/ml leupeptin with or without 0.5mg/ml phosphatidyl serine (PS; obtained from UBI, Lake Placid, NY) then immediately mixed with Laemmli buffer on ice; the entire mixing process took less than 1 min.

Stock solutions of phosphatidyl serine (PS), phosphatidyl choline (PC) and phosphatidyl inositol (PI), commercially-obtained in chloroform:methanol, were treated as described previously (Baudier and Cole, 1987; Baudier et al., 1987). Chloroform:methanol was slowly evaporated by blowing a gentle stream of nitrogen gas across the surface of the solution. The resulting phospholipid-containing, dry sediment was resuspended in 50mM Tris-HCl (pH7.4) to yield a concentration of 10mg/ml phospholipid, which was then sonicated and added directly to tau incubation mixtures. To control for the potential of residual chloroform:methanol inducing tau denaturation, 1–4x of the corresponding amount of chloroform:methanol lacking phospholipids was evaporated as described above, Tris buffer added, the tube sonicated, and added to tau incubation mixtures.

2.2. Cell-Free MT Assembly Assays

Twice-cycled bovine brain tubulin free of MAPs (Shea and Fischer, 1996) was incubated for 30min at 37°C with and without 5ng/ml purified tau and 0.5μg/ml PS in 100mM PIPES (pH6.6) containing 1mM EGTA and 1mM $MgCl_2$, then centrifuged at 100,000 x g for 30min at 37°C. Resultant pellet and supernatant fractions were subjected to immunoblot analysis with 5E2 and a monoclonal antibody (Sigma Immunochemicals, Inc.) directed against α-tubulin (Shea and Fischer, 1996).

2.3. Cell-Free Calpain-Mediated Proteolysis

Purified tau (5ng/ml) in 20 mM Tris (pH 7.5) was incubated at 37°C with m calpain (Sigma) and 2mM $CaCl_2$ as described previously (Shea et al., 1996) in the presence and absence of 0.5mg/ml PS. Samples were withdrawn at 0–30min and added to concentration Laemmli buffer on ice.

2.4. Electrophoresis and Immunoblot Analysis

Samples were subjected to SDS-gel electrophoresis and either dried and placed against X-ray film to generate autoradiographs or transferred to nitrocellulose in a Hoefer Transphor apparatus. Nitrocellulose replicas were blocked with normal goat serum, then incubated overnight at 4°C with one or more of the following phospho-dependent and -independent monoclonal antibodies (below) that collectively span the length of the tau molecule. Replicas were rinsed and incubated for 2hr at room temp with a 1:5000 dilution of alkaline phosphatase-conjugated goat anti-mouse IgG and visualized as described (Shea et al., 1996). Semi-quantitative densitometric analyses was carried out using NIH Image (version 1.57; available by anonymous FTP from *zippy.nimh.nih.gov*)) following digitization of immunoblots and autoradiographs via a UMAX scanner (with transparancy adaptor) connected to a Macintosh PowerPC 7100 (e.g., Shea et al., 1996). Mean and standard error of the mean were calculated following digitization and analyses of multiple samples. Error bars are not reported for graphs expressing ratios of samples +/– PS since relative, rather than absolute, values are generated by comparative densitometry; values presented are in all cases representative. All reagents unless indicated otherwise were obtained from Sigma Chemical Co., Inc. (St. Louis, MO).

2.5. Antibodies Utilized in This Study

Multiple antibodies were utilized that collectively span the entire tau molecule and react with the following epitopes: PHF-1 (generous gift of Dr. S. Greenburg), which recognizes ser 396 and 404 when phosphorylated; AT-8 (Immunogenetics) and Tau-1 (Sigma) which recognize ser-202 when phosphorylated and nonphosphorylated, respectively; ALZ-50 (generous gift of Dr. P. Davies), a conformation-dependent antibody that recognizes the N-terminus; 5E2 (generous gift of Dr. Ken Kosik), N- and C-terminal monoclonal antibodies T14 and T46 (respecitively; generous gifts of Dr. Virginia Lee); 5A6 (N-terminal to exon 2); 10C4 (C terminal to exon 11) and 1B3 (which recognizes an undefined epitope within the central portion of the tau molecule; monoclonal antibodies 5A6, 10C4 and 1B3 were generous gifts of Dr. Gail Johnson); and JM (generous gift of Dr. Marc Mercken), a rabbit polyclonal antibody prepared against the purified human tau utilized in this study.

2.6. Individual and Multiple Kinase Incubation

Tau (5ng/ml) was incubated at 37°C for 24hr in 20mM Tris (pH 7.5) containing 100µM ATP, 5mM $CaCl_2$, 100mM $MgCl_2$ and 0.5mg/ml phosphatidyl serine (PS). Stock solutions of PS, commercially-obtained in chloroform:methanol, were treated as described previously (Baudier and Cole, 1987; Baudier et al., 1987). Chloroform:methanol was slowly evaporated by blowing a gentle stream of nitrogen gas across the surface of the solution. The resulting phospholipid-containing, dry sediment was resuspended in 50mM Tris-HCl (pH7.4) to yield a concentration of 10mg/ml PS, which was then sonicated and aliquoted into the above buffer. Incubation mixtures included 0.75–7.5 activity units (au) of PKC (UBI Lake Placid, NY), 0.2–20au of PKA (Sigma Chemical Co), or 75au of MAP kinase (UBI); the appropriate range of concentrations for each kinase and length of incubation was determined in preliminary studies. For example, previous studies indicated that as little as 2hr of incubation with PKC resulted in the incorporation of phosphate groups by tau, but that 20–24hr of incubation were required for PKC to confer AD-like alterations in tau antigenicity and electrophoretic migration (Cressman et al., 1995a). For autoradiographic analyses, 250nCi g-AT[P^{32}] (DuPont, Boston, MA) was included in the above reaction. Reactions were terminated by addition of concentrated sample buffer on ice, and samples were subjected to gel electrophoresis followed by autoradiographic or immunoblot analysis.

For sequential kinase treatments, samples were boiled for 5min following initial kinase incubations to inactivate the first kinase. Dephosphorylation of tau was carried out via 18hr incubation with 500unit/ml E.coli alkaline phosphatase (Sigma) in the presence of 50µg/ml leupeptin in 20mM Tris-HCl (pH8.0) followed by boiling (to inactivate and denature alkaline phosphatase) for 5min and centrifugation at 15,000 x g for 15min. Tau was recovered in the resulting heat-stable supernatant (Shea et al., 1996). Reactions were terminated by addition of concentrated Laemmli buffer on ice, and samples were subjected to gel electrophoresis followed by autoradiographic or immunoblot analysis. Samples were subjected to SDS gel electrophoresis and either dried and subjected to autoradiography or transferred to nitrocellulose for immunoblot analysis. Replicas were immunostained with PHF-1 (generous gift of Dr. S. Greenburg), which recognizes ser 396 and 404 when phosphorylated; AT-8 (Immunogenetics) and Tau-1 (Sigma) which recognize ser-202 when phosphorylated and nonphosphorylated, respectively; and ALZ-50 (generous gift of Dr. P. Davies), a conformation-dependent antibody that recognizes the N-terminus of tau.

Immunoblots were digitized and subjected to densitometric analyses as described above. Densitometric values for each incubation condition were expressed as the fold change relative to the respective densitometric values obtained for control samples for each antibody. Dephosphorylated samples were normalized according to their respective dephosphorylated control samples, and no attempt was made to normalize dephosphorylated and non-dephosphorylated densitometric values.

3. PHOSPHOLIPIDS ALTER TAU CONFORMATION, PHOSPHORYLATION, PROTEOLYSIS AND ASSOCIATION WITH MICROTUBULES

3.1. Phospholipids Alter Tau Antigenicity

Mixing of PS with purified human brain tau at a PS:tau ratio of 100:1 for approximately 30 sec prior to SDS gel-electrophoresis altered tau immunoreactivity towards epitopes spanning the entire tau molecule (Fig. 1). Notably, certain "AD-like" phospho-epitopes (Alz-50, PHF-1 and AT-8) were markedly reduced by PS. By contrast, immunoreactivity towards an antibody that recognizes a nonphosphorylated epitcpe (Tau-1) and the phosphate-independent epitope 5E2 were increased following incubation with PS (Fig. 1). Several other phospho-independent epitopes were unchanged by PS; these latter findings confirm that the reduction of "AD-like" tau immunoreactivity was not due to prevention of entry of tau into gels or a generalized interference with tau immunoreactivity. Autoradiographic analysis of samples incubated with g-AT[P^{32}] did not reveal any residual/contaminating kinase activity (Fig. 1). Increasing the amount of tau relative to PS to yield a PS:tau ratio of 1.7:1 under these conditions eliminated the influence of PS on tau antigenicity with most antibodies, and markedly reduced the impact of PS on 5E2 immunoreactivity (Fig. 2). All subsequent analyses of the influence of PS on tau were carried out at the above higher phospholipid:tau ratio (100:1::phospholipid:tau).

The possibility remained that the apparent influence of PS and PC on tau antigenicity was actually a consequence of tau denaturation due to the introduction of residual chloroform:methanol (the initial phospholipid solvent) that may have persisted despite nitrogen evaporation of stock phospholipid solutions to dryness. This possibility was tested and eliminated by exposing tau to the putative residue resulting from phospholipid-free chloroform:methanol. Antigenic changes were not induced when tau was exposed to the putative residual chloroform:methanol resulting from nitrogen evaporation of up to 4 volumes of the original amount of phospholipid-free chloroform:methanol (Fig. 3). These findings confirmed that the observed alterations in tau antigenicity were mediated by phospholipids themselves and not any putative chloroform:methanol residue. Increasing the concentration of SDS utilized in sample preparation for electrophoresis from 1% to 5% did not reverse the effects of PS on tau as ascertained by maintenance of an increase in 5E2 immunoreactivity (Fig. 4). Effects similar to those observed with PS were also induced by PC but not by PI (Fig. 5).

We interpret these findings to indicate that even a brief (<1min) exposure to excess PS induces an-SDS-resistant conformational alteration in the tau molecule that renders certain antigenic sites cryptic, renders others more prominent, and leaves additional sites unaltered. The presence of PS also impaired the ability of MAP kinase to phosphorylate tau, and in doing so, to confer PHF-1, AT-8 and ALZ-50 immunoreactivity upon tau; this finding, along with the apparent SDS-resistance of the influence of PS on tau herein and previous studies (Baudier and Cole, 1987; Baudier et al., 1987), suggests that any putative phospholipid-induced alteration in tau conformation be relatively stable. In these earlier studies, only 50% of the total tau was readily phosphorylated in the absence of PS, while 100% was able to be phosphorylated in the presence of PS, leading the investigators to conclude that PS disrupted tau-tau dimers (Baudier and Cole, 1987). Moreover, in these earlier analyses, certain phospholipids also altered the electrophoretic migration of tau, indicative of a conformational change; in this regard, we note that the electrophoretic migra-

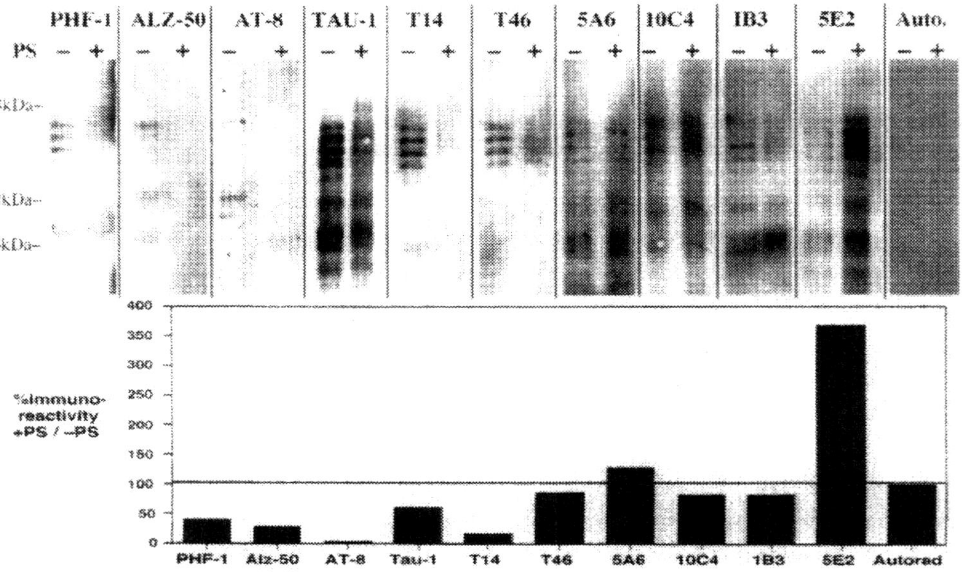

Figure 1. Phosphatidyl serine alters tau antigenicity in the absence of detectable kinase activity. Immunoblot analyses of human brain tau incubated for approximately 30 sec in the presence (+PS) or absence (–PS) of 0.5mg/ml PS prior to SDS gel-electrophoresis and immunoblot analysis with various tau antibodies as indicated. An additional sample incubated with g-AT[P^{32}] was subjected to autoradiography ("Auto."). The accompanying graph presents the ratio of values obtained from these immunoblots and autoradiograph of samples incubated in the presence (+) and absence (–) of PS. Note that PS markedly reduced the reactivity of tau with some antibodies, including, but not limited to, the reduction of so-called "AD-like" immunoreactivities PHF-1, ALZ-50 and AT-8, only slightly altered others (e.g., 5A6, 10C4, 1B3, and markedly increased immunoreactivity with the phospho-independent antibody 5E2). Note further that autoradiographic analyses following incubation with g-AT[P^{32}] failed to indicate the presence of any contaminating kinase activity (compare with autoradiograph from Fig. 3).

tion of Tau-1 immunoreactive tau species were altered (e.g., Fig. 1). Antigenicity, however, was not examined in the earlier study of Baudier and Cole (1987).

3.2. Phosphatidyl Serine Alters the Ability of Tau to Undergo Calpain-Mediated Proteolysis

Consistent with previous studies (Yang and Ksiezak-Reding, 1995; Litersky et al., 1993; Johnson et al., 1989; Shea et al., 1996), purified m calpain rapidly degraded tau under cell free conditions (Fig. 6). PS significantly decreased the rate of calpain-mediated tau degradation. This was apparently not the result of a general inhibitory effect of PS on calpain, since we have previously demonstrated that calpain-mediated degradation of azocasein is not affected by PS (Cressman et al., 1995b; Lang et al., 1995).

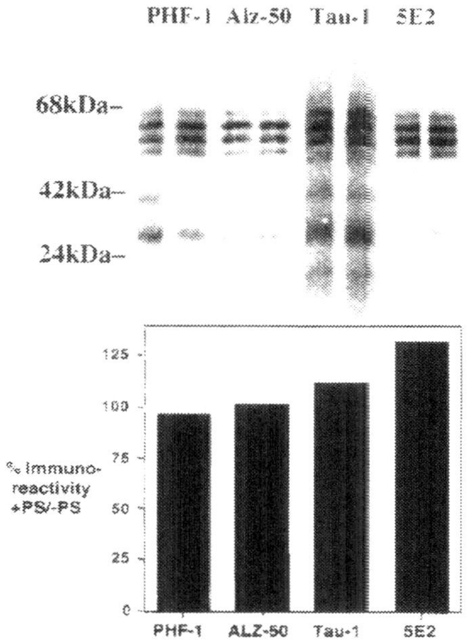

Figure 2. PS-mediated alteration in tau antigenicity is dependent upon a high PS:tau ratio. Immunoblot analyses of 300ng/ml human brain tau incubated for 30 sec with 0.5mg/ml PS prior to SDS gel-electrophoresis and immunoblot analysis with various tau antibodies as indicated. The accompanying graph presents the ratio of values obtained from these immunoblots and autoradiograph of samples incubated in the presence (+) and absence (–) of PS. Note that inclusion of this higher tau concentration relative to PS eliminated the influence of PS on PHF-1, ALZ-50 and Tau-1 immunoreactivity, and markedly diminished its influence on 5E2 immunoreactivity.

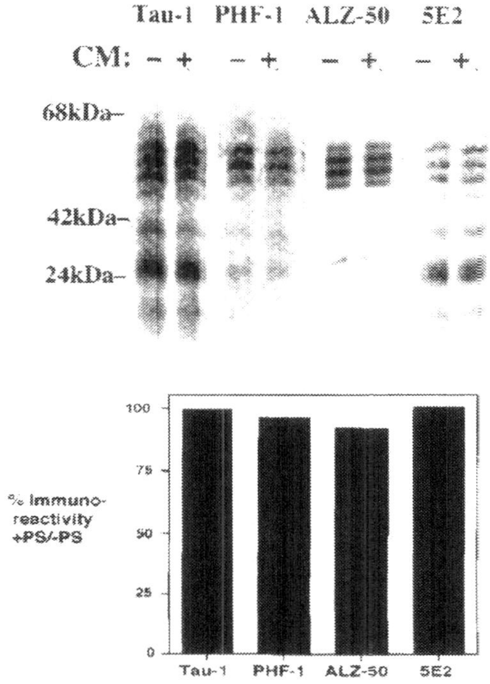

Figure 3. The PS-induced alteration in tau antigenicity cannot be attributed to tau denaturation by residual chloroform-methanol (the initial phospholipid solvent). Phospholipid-free chloroform:methanol ("CM") at 4x the amount of phospholipid stock solutions used in all experiments was dried via nitrogen evaporation. Tau in buffer was added and the solution sonicated in an identical manner as carried out for the phospholipid stock solution. Approximately 30 sec later, samples were subjected to SDS gel-electrophoresis and immunoblot ana|lysis with various tau antibodies as indicated. The accompanying graph presents the ratio of values obtained from these immunoblots and autoradiograph of samples incubated in the presence (+) and absence (–) of CM. Note that inclusion of higher tau concentrations relative to PS eliminated the influence of PS on PHF-1, ALZ-50 and Tau-1 immunoreactivity, and markedly diminished its influence on 5E2 immunoreactivity.

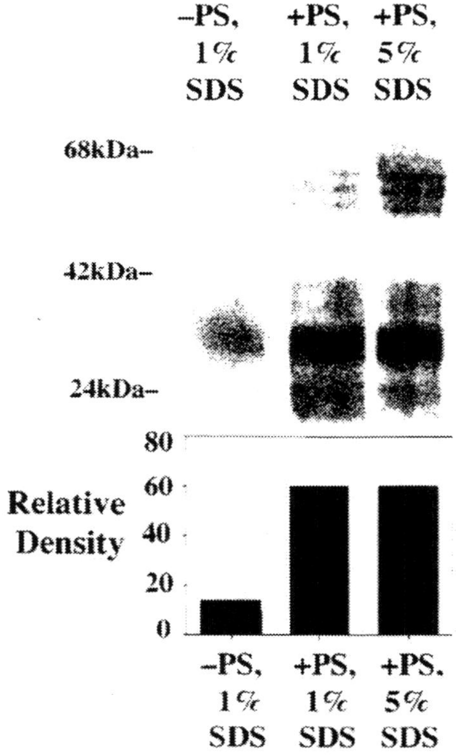

Figure 4. Excess SDS does not reverse the PS-induced alteration in tau antigenicity. Immunoblot analyses with monoclonal antibody 5E2 of 5ng/ml human brain tau incubated for approximately 30 sec in the absence (-PS) or presence (+PS) of 0.5mg/ml PS followed by boining with 1% SDS (the normal level of SDS in Laemlli treatment buffer) or 5%SDS for 1min prior to SDS gel-electrophoresis as indicated. The accompanying graph presents the relative density in arbitrary densitometric units of values obtained from resulting immunoblots. Note that 5% SDS does not attenuate the PS-induced increase in 5E2 immunoreactivity.

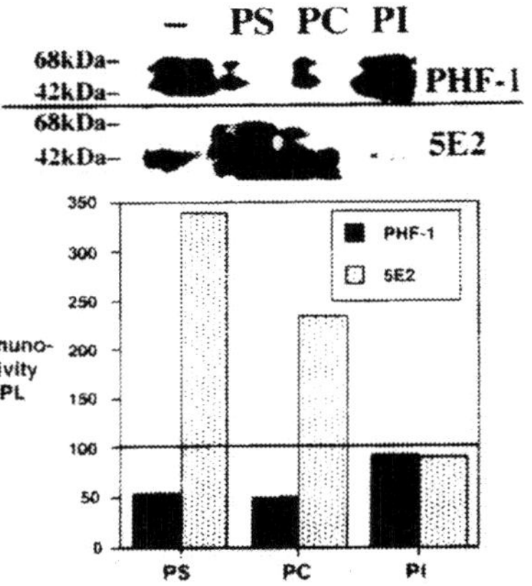

Figure 5. Multiple phospholipids alter tau antigenicity. Immunoblot analyses of 5ng/ml human brain tau incubated for approximately 30 sec in the absence (–) or presence of 0.5mg/ml PS, PC or PI, prior to SDS gel-electrophoresis and immunoblot analysis with PHF-1 or 5E2 as indicated. The accompanying graph presents the ratio of values obtained from these immunoblots and autoradiograph of samples incubated in the absence (–) and presence (+) of each phospholipid species as indicated. Note that PC, like PS, reduces PHF-1 immunoreactivity and increases 5E2 immunoreactivity (see also Fig. 1), while PI does not appreciably alter either of these immunoreactivities from phospholipid-free control samples.

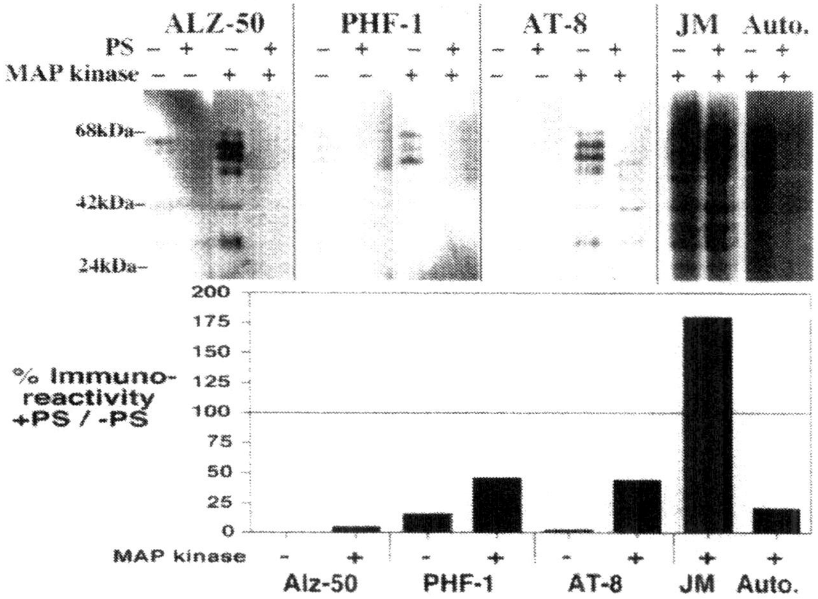

Figure 6. PS facilitates calpain-mediated proteolysis of tau. Purified tau (5ng/ml) in 20 mM Tris (pH 7.5) was incubated at 37°C with m calpain and 2mM $CaCl_2$ in the presence (+PS) and absence (–PS) of 0.5mg/ml PS. Samples were withdrawn at 0–30min and added to concentrated Laemmli buffer on ice to stop the reaction, followed by immunoblot analysis with 5E2. To facilitate these comparitive analyses of tau degradation by calpain, digitized images of 5E2-reacted immunoblots were normalized to yield identical total tau levels in the presence and absence of PS at t = 0min; representative normalized images are presented in this figure. Values present the mean ± standard error of the mean of the % of 5E2 immunoreactivity remaining at 5–15min versus that at time 0 for 3 independent, normalized experiments. Note that tau is more rapidly degraded in the presence of PS than in its absence.

3.3. Phosphatidyl Serine Alters the Ability of Tau to Be Phosphorylated by a Non-Phospholipid-Dependent Kinase

MAP kinase phosphorylates tau under cell-free conditions and, in doing so, alter its antigenicity (see refs. in Introduction). We examined the influence of PS on the ability of tau to serve as a substrate for MAP kinase under cell-free conditions. Immunoblot analysis of samples incubated in the absence of PS confirmed a MAP kinase-mediated increase in PHF-1, ALZ-50 and AT-8 immunoreactivity, while corresponding autoradiographic analysis of samples incubated with g-AT[P^{32}] confirmed MAP kinase-mediated tau phosphorylation (Fig. 7). However, the ability of MAP kinase to confer PHF-1, ALZ-50 and AT-8 immunoreactivity to tau, and to phosphoryate tau, was markedly impaired by co-incubation with PS (Fig. 7). Notably, the activity of MAP kinase (or, at the very least, the commercially-available isoform used in this study) is not directly affected by phospholipids (S. Pelech, personal communication).

It should be noted that the diminished ability of tau to be phosphorylated by MAP kinase or to associate with MTs is unlikely to be derived from simple steric prevention of these phenomena by the presence of excess PS, since this same concentration of PS enhanced tau proteolysis by calpain.

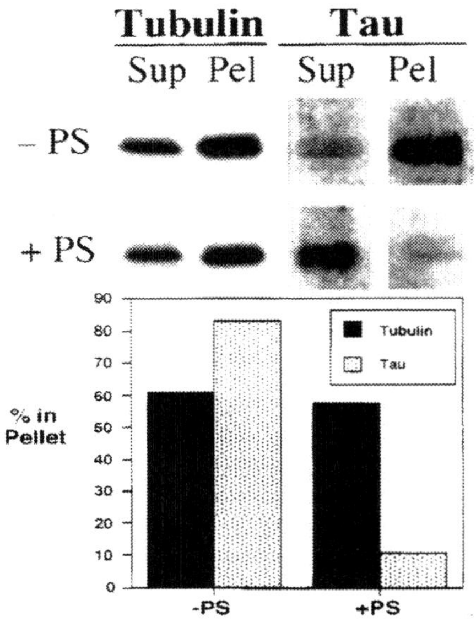

Figure 7. Phosphatidyl serine alters the ability of tau to be phosphorylated by MAP kinase. Immunoblot analyses of 5ng/ml human brain tau incubated with (+) or without (–) 0.5μg/ml PS was incubated with (+) or without (–) MAP kinase for 24hr in the presence of g-AT[P^{32}] prior to SDS gel-electrophoresis followed by transfer to nitrocellulose, immunoblot analysis with various tau antibodies, and subsequent autoradiography of the dried immunoblot. The portion of the immunoblot reacted with polyclonal antibody JM was used to generate the corresponding autoradiograph ("Auto."). The accompanying graph presents the % of immunoreactive or autoradiographic density generated by MAP kinase in the presence of PS versus that in the absence of PS. Note that in the absence of PS, MAP kinase increases tau immunoreactivity with PHF-1, Alz-50 and AT-8, and phosphorylates tau as evidenced by the incorporation onto tau of radiolabeled phosphate groups, while PS markedly attenutates these phenomena. Note further that, like monoclonal antibody 5E2, immunoreactivity of tau with polyclonal JM is markedly enhanced by PS.

3.4. Phosphatidyl Serine Alters the Ability of Tau to Associate with MTs under Cell-Free Conditions

Tau was incubated with purified bovine brain tubulin under conditions that promote MT assembly, after which MTs were sedimented by high-speed centrifugation (100,000 x g, at 37°C) for 15min. As previously demonstrated (Shea and Fischer, 1996), a significant amount of tau was co-sedimented with these newly-formed MTs (Fig. 8). Inclusion of PS during co-incubation of tau and tubulin did not significantly reduce the extent of MT formation, but did reduce the extent of co-sedimentation of tau with newly-formed MTs (Fig. 8). PS did not induce tau sedimentation in the absence of tubulin under these conditions (not shown); in this regard, while PS has been shown to promote polymerization of tau into filaments in previous studies (Binder et al., 1996; Kuret et al., 1996) as well as in our hands (to be published), the length of incubation required for significant tau filament formation significantly exceeds that utilized in the present study to assay for co-sedimentation of tau with MTs (not shown, see Binder et al., 1996).

Diminished ability of tau to associate with MTs in the presence of PS may indicate that tau *in situ* is associated with either membranes or MTs, and does not simultaneously associate with both. Our demonstration of diminished calpain-mediated proteolysis in the absence of PS may relate to a previous study indicating that phosphorylation alone may not be fully responsible for conferring resistance of tau to calpain-mediated proteolysis (Yang and Ksiezak-Reding, 1995). One interpretation of reduced tau proteolysis in the absence of PS is that cytosolic tau (i.e., not associated with phospholipids or MTs) may demonstrate an increased potential for inappropriate accumulation; such accumulation, perhaps coupled with key phosphorylation events, may, under certain conditions, contribute towards PHF formation (Bancher et al., 1989; Shea et al., 1996).

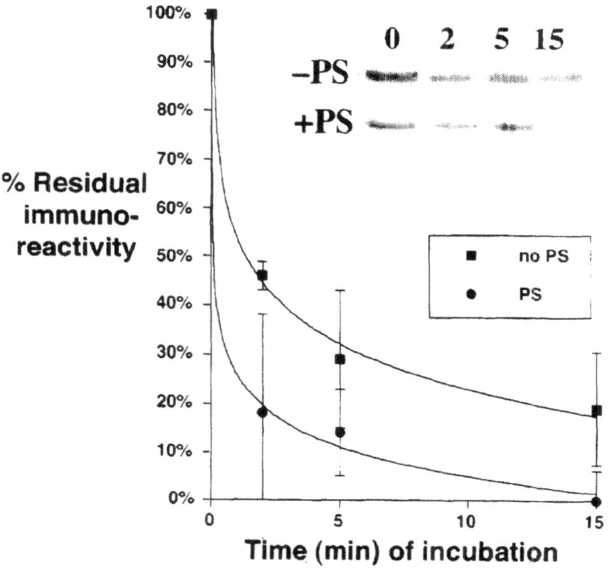

Figure 8. PS reduces the ability of tau to co-sediment with microtubules. Twice-cycled bovine brain tubulin was incubated for 30min at 37°C with purified tau and 0.5μg/ml PS in 100mM PIPES (pH6.6) containing 1mM EGTA and 1mM $MgCl_2$, then centrifuged at 100,000 x g for 30min at 37°C. Resultant pellet ("Pel") and supernatant ("Sup") fractions were subjected to immunoblot analysis with 5E2 and a monoclonal antibody directed against α-tubulin. To facilitate these comparitive analyses of tau distribution among supernatant and pellet fractions, digitized images of 5E2-reacted immunoblots were normalized to yield identical total tau levels before densitometry; representative normalized images are presented in this figure. The accompanying graph presents the % of tau or tubulin sedimented in the absence (−PS) or presence (+PS) of PS. Note the apparent lack of an influence of PS on polymerization of tubulin, but a marked attenuation by PS of the amount of tau that co-sediments with these newly-formed microtubules.

Recent studies have indicated that incubation with phospholipids can induce tau polymerization (Binder et al., 1996; Kuret et al., 1996), leaving open the possibility that our observed reduction in certain phosphate-dependent immunoreactivities was derived from selective failure of a portion of tau, or particular tau isoforms, to enter the gel due to SDS-resistant polymerization. However, the observation of identical, and in some cases, increased levels of immunoreactivity with certain monoclonals indicates that similar levels of tau entered gels. These latter findings also confirm that PS did not induce generalized interference with all antibody interactions. While tau antigenic alteration in the presence of PS is evident in immunoblot analyses, it remains unclear whether or not similar alterations in tau antigenicity are reflected in immunocytochemical and immunofluorescent analyses; i.e., whether or not certain epitopes are diminished or rendered cryptic in phospholipid-rich cellular subregions and vice-versa.

Previous analyses suggest that the N-terminal projection domain of tau mediates its association with the plasma membrane in intact cells (Brandt et al., 1995). While it is reasonable to conclude that the N-terminal region also mediates tau-PS associations under cell-free conditions, our antibody library revealed that PS induced antigenic changes along the entire length of the molecule, indicative of overall conformational change. Thus, we cannot be certain of the nature and extent of tau-PS associations along the molecule. This PS-induced conformational change is reminiscent of the previous demonstration of marked conformation alterations in tau molecule following phosphorylation (Hagestedt et al., 1989). This putative PS-induced conformational change exerts a major physiological impact since the ability of tau to undergo phosphorylation, proteolysis and to associate with microtubules were altered by PS. Detailed studies of full-length tau and various fragments by techniques such as circular dichroism will be required to ascertain the nature and extent of alteration in tau configuration by phospholipids (e.g., Willis, 1994).

Tau has previously been shown to associate with phospholipid vesicles (Surridge and Burns, 1994). It was further noted in these analyses that MAP2 associates with PI via its N-terminal projection domain and that this association altered the configuration of the MAP2 MT-binding domain.(Surridge and Burns, 1994; Burns and Surridge, 1995). Conversely, tau did not associate with PI (Surridge and Burns, 1994; Burns and Surridge, 1995); this latter finding is consistent with our failure to observe antigenic alterations in tau following incubation with PI. The possibility exists that differential interaction of tau and MAP2 with axonal and dendritic membranes and/or membrane transport systems, respectively, may contribute to the differential segregation of tau and MAP2 within axons and dendrites during the development of polarity (Kosik and Finch, 1987).

The ability of tau to associate with the plasma membrane via its N-terminal projection domain (Brandt et al., 1995), while it associates with MTs via its C-terminus (Aizawa et al., 1988; Lee et al., 1989; Himmler et al., 1989), leaves open the additional possibility that tau *in situ* may bridge the MT-membrane distance; this possibility is supported by the ability of tau to form polymers under physiological conditions in the presence of lipids (Binder et al., 1996; Kuret et al., 1996). In light of these recent considerations, it remains possible that the inhibitory effect on neuritogenesis observed following antisense oligonucleotide-mediated tau inhibition was not directly related to effects on MT polymerization, as orignially hypothesized (Caceres and Kosik, 1990; Caceres et al., 1991; Shea et al., 1992a), but rather due at least in part to impairment of membrane-tau associations critical for neuritogenesis.

It has been hypothesized that (inappropriate) dissociation of tau from MT may foster tau hyperphosphorylation (Brambett et al., 1993). One interpretation of the findings in the present study suggest that tau may also be potentially at risk for inappropriate phosphorylation/conformational change following (inappropriate) dissociation from phospholidpids *in situ*. It must further be considered that some degree of antigenicity previously attributed only to changes in tau phosphorylation state may actually reflect conformational changes brought upon by the presence or absence of phospholipids, perhaps in combination with key phosphorylation events. This latter consideration may indicate why certain epitopes, such as that recognized by ALZ-50, can be increased by phosphorylation in cell-free analyses yet are clearly not entirely phospho-dependent (Goedert et al., 1993). Notably, PHFs, which are comprised chiefly if not entirely of tau, have been observed arising from the surface of cytomembranes (Gray et al., 1987).

Finally, although induction of epitopes such as PHF-1, ALZ-50 and AT-8 are often referred to as "AD-like" changes due to their similarity with tau isolated from PHFs, it must be kept in mind that such immunoreactivities are detected during normal development (Goedert et al., 1993; Pope et al., 1993, 1994; Rebhan et al., 1995) Furthermore, these "abnormal" sites are detectable in normal adult brain biopsy tissue, autopsy-derived fetal tissue and rapidly-processed rat tissue (Garver et al., 1994; Matsuo et al., 1994). Our findings further suggest that differential association of tau with phospholipids during development, and any changes in such interactions during the postmortem interval and/or isolation, may contribute to the observed tau antigenic profile.

4. THE ORDER OF EXPOSURE OF TAU TO VARIOUS KINASES ALTERS THE GENERATION OF "AD-LIKE" PHOSPHO-EPITOPES

4. 1. PKC-Mediated Phosphorylation Can Generate and Mask Tau Phospho-Epitopes

Incubation of tau with 0.75–7.5au of PKC reduced Tau-1 immunoreactivity and increased AT-8 immunoreactivity in a concentration-dependent manner. Incubation with a 75 PKCau further reduced Tau-1, yet, by contrast, generated substantially less AT-8 reactivity than did 7.5au (Fig. 9). PKC induced a similar biphasic pattern of ALZ-50 immunoreactivity, in which ALZ-50 immunoreactivity was induced by 0.75–7.5au of PKC but not by 75au. These data suggest that PKC phosphorylates tau at multiple sites, some of which apparently demonstrate less affinity for PKC and therefore require higher PKC concentrations, and that this latter site/group of sites either eliminated or masked both the AT-8 and ALZ-50 epitope. PKA over a 100-fold range of activity reduced Tau-1 immunoreactivity without generating AT-8 immunoreactivity. PKA did not induce ALZ-50 reactivity (Fig. 9).

This phenomenon is likely to reflect conformational changes brought about by sequential phosphorylation of epitopes of differing affinity for PKC. This possibility is supported by the observation of a similar biphasic generation of immunoreactivity towards the conformation-dependent ALZ-50 epitope (Goedert et al., 1993). In this regard, the Tau-2 epitope (not investigated herein) is also conformation-dependent, and this conformation is regulated by phosphorylation (Lang and Otvos, 1992). Concentration-dependence of phospho-epitope generation by PKC in the present study may also provide a partial explanation as to why PKC was able to confer "AD-like" alteration in tau antigenicity and electrophoretic migration in some (Cressman and Shea, 1995; Cressman et al., 1995a; Boyce and Shea, 1997) but not all (e.g., Steiner et al., 1990) previous studies. Indeed, in our previous studies, we were unable to deplete the Tau-1 epitope, nor to generate the AT-8 epitope, using a concentration of PKC that did foster PHF-1 and ALZ-50 immunoreactivity to purified tau (Cressman and Shea, 1995; Cressman et al., 1995a). The data of the present study also provide evidence, as previously suggested (Szendrei et al., 1993), that phospho-dependent loss of the Tau-1 epitope is not synonymous with generation of the so-called "complementary" AT-8 epitope, and that both the loss of Tau-1 and the generation of AT-8 may require multiple phosphoryation events in specific order (see also Biernat et al., 1992; Liu et al., 1993).

4.2. PKC-Mediated Tau Phosphorylation Inhibits Site-Specific Tau Phosphorylation by PKA

Incubation of tau with 0.75au of PKC did not appreciably reduce Tau-1 but impaired reduction of reactivity with this epitope by PKA (Fig. 10). When tau was first incubated with PKA, however, subsequent incubation with PKC did not reverse the PKA-mediated reduction in Tau-1 (Fig. 10). These findings demonstrate that PKC-mediated phosphorylation could influence subsequent PKA-mediated tau phosphorylation, and that the order of kinase treatment was important in such effects. Since tau may have retained some phosphate groups during isolation from brain, we next dephosphorylated tau, in order to determine whether or not any such phosphates could influence subsequent cell-free phosphorylation. Dephosphorylated tau exhibited increased activity with Tau-1 (Fig. 10),

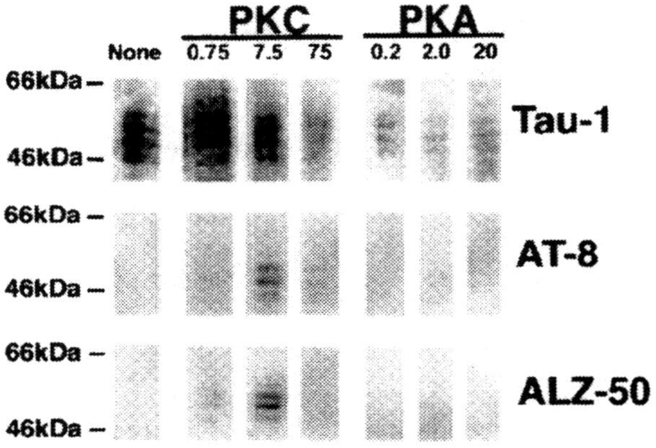

Figure 9. Alteration in tau antigenicity by PKC and PKA. Purified human brain tau was incubated with PKC and PKA and subjected to electrophoresis and immunoblot analysis with Tau-1, AT-8 and ALZ-50 as described in Materials and Methods. Note that 0.75 -75 activity units (au) of PKC progresssively reduced Tau-1 immunoreactivity. PKC at 0.75 and 7.5au progressively generated AT-8 immunoreactivity, but 75au generated less AT-8 immunoreactivity than that generated by 7.5au. A similar bi-phasic pattern of ALZ-50 immunoreactivity was generated by these PKC concentrations. PKA over a 100-fold range of au reduced Tau-1 immunoreactivity but did not generate detectable AT-8 immunoreactivity. PKA also did not generate detectable ALZ-50 immunoreactivity.

suggesting that tau indeed had retained some phosphates during isolation that were removed by alkaline phosphatase. Moreover, dephosphorylation inhibited the ability of PKA to reduce Tau-1 immunoreactivity (Fig. 10), indicating that *in situ* phosphorylation at one or more sites rendered tau more susceptible to PKA-mediated antigenic alterations at least at the Tau-1 site.

4.3. PKC-Mediated Phosphorylation Enhamces Subusequent Phosphorylation by MAP Kinase

As shown above (Fig. 7), incubation of tau with PS reduced the ability of MAP kinase to phosphorylate tau under these conditions. Nevertheless, prior phosphorylation of tau by 0.75au PKC was able to facilitate MAP kinase-mediated reduction of Tau-1, and induction of AT-8, immunoreactivity (Fig. 11). Since PS is a necesseasry co-factor for PKC activity, we were unable to ascertain whether or not this effect would be observed in the absence of PS, nor were we able to do so for studies involving PKA (above). By contrast, prior PKC-mediated phosphorylation did not facilitate MAP kinase-induced alteration in tau antigenicity towards PHF-1 or ALZ-50 (Fig. 11).

4.4. Prior in Situ Phosphorylation by Undisclosed Kinase(S) Influences Phosphorylation in Vitro

Dephosphorylation of tau induced opposing effects on phosphorylation by MAP kinase and PKC. Dephosphoryation of tau did not alter the influence of MAP kinase on Tau-1, AT-8 or PHF-1 immunoreactivity, but reduced the ability of MAP kinase to generate ALZ-50 immunoreactivity (Fig. 11). Conversely, dephosphorylation inhibited the ability

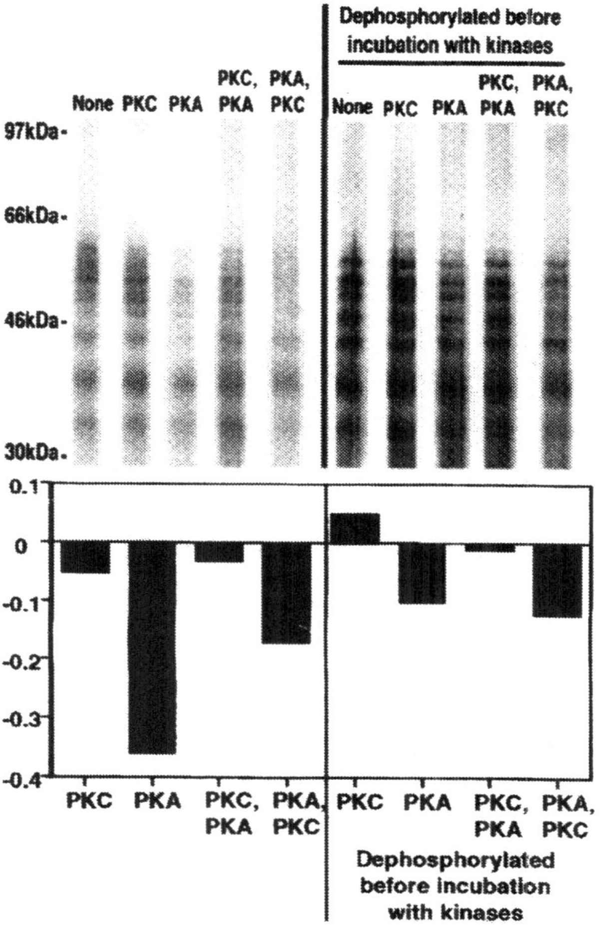

Figure 10. Prior phosphorylation of tau by PKC under cell-free conditions and by unknown kinases in situ alters the ability of PKA to phosphorylate tau. The top panel presents representative immunoblot analysis of tau before ("none") and following incubation with PKC and/or PKA, with or without prior dephosphorylation as indicated, and probed with Tau-1. The order of kinase incubation in sequential treatments corresponds to the order listed above individual lanes. The lower panel presents densitometric analyses of the above immunoblot, with values expressed as fold change of respective densitometric values obtained for the control sample ("none"). Dephosphorylated samples ("dephosphorylated before incubation") were normalized according to their respective dephosphorylated control sample, and no attempt was made to normalize dephosphorylated and non-dephosphorylated densitometric values. Control values are not included in the graph.

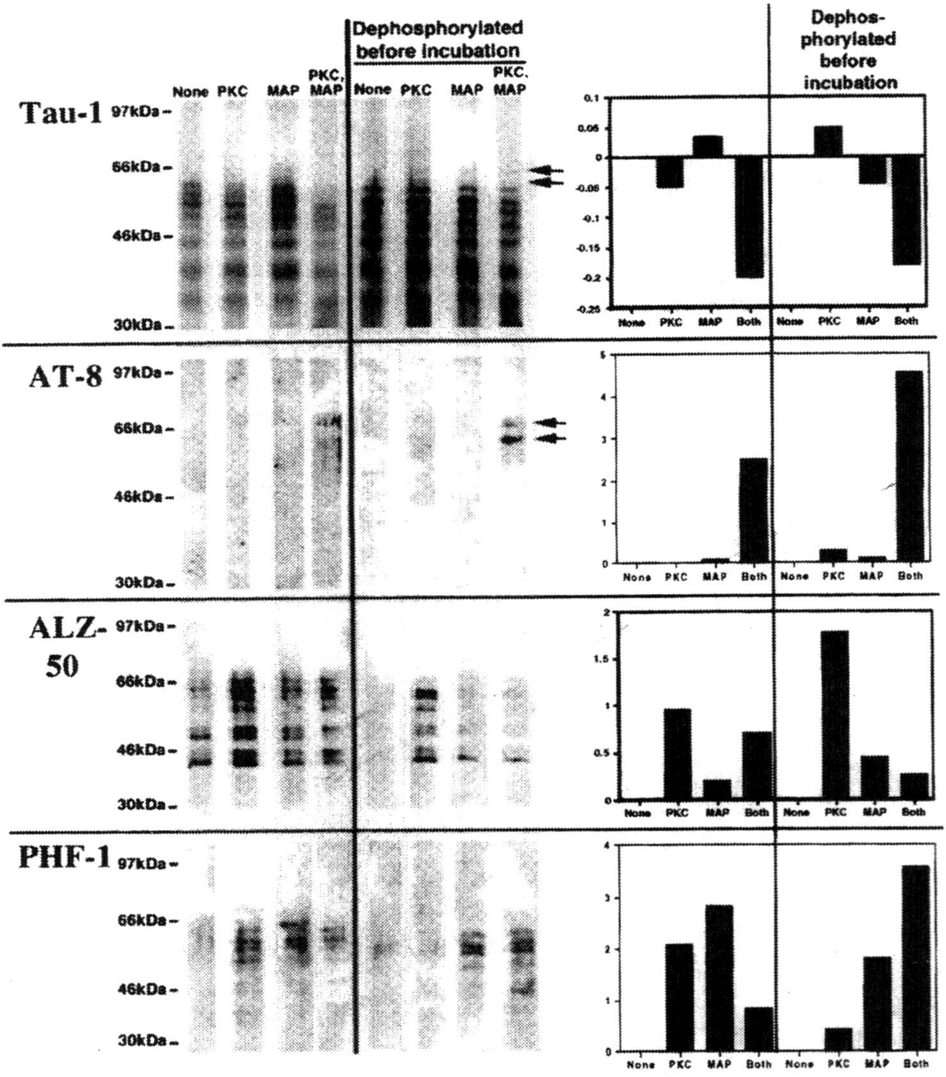

Figure 11. Prior PKC-mediated phosphorylation of tau influences subsequent MAP kinase-mediated tau phosphorylation. Panels on the left present representative immunoblot analysis of tau before ("none") and following incubation with PKC and/or MAP kinase, with or without prior dephosphorylation, and probed with various antibodies as indicated. Arrows to the right of immunoblots probed with Tau-1 and AT-8 indicate tau isoforms migrating at approximately 68 and 62kDa. Panels on the right present densitometric analyses of respective immunoblots, with values expressed as fold change of respective densitometric values obtained for control samples ("none"). Dephosphorylated samples ("dephosphorylated before incubation") were normalized according to their respective dephosphorylated control samples, and no attempt was made to normalize dephosphorylated and non-dephosphorylated densitometric values.

of PKC to generate PHF-1 immunoreactivity, but did not alter the influence of PKC on Tau-1, AT-8 or ALZ-50 immunoreactivity (Fig. 11).

Dephosphorylation of tau altered the generation of certain epitopes by kinases in individual and sequential incubations, and, moreover, did so in a manner unique for each incubation regimen. These latter findings further underscore the possible importance of hierarchical phosphorylation of tau *in situ* and leave open the possibility that the initial accumulation of so-called "hyperphosphorylated" tau may involve displacement of one or more "protective" regulatory phosphorylation events due to hyperactivation of an inappropriate kinase (see also Kosik and Greenberg, 1994). Notably, a similar effect could result from a deficiency in one or more of the putative protective kinases. In this regard, we have previously shown that increasing intracellular PKC activity by phorbol ester increased certain "AD-like" tau epitopes (Cressman and Shea, 1995; Shea et al., 1996), and that inhibition of MAP kinase activity prevents TPA-induced PHF-1 immunoreactivity in human neuroblastoma cells (Pundreddy and Shea, 1997). While we had previously considered that this sequential effect was exclusively due to PKC-mediated activation of the MAP kinase pathway (e.g., Pelech, 1995), it must also now be considered that PKC-mediated phosphorylation of tau may render tau more susceptible to subsequent phosphorylation by MAP kinase (see also Singh et al., 1995a,b). Such facilitation within intact cells may rely on displacement of tau from microtubules and/or disruption of segregation of tau within the axonal compartment of neurons. For example, phosphorylation within the microtubule-binding region of tau, by lessening the ability of tau to associate with microtubules, may foster subsequent hyperphosphorylation (for review, see Mandelkow et al., 1995). Conflicting results have appeared regarding the ability of MAP kinase to phosphorylate tau in intact cells (Cressman et al., 1996; Lu et al., 1993; Latimer et al., 1995). The data of the present study suggest that one factor contributing to this discrepancy may be prior *in situ* phosphorylation of tau by one or more "protective" kinases.

Hyperphosphorylation of tau following PKC hyperactivation within human neuroblastoma (Cressman and Shea, 1995; Shea et al., 1996), coupled with the previous demonstration that prior tau phosphorylation by PKA reduced subsequent tau phosphorylation by MAP kinase (Blanchard et al., 1994) leaves open the possibility that tau may be rendered more or less susceptible to "AD-like" hyperphosphorylation *in situ* by intial phosphorylation by PKC or PKA, respectively. The possibility of divergent influence of these kinases on tau is supported by the combined observations that neuritogenesis in neuroblastoma cells is microtubule-dependent, is inhibited by intracellular delivery of antibodies and antisense oligonucleotides directed against tau (Shea et al., 1992a; Shea and Beermann, 1994), and is promoted by PKA activation yet inhibited by PKC activation (Shea et al., 1992b). Further studies will be required to clarify whether or not these kinases exert opposing influences on tau behavior *in situ*.

Signal transduction kinases such as PKC, PKA and MAP kinase are known to interact at multiple levels in neuronal homeostasis (e.g., Hollingsworth et al., 1986; Kosik and Greenberg, 1994; Peraldi et al., 1995; Pelech and Sangria, 1992; Pelech, 1995; Shea et al., 1992b). These findings further suggest that such interaction may influence the net phosphorylation state of tau not only by interaction among kinase pathways themselves, but also by hierarachical phosphorylation of tau protein.

5. CONCLUSION

The studies presented herein demonstrate a profound influence of phospholipids on tau biology and further studies indicating the importance of sequential kinase activities on tau phosphorylation state. Whether or not the accumulation of hyperphosphorylated tau and/or actual PHF formation are causative in AD neuropathology or instead represent end-stage "by-products" remain controversial. Nevertheless, the accumulation of hyperphosphorylated tau is clearly indicative of previous and/or ongoing disregulation in signal transduction pathways affecting tau. Continued elucidation of such factors and intracellular pathways that influence tau steady-state levels and phosphorylation state may uncover avenues for therapeutic intervention to prevent or delay neuronal loss in Alzheimer's disease.

ACKNOWLEDGMENT

This research was supported by the National Institute for Aging.

REFERENCES

Aizawa H, Kawasaki H, Murofushi S, Kotani K, Suzuki and Sakai H (1988) Microtubule-binding domain of tau proteins. J Biol Chem 263:7703–7707.

Arioka M, Tsukamoto M, Ishiguro K, Kato R, Sato K, Imahori K and Uchida T (1993) Tau protein kinase II is involved in the regulation of the normal phosphorylation state of tau protein. J Neurochem 60:461–468.

Baas PW, Pienkowski TP, Cimbalnik KA, Toyama K, Bakalis S, Ahmand FJ, and Kosik KS (1994) Tau confers drug-stability but cold-stability to microtubules in living cells. J Cell Sci 107: 135–143.

Bancher C, Brunner C, Lassmann H, Budka H, Jellinger K, Wiche G, Seiteberger F, Grundke-Iqbal I, Iqbal I and Wisniewski HM (1989) Accumulation of abnormally phosphorylated tau precedes the formation of neurofibrillary tangles in Alzheimer's disease. Brain Res 477: 90–99.

Baudier J and Cole RD (1987) Phosphorylation of tau proteins to a state like that in Alzheimer's brain in catalyzed by a calcium/calmodulin-dependent kinase and modulated by phospholipids. J Biol Chem 262:17577–17583.

Baudier J, Lee S-H and Cole RD (1987) Separation of the different microtubule-associated tau proteins from bovine brain and their mode II phosphorylation by Ca2+/phospholipid-dependent protein kinase C. J Biol Chem 262:17584--17590.

Baumann. K, Mandelkow, E-M, Biernat, J, Piwnica-Worms, H, and Mandelkow, E. (1993) Abnormal Alzheimer-like phosphorylation of tau-protein by cyclin-dependent kinases cdk2 and cdk5. FEBS Lett 336:417–424.

Biernat J, Mandelkow E-M, Schroter, Lichtenberg-Kraag B, Steiner B, Berling B, Meyer H, Mercken M, Vandermeeren A. Goedert M, et al (1992) The switch of tau protein to an Alzheimer-like state includes the phosphorylation of two serine-proline motifs upstream of the microtubule-binding region. EMBO J 11:1593–1597.

Binder LI, Wilson DW and Kuret J (1996) Lipid stimulation of tau polymerization into Alzheimer straight filament-like structures. J Neurochem 66 Supp: S95

Black MM, Slaughter T, Moshiach M, Obrocka M and Fischer I (1996) Tau is enriched on dynamic microtubules in the distal region of growing axons. J Neurosci 16:3601–3619.

Blanchard BJ, Raghnandan R, Roder HM and Ingram VM (1994) Hyperphosphorylation of human tau by brain kinase PK40erk beyond phosphorylation by cAMP-dependent PKA: Relation to Alzheimer's disease. Biochem Biophys Res Comm 200:187–194.

Boyce JJ and Shea TB (1997) Phosphorylation events mediated by protein kinase Cα and ε participate in regulation of tau steady-state levels and generation of certain "Alzheimer-like" phospho-epitopes. Int J Dev Neurosci: 15:295-307.

Brambett GT, Goedert M, Jakes R, Merrick SE, Trojanowski JQ and Lee V M-Y (1993) Abnormal tau phosphorylation at ser-396 in Alzheimer's disease recapitulates development and contributes to reduced microtubule binding. Neuron 10: 1089–1099.

Brandt R, Leger J and Lee G (1995) Interaction of tau with the neural plasma membrane mediated by tau's amino-terminal projection domain. J Cell Biol 131:1327–1340.

Burns RG and Surridge CD (1995) The phosphatidylinositol-binding site of microtubule-associated protein MAP2. Biochem Soc Trans 23: 41–46.

Caceres A and Kosik KS (1990) Inhibition of neuronal polarity by tau antisense oligonucleotides in primary cerebellar neurons. Nature 343:461–463.

Caceres A, Pottrebic S and Kosik KS (1991) The effect of antisense oligonucleotides on neurite formation of cultured cerebellar macroneurons. J Neurosci 11: 1515–1523.

Cleveland DW, Hwo SY and Kirschner MW (1977) Purification of tau, a microtubule-associated protein that induces assembly of microtubules from purified tubulin. J Mol Biol 116:227–247.

Cressman CM and Shea TB (1995) Hyperphosphorylation of tau and filopodial retraction following microinjection of protein kinase C catalytic subunits. J Neurosci Res 42: 648–656.

Cressman CM, Mercken MM and Shea TB (1995a) Alteration in tau antigenicity and electrophoretic migration by PKCα under cell-free conditions. Neurosci Res Commun 17: 61–64.

Cressman CM, Mohan, PS, Nixon RA, Griffin WR and Shea TB (1995b) Proteolysis of protein kinase C: mM and μM calcium-requiring calpains have different abilities to generate, and degrade, the free catalytic subunit, protein kinase M. FEBS Lett 367: 223–227.

Dreschel DN, Hyman AA, Cobb MH and Kirschner MW (1992) Modulation of the dynamic instability of tubulin assembly by the microtubule-associated protein tau. Mol Biol Cell 3:1141–1154.

Drewes, G., Lichtenberg-Kragg, B., Doring, F., Mandelkow, E.-M., Bienart, J., Doree, M., and Mandelkow, E. (1992) Mitogen activated protein (MAP) kinase transform tau protein into an Alzheimer-like state. EMBO J. 11:2131–2138.

Drubin DG and Kirschner MW (1986) Tau protein function in living cells. J Cell Biol 103: 2739–2746.

Garver TD. Harris KA. Lehman RAW, Lee Vm-Y, Trojanowski JQ and Billingsley ML (1994) Tau phosphorylation in human, primate and rat brain: Evidence that a pool of tau is highly phosphorylated in vivo and is rapidly dephosphorylated in vitro. J Neurochem 63:2279–2287.

Goedert, M., Cohen, E. S., Jakes, R., and Cohen, P. (1992) p42 MAP kinase phosphorylation sites in microtubule-associated protein tau are dephosphorylated by protein phosphatase 2A1. FEBS Lett. 312:95–99.

Goedert M (1993) Tau protein and the neurofibrillary pathology of Alzheimer's disease. Trends Neurosci 16:460–465.

Goedert M, Jakes R, Crowther RA, Six J, Lubke U, Vandermeeren M, Cras P, Trojanowski JQ and Lee V M-Y (1993) The abnormal phosphorylation of tau proteins at ser-202 in Alzheimer's disease recapitulates phosphorylation during development. Proc Natl Acad Sci USA 90:5066–5070.

Goedert M, Jakes R, Crowther RA, Cohen P, Vanmechelen E, Vandermeeren M and Cras P (1994) Epitope mapping of monoclonal antibodies to the paired helical filaments of Alzheimer's disease: identification of phosphorylation sites in tau protein. Biochem J 301:871–877.

Gray EG, Paula-Barbosa M and Roher (1987) Alzheimer's disease: paired helical filaments and cytomembranes. Neuropath Appl Neurobiol 13:91–110.

Hagestedt T, Lichtenberg B, Wille H, Mandelkow E-M and Mandelkow E (1989) Tau protein becomes long and stiff upon phosphorylation: Correlation between paracrystalline structure and degree of phosphorylation. J Cell Biol 109:1643–1651.

Hanger, D P, Hughes, K, Woodgett, J R, Brion, J P, and Anderton, B H (1992) Glycogen synthase kinase-3 induces Alzheimer's disease-like phosphorylation of tau: generation of paired helical filament epitopes and neuronal localization of the kinase. Neurosci. Lett. 147:58–62.

Himmler A, Dreschel D, Kirschner MW and Martin DW (1989) Tau consists of a set of proteins with repeated C-terminal microtubule-binding domains and variable N-terminal domains. Mol Cell Biol 9:1381–1388.

Hollingsworth EB, Ukena D and Daly JW (1986) The protein kinase C activator phorbol-12-myristate-13-acetate enhances cyclic AMP accumulation in pheochromocytoma cells. FEBS Lett 196:131–134.

Johnson GVW, Jope RS and Binder LI (1989) Proteolysis of tau by calpain. Biochem Biophys Res Commun 163: 1505–1511.

Kempf M, Clement A, Faissner A, Lee G and Brandt R (1996) Tau binds to the distal axon early in development of polarity in a microtubule- and microfilament-dependent manner. J Neurosci 16:5583–5592.

Kobayashi, S, Ishiguro, K, Omori, A, Takamatsu, M, Arioka, M, Imahora K, and Uchida, T (1993) A cdc-related kinase PSSALRE/cdk5 is homologous with the 30kDa subunit of tau protein kinase II, a proline-directed kinase associated with microtubules. FEBS Lett. 335: 171–175.

Kosik KS (1993) The molecular and cellular biology of tau. Brain Pathol 3:39–43.

Kosik KS and Finch EA (1987) MAP2 and tau segregate into dendritic and axonal domains after the elaboration of morphologically distinct neurites: An immunochemical study of cultured rat cerebellum. J Neurosci 7:3142–3153.

Kosik KS and Greenberg SM (1994) Tau protein and Alzheimer's disease. In: Alzheimer's disease (Terry RD, Katzman R and Bick KL, ed) Raven Press, NY pp 335–344.

Kuret J, Hantash B, Cha D, Wilson DM and Binder LI (1996) Morphological and structural charactierization of lipid-induced tau polymers. Mol Biol Cell 7:569a.

Lang D, Beermann ML, Hauser G, Cressman CM and Shea TB (1995) Phospholipids inhibit proteolysis of protein kinase Cα by mM calcium-requiring calpain. Neurochem Res 20: 1361–1364.

Lang E and Otvos L (1992) A serine-proline change in the Alzheimer's disease-associated epitope Tau-2 results in altered secondary structure, but phosphorylation overcomes the conformational gap. Biochem Biophys Res Comm 188:162–169.

Latimer DA, Gallo J-M, Lovestone S, Miller CCJ, Reynolds CH, Marquardt B, Stable S, Woodgett JR and Anderton BH (1995) Stimulation of MAP kinase by v-raf transformation of fibroblasts fails to induce hyerphosphorylation of transfected tau. FEBS Lett 365:42–46.

Ledesma, M D, Correas, L, Avila, J, and Diaz-Nido, J (1992) Implication of brain cdc2 and MAP kinases in the phosphorylation of tau protein in Alzheimer's disease, FEBS Lett. 308: 218–224.

Lee G, Neve RL and Kosik KS (1989) The microtubule binding domain of tau protein. Neuron 2:1615–1624.

Lee G, Kwei SL, Newman ST, Lu M and Yiu Y (1996a) A new molecular interactor for tau protein. Trans Am Soc Neurosci 22:975.

Lee G, Kwei SL, Newman ST, Olmstead J and Yiu Y (1996b) A new molecular interactor for tau protein. Mol Biol Cell 7:570a.

Litersky JM, Scott CW and Johnson GVW (1993) Phosphorylation, calpain proteolysis and tubulin binding of recombinant tau isoforms. Brain Res 604:32–40.

Liu W-K, Moore WT, Williams RT, Hall FL and Yen S-H (1993) Application of synthetic phospho- and unphospho- peptides to identify phosphorylation sites in a subregion of the tau molecule, which is modified in Alzheimer's disease. J Neurosci Res 34:371–376.

Lu, Q, Soria, J P, and Wood, J G (1993) p44mpk MAP kinase induces Alzheimer type alterations in tau function and in primary hippocampal neurons. J. Neurosci. Res.. 35: 439–444.

Mandelkow EM and Mandelkow E (1993) Tau as a marker for Alzheimer's disease. Trends Biochem Sci 18: 480–483.

Mandelkow, E-M, Drewes, G, Biernat, J, Gustke, N, Van Lint, J, Vandenheede, J R, and Mandelkow, E (1992) Glycogen synthase kinase 3 and the Alzheimer's disease-like state of microtubule-associated protein tau. FEBS Lett. 314:315–321.

Mandelkow E-M, Biernat J, Drewes G, Gustke N, Trinczek B and Mandelkow E (1995) Tau domains, phosphorylation, and interactions with microtubules. Neurobiol Aging 16:355–362.

Mandell JW and Banker GA (1996) A spatial gradient of tau protein phosphorylation in nascent axons. J Neurosci 16:5727–5740.

Matsuo, E S, Shin, R-W, Billingsley, M L, Van deVoorde, A, O'Connor, M, Trojanowski, J Q and Lee, V M-Y (1994) Biopsy-derived adult human brain tau is phosphorylated at many of the same sites as Alzheimer's disease paired helical filament tau. Neuron 13: 989–1002.

Morishima-Kawashima M, Hasegawa M, Takio K, Suzuki M, Yoshida H, Titani K and Ihara Y (1995) Prolin-directed and non-proline-directed phosphorylation of PHF-tau. J Biol Chem 270:823–829.

Mulot, S F C, Hughes, K, Woodgett, J R, Anderton, B H, and Hanger, D P (1994) PHF-tau a from Alzheimer's brain comprises four species on SDS-PAGE which can be mimicked by in vitro phosphorylation of human brain tau by glycogen synthase kinase-3β. FEBS Lett. 349: 359–364.

Paudel HK, Lew J, ALi Z, Wang JH (1993) Brain proline-directed protein kinase phosphorylates tau on sites that are abnormally phosphorylated in tau associated with Alzheimer's paired helical filaments. J Biol Chem 268:23512–23518.

Pelech SL (1995) Networking with proline-directed protein kinases implicated in tau phosphorylation. Neurobiol Aging 16: 247–261.

Pelech SL and Sangria JS (1992) Mitogen-activated protein kinases: versatile transducers for cell signalling. Trends Biochem Soc 17: 233–238.

Peraldi P, Frodin M, Barnier JV, Calleja V, Scimeca J-C, Filloux C, Calothy G and Van Obberghen E (1995) Regulation of the MAP kinase cascade in PC12 cells: B-Raf activates MEK-1 (MAP kinase or ERK kinase) and is inhibited by cAMP. FEBS Lett 357:290–296.

Pope, W B, Enam, S A, Bawa, N, Miller, B E, Ghanbari, H.A, and Klein, W L (1993) Phosphorylated tau epitope of Alzheimer's disease is coupled to axon development in the avian central nervous system. Exp. Neurol. 120: 106–113.

Pope, W B, Lambert, M P, Leypold, B, Seupaul, R, Sletten, L, Krafft, G, and Klein, W L (1994) Microtubule-associated protein tau is hyperphosphorylated during mitosis in the human neuroblastoma cell line SH-SY-5Y. Exp. Neurol. 126: 185–194.

Pundreddy S and Shea TB (1997) AD-like tau phosphorylation in human neuroblastoma cells following PKC hyperactivation is mediated by MAP kinase. Neurosci Res Commun. 21:173–177.

Raghunandan R and Ingram VM (1995) Hyperphosphorylation of the cytoskeletal protein tau by the MAP-kinase PK40erk: Regulation by prior phosphorylation with cAMP-dependent protein kinase A. Biochem Biophys Res Comm 215: 1056–1066.

Rebhan M, Vacun G and Rösner H (1995) Complementary distribution of tau proteins in different phosphorylation states within growing axons. NeuroReport 6:429–432.

Sengupta A, Wu Q, Grundkle-Iqbal I, Iqbal K and Singh TJ (1997) Potentiation of GSK-3-catalyzed Alzheimer-like phosphorylation of human tau by cdk5. Mol Cell Biochem 167:99–105.

Shea TB, Beermann ML, Nixon RA and Fischer I (1992a) Microtubule-associated protein tau is required for axonal neurite elaboration by neuroblastoma cells. J Neurosci Res 32:363–374.

Shea TB, Beermann ML, Leli U and Nixon RA (1992b) Opposing influences of protein kinase activities on neurite outgrowth in human neuroblastoma cells: Initiation by kinase A and restriction by kinase C. J Neurosci Res 33:398–407.

Shea TB and Beermann ML (1994) Respective roles of neurofilaments, microtubules, MAP1B and tau in the outgrowth and stabilization of axonal neurites. Mol Biol Cell 5:863–875.

Shea TB and Fischer I (1996) Phosphatase inhibition in human neuroblastoma cells alters tau antigenicity and renders it incompetent to associate with exogenous microtubules. FEBS Lett 380:63–67.

Shea TB, Spencer MJ, Beermann ML, Cressman CM, Nixon RA (1996) Calcium influx into human neuroblastoma cells induces ALZ-50 immunoreactivity: Involvement of calpain-mediated hydrolysis of protein kinase C. J Neurochem 66:1539–1549.

Singh TJ, Zaidi T, Grundke-Iqbal I and Iqbal K (1994a) Modulation of GSK-3-catalyzed phosphorylation of microtubule-associated protein tau by non-proline-dependent protein kinases. FEBS 358: 4–8.

Singh TJ, Haque N, Grundke-Iqbal I and Iqbal K (1994b) Rapid Alzheimer-like phosphorylation of tau by the synergistic actions of non-proline dependent kinases and GSK-3. FEBS Lett 358: 267–272.

Steiner, B, Mandelkow, E-M, Biernat, J, Gustke, N, Meyer, H E, Schmidt, B, Mieskes, G, Söling, H D, Dreschsel, D, Kirschner, M W, Goedert, M, and Mandelkow, E (1990) Phosphorylation of microtubule-associated protein tau: identification of the site for Ca^{2+}-calmodulin dependent kinase and relationship with tau phosphorylation in Alzheimer tangles. EMBO J. 9:3539–3544.

Szendrei GI, Lee V M-Y and Otvos L (1993) Recognition of the minimal epitope of monoclonal antibody Tau-1 depends upon the presence of a phosphate group but not its location. J Neurosci Res 34:243-249

Surridge CD and Burns RG (1994) The difference in the binding of phosphatidylinositol distinguishes MAP2 from MAP2C and tau. Biochem 33:8051–8057.

Takemura R, Okabe S, Umeyama T, Kanai Y, Cowan NI and Hirokawa N (1992) Increased microtubule stability and alpha-tubulin acetylation in cells transfected with microtubule-associated proteins MAP1B, MAP2 or tau. J Cell Sci 103:953–964.

Trojanowski, J Q, Schmidt, M L, Shin, R-W, Bramblett, G T, Rao, D, and Lee, V M-Y (1993a) Altered tau and neurofilament proteins in neurodegenative diseases: diagnostic implications for Alzheimer's disease and Lewy body dementias. Brain Pathol. 3:45–54.

Trojanowski, J Q, Schmidt, M L, Shin, R-W, Bramblett, G T, Goedert, M, and Lee, V M-Y (1993b) PHF tau (A68): from pathological marker to potential mediator of neuronal dysfunction and degeneration in Alzheimer's disease. Clin. Neurosci. 1:184–191.

Vulliet, R, Halloran, S M, Braun, R K, Smith, A J, and Lee, G (1992) Proline-directed phosphorylation of human tau protein. J. Biol. Chem. 267: 22570–22574.

Willis KJ (1994) Interaction with model membrane system induces secondary structure in amino-terminal fragments of parathyroid hormone-related protein. Int J Peptide Protein Res 43:23–28.

Yang L-S and Ksiezak-Reding H (1995) Calpain-induced proteolysis of normal human tau and tau associated with paired helical filaments. Eur J Biochem 233: 9–17.

12

CELL CULTURE MODELS OF NEURONAL DEGENERATION AND NEUROPROTECTION

Implications for Parkinson's Disease

Efthimia T. Kokotos Leonardi and Catherine Mytilineou

Department of Neurology
Mount Sinai School of Medicine
New York, New York 10029

Parkinson's disease (PD) is a neurodegenerative disorder characterized by a progressive loss of dopaminergic neurons in the substantia nigra (McGeer, et al., 1988). The disease has a late onset, and is fairly prevalent, affecting 1% of the general population over the age of 50 (Polymeropoulos, et al., 1996). Treatment for the disorder is largely symptomatic, usually beginning with dopaminergic agonist therapy, or with a combination of levo-dopa (L-DOPA) and dopaminergic receptor agonists (see Rabey, 1995 and Corboy et al., 1995 for review).

Recently, a gene has been discovered that is linked to PD (Polymeropoulos, et al., 1996). However, the gene is only linked to a small percentage of PD patients; the underlying etiology of idiopathic PD still remains unclear. Studies using mitochondria isolated from platelets (Parker et al., 1989), muscle tissue (Shoffner et al., 1991) and fibroblasts (Mytilineou et al., 1994) of PD patients show a significant reduction of complex I activity. A defect in complex I has also been observed in the substantia nigra of PD patients by Schapira et al., (1992). These findings support the hypothesis that one underlying cause of PD may be a defect or reduction in complex I in the tissues of these patients.

1-Methyl-4-phenyl-1,2,3,6-tetrahydropyridine (MPTP), is a neurotoxin that produces parkinsonian symptoms in humans, monkeys and mice (Langston et al, 1983; Langston et al., 1984; Burns et al., 1983; Heikkila et al., 1984). MPTP is oxidized by monoamine oxidase type B (MAO-B) to form MPDP+ (Chiba et al., 1984); MPDP+ is subsequently converted to the pyridinium ion, MPP+ (Chiba et al., 1984; Castagnoli et al, 1985). MPP+ enters dopaminergic neurons from the extracellular space through the high-affinity transporter protein (Chiba et al., 1985) and accumulates within these cells in millimolar concentrations (Nicklas et al., 1985). Cell death ensues presumably via MPP+-mediated inhibition of complex I of the mitochondrial respiration chain (Nicklas, et

Molecular and Cellular Mechanisms of Neuronal Plasticity, edited by
Ehrlich. Plenum Press, New York, 1998.

al. 1985; Nicklas et al., 1987). In addition, a redox reaction that occurs between MPDP+ and MPP+ generates superoxide radicals (Zang and Misra, 1993), which in turn, cause lipid perodixation and damage to the dopaminergic cell membrane.

Interestingly, complex I is also vulnerable to oxidative stress (Glinka et al., 1996). Evidence of oxidative stress, such as increased lipid peroxidation (Dexter et al., 1994), decreased levels of reduced glutathione (Jenner, 1993) and increased superoxide dismutase activity (Saggu et al, 1989) is present in the substantia nigra of PD patients. Oxidative stress in the substantia nigra may have several origins, including: a) hydrogen peroxide formation as a by-product of dopamine catabolism by MAO-B, b) a conversion of hydrogen peroxide into hydroxyl free radicals via iron-dependent Fenton reaction and c) the auto-oxidation of dopamine into melanin and its concomitant free radical production (Owen et al, 1996; Cohen, 1985).

Parkinson's disease may also arise from excitatory amino acid (EAA) toxicity to the substantia nigra following an overstimulation of glutamatergic receptors; receptors of the NMDA subtype are localized on the substantia nigra pars compacta (Krebs et al., 1991; Mereu et al., 1991). Activation of ionotropic EAA receptors causes an initial entry of Ca^{+2} into the cell (Choi, 1990), followed by a massive disruption in intracellular Ca^{+2} homeostasis (Carillo et al., 1992; Fransden and Shousboe, 1993). These processes ultimately lead to cell death (Choi et al., 1987; Kato et al., 1991), possibly by an uncoupling of the mitochondrial proton gradient (Lafon-Cazal et al., 1993).

Our laboratory has used dissociated fetal rat mesencephalic cell cultures to study dopaminergic cell toxicity and degeneration arising from 1) MPTP, 2) oxidative stress and 3) excitotoxicity. As these systems represent models of some of the neurodegenerative processes that may underlie Parkinson's disease, it is hoped that an understanding of dopaminergic cell degeneration and neuroprotection in a closed, reproducible system can provide some insights into the disease mechanism.

Neurodegenerative effects to dopaminergic cells in these studies were determined by measuring the high-affinity uptake capacity of cultures for [^{3}H]-dopamine (DA) or by the use of antibodies directed against tyrosine hydroxylase (TH), the rate-limiting enzyme in catecholamine synthesis. Since the dopamine transporter is localized on dopaminergic fibers, a reduction in the uptake capacity for [^{3}H]-DA by these cells indicates a fiber degeneration or loss. The use of both functional and morphological indices of dopamine neurons provide an accurate measure of the extent and localization of damage to cultured dopaminergic cells. Descriptions of methodologies used for the establishment of cultures, measurement of [^{3}H]-DA uptake and TH-immunocytochemistry are provided in previous publications (Hou et al, 1996; Mytilineou et al., 1997).

1. NEURODEGENERATIVE EFFECTS OF MPTP ON FETAL RAT MESENCEPHALIC CULTURES

Because MPTP causes parkinsonian symptoms in humans, mice, and monkey, data obtained from *in vitro* studies on the mechanisms of MPTP toxicity are useful toward furthering the understanding of the neurodegenerative processes that take place in Parkinson's disease. The effects of MPTP have been characterized in rat mesencephalic explant and dissociated cell cultures. Although rats are not sensitive to MPTP toxicity *in vivo* our studies showed that dopamine neurons from embryonic rat brain are sensitive to MPTP *in vitro* (Mytilineou and Cohen, 1984).

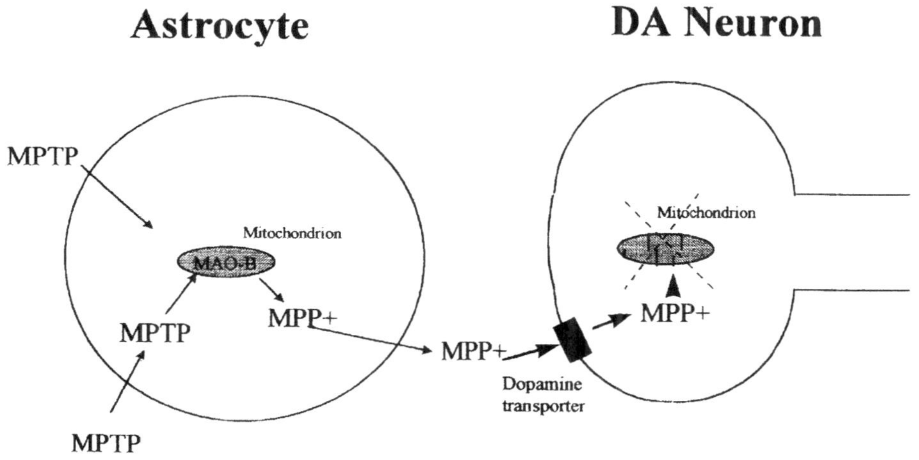

Figure 1. Highly lipophilic MPTP enters the astrocytes where it is metabolized to MPP+ in a two-step process which involves MAO-B. MPP+ is then released into the extracellular space where it is actively taken up into the dopaminergic neuron by the dopamine transporter. MPP+ inhibits complex I of mitochondrial respiration, which leads to the death of the dopaminergic neuron.

MPTP is toxic to cultured dopamine neurons in a concentration-dependent manner (Friedman and Mytilineou, 1987). The concentrations of MPTP required to produce deleterious effects in the *in vitro* paradigm correspond well to the toxic doses for monkeys *in vivo* (Langston et al., 1984; Mytilineou et al, 1988), lending support to the use of a tissue culture model as an effective tool in the study of the cellular actions of neurotoxic agents.

MPTP toxicity to mesencephalic cultures is dependent upon MAO oxidation of MPTP, indicating that MPTP is a protoxin which requires oxidation by MAO to form a toxic metabolite (Mytilineou and Cohen, 1984). In support of this hypothesis, MPP+, the oxidative product of MAO deamination of MPTP, was found to be more toxic to mesencephalic cultures than MPTP (Mytilineou et al., 1985). The glial localization of MAO-B in the brain (Levitt et al., 1982) led to the formulation of a hypothesis for MPTP toxicity to dopamine neurons by Javitch et al (1985, see Figure 1). They suggested that MPTP is metabolized by glial cells to MPP+, which is then taken up by dopaminergic neurons (Chiba et al., 1985; Javitch et al., 1985; Javitch and Snyder, 1985) where it accumulates at high concentrations and causes cell death. Our cell culture studies support the hypothesis that MPTP is metabolized by glial cells and that MPP+ can diffuse into the extracellular space, from where it can be taken up by catecholaminergic neurons (Mytilineou and Friedman, 1988). We showed that the metabolism [^{3}H]-MPTP and ensuing toxicity to dopaminergic neurons were dependent upon the presence of glial cells (Mytilineou and Friedman, 1988). Treatment of cultures with the anti-mitotic agent 3-fluoro-5-deoxyuridine (FuDR) significantly reduced the number of glial cells and the amount of MPTP metabolite produced (Mytilineou and Friedman, 1988). These observations suggested that neuronal-glial interactions may be important in the pathophysiology of Parkinson's disease.

One way MPP+ can exert its toxic effects to dopaminergic cells is by inhibiting complex I of the mitochondrial respiratory chain (Nicklas et al., 1987). Our studies showed that dependence on oxidative metabolism sensitized cells to MPP+ toxicity and a switch to glycolysis in older cultures rendered dopamine neurons resistant to MPTP and MPP+ toxicity (Danias et al., 1989). Thus alterations in mitochondrial function is a crucial component in the neurotoxic mechanism of MPP+ toxicity and possibly, Parkinson's disease.

2. PROTECTION OF DOPAMINERGIC NEURONS AGAINST MPTP TOXICITY BY GROWTH FACTORS

A MPTP model of Parkinson's disease *in vitro* is helpful not only in the determination of degenerative processes occurring in PD, but also in the screening of agents that may be useful in its treatment and management. Identification of dopaminergic growth factors that promote survival after exposure to specific dopaminergic neurotoxins *in vitro* can provide useful information in the development of clinical treatments that may delay the progression of Parkinson's disease. Several peptide growth factors with neurotrophic properties protect dopaminergic cells against MPTP toxicity *in vitro* and *in vivo*.

Basic fibroblast growth factor (bFGF) and epidermal growth factor (EGF) promote the survival of dopaminergic neurons in mesencephalic cell cultures and enhance the uptake capacity of these cells for [^{3}H]-DA (Engele and Bohn, 1991; Ferrari et al., 1989; Casper et al., 1991). [^{3}H]-DA uptake in cultures treated with these growth factors was approximately 2–3 fold higher after one week, and by the end of the third week, reached levels 10–15 fold higher than controls (Casper, et al., 1991; Park and Mytilineou, 1992). In addition, dopaminergic neurons visualized by immunocytochemical staining against TH showed more elaborate processes in EGF and bFGF treated cultures. The time course of enhanced neuritic outgrowth paralleled the increase in [^{3}H]-DA uptake by the cultures (Park and Mytilineou, 1992). Immunocytochemistry for glial fibrillary acidic protein (GFAP) showed that there was a dramatic increase in the density of GFAP+ astrocytes in the cultures treated with growth factors, suggesting the involvement of glial cells in the neurotrophic activity (Park and Mytilineou, 1992). The elimination of the trophic activity by inhibition of glial cell proliferation further confirmed the involvement of glial cells in the neurotrophic actions of EGF and bFGF (Park and Mytilineou, 1992; Engele and Bohn, 1991).

The effects of EGF or bFGF were examined to determine whether they modify the extent of damage caused by MPP+ to dopaminergic neurons (Park and Mytilineou, 1992). The growth factor concentrations used in these experiments were those shown to have maximal trophic effects in primary neuronal cultures (Engele and Bohn, 1991; Ferrari et al., 1989; Knusel et al., 1990). The concentration of MPP+ used (10 μM) has been shown to affect only dopaminergic neurons in mesencephalic cell culture (Michel et al., 1990; Sanchez-Ramos et al., 1986).

First the short-term effects of bFGF and EGF on MPP+ toxicity were examined. Cultures maintained with or without EGF or bFGF for 4 days were exposed to 10 μM MPP+ for 24h followed by a 24h incubation in a toxin-free medium. Short-term treatment of cultures with bFGF or EGF did not afford dopamine neurons protection against MPP+ (Park and Mytilineou, 1992).

The effects of growth factors on damaged dopaminergic neurons were also investigated. A 24h exposure of DIV4 cultures to 10 μM MPP+ in control, bFGF and EGF treated cultures was used. Dopaminergic neurons were subsequently observed for two weeks (Figure 2). While MPP+ treated cultures showed a gradual slow increase in [^{3}H]-DA uptake capacity between days 6 and 18 *in vitro*, MPP+ treated cultures exposed to

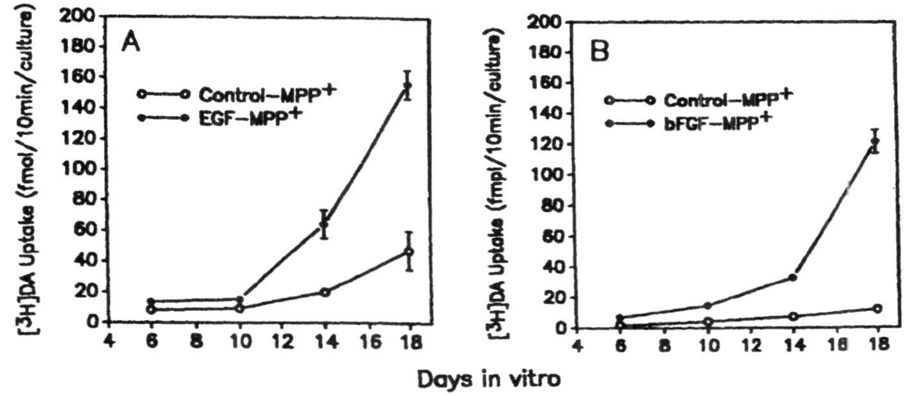

Figure 2. EGF and bFGF stimulate the recovery in [^{3}H]DA uptake following MPP+ treatment. Mesencephalic cultures grown in the absence or presence of EGF (10 ng/ml, A) or bFGF (10 ng/ml, B) were treated with 10 μM MPP+ for 24 h on DIV 4. [^{3}H]DA uptake was measured at various times following MPP+ treatment. Growth factors were present in the appropriate groups throughout the experiment. 24 h after MPP+ treatment uptake in all groups was reduced by 60–70%. Although some growth occured in control cultures after MPP+, EGF and bFGF stimulated a much larger increase in DA uptake (from Park and Mytilineou, 1992).

EGF or bFGF showed a dramatic increase in DA uptake beginning on DIV 12 with a sharp rise between DIV 14–18, demonstrating that exposure to bFGF and EGF promotes regrowth of damaged dopaminergic processes (Park and Mytilineou, 1992). TH immunocytochemistry of MPP+ treated cultures showed a re-growth of dopaminergic fibers over time, consistent with observations of increased [^{3}H]-DA uptake capacity (Park and Mytilineou, 1992).

Long-term treatment with growth factors (10 days) also protected cultured dopaminergic neurons against 48h exposure to MPP+. At 5 μM MPP+, control cultures had an 86% loss in uptake capacity, while after EGF or bFGF treatment, the same concentration of MPP+ caused a 56% and 48% loss in uptake, respectively (Park and Mytilineou, 1992).

The role of glial cells in the long-term protective effects afforded to MPP+ treated cultures by EGF and bFGF was investigated. Treatment with the mitotic inhibitor, FuDR, significantly reduced the amount of GFAP+ labeled cells in all cultures, especially those treated with growth factors (77%, 91% and 98% reductions in control, EGF and bFGF cultures, respectively). The protection of bFGF and EGF against MPP+ was completely abolished by FuDR treatment, suggesting that these growth factors exert their neuroprotective effects by promoting glial proliferation.

The neuroprotective effects of bFGF against MPTP have also been examined *in vivo* by Otto and Unsicker (1990), who showed that bFGF treatment reverses the morphologic and biochemical deficits following MPTP lesions of the nigrostriatal system in mice.

Other agents with neuroprotective properties against MPP+ include brain derived neurotrophic factor (BDNF) (Hyman et al., 1991; Spina et al., 1992; Beck et al., 1992) and neurotrophic factor 4/5 (Hynes et al., 1994).

Recently, a glial cell-line derived neurotrophic factor (GDNF) was purified and cloned (Lin et al., 1993, 1994). GDNF is relatively specific, glial cell independent trophic agent for dopaminergic neurons (EC_{50} ~ 40 pg/mL) (Lin et al., 1993, 1994). Since other dopaminergic growth factors have been shown to protect against MPP+ toxicity to cultured fetal dopaminergic neurons, it was of interest to investigate whether GDNF can also protect dopaminergic cells against MPP+.

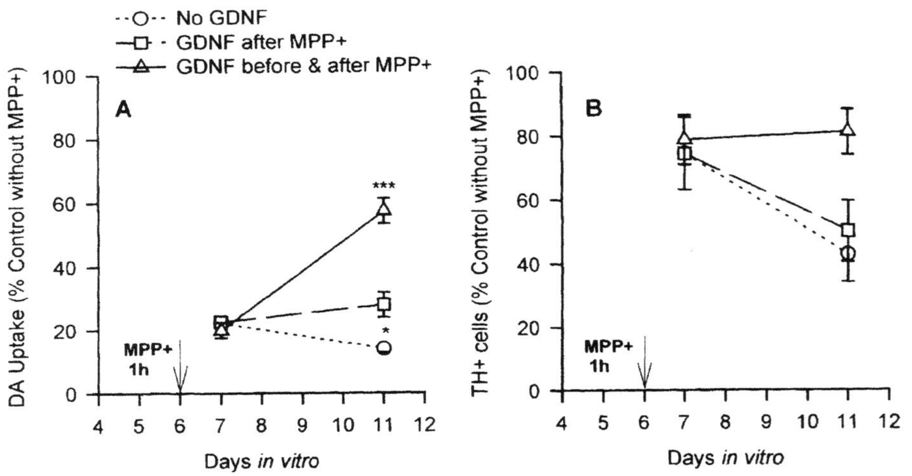

Figure 3. GDNF stimulation of survival and regrowth of DA neurons after MPP+ treatment. Mesencephalic cultures maintained in the absence or presence of GDNF (1 ng/ml) were treated with MPP+ (100 μM) for 1 h on DIV 6. Following MPP+ treatment, control cultures were subdivided into two groups: one that received GDNF and one that did not. Cultures pretreated with GDNF were given GDNF after MPP+ treatment. [^{3}H]DA uptake and TH immunocytochemistry were performed on DIV 7 and DIV 11. Data are mean ± SEM expressed as percent of corresponding controls (adapted from Hou et al., 1996).

The effects of GDNF were determined at first in control cultures exposed to a chemically defined medium (Hou et al, 1996). Growth stimulating effects of GDNF treatment to dopaminergic neurons were determined by the measurement of [^{3}H]-DA uptake and by counting the number of TH+ cells per cm^2. Optimal effects of GDNF were observed in both uptake and TH+ cell number at 1 ng/mL; there was no effect of GDNF on GFAP positive cells (Hou et al., 1996).

Neuroprotective actions of GDNF against MPP+ in fetal mesencephalic cultures were determined by treatment with GDNF (0, 0.1, 1 and 10 ng/mL) in defined medium on DIV 0 followed by a 1h exposure to 100 μM MPP+ on DIV 5. [^{3}H]-DA uptake was measured 24h after MPP+ removal from cultures. While GDNF itself increased [^{3}H]-DA uptake at 0.1 and 1 ng/mL, it did not significantly improve the reduction in uptake by MPP+ at any concentration (Hou et al., 1996). These findings demonstrate that GDNF cannot prevent MPP+ toxicity from taking place.

We also examined whether GDNF can promote the regrowth of dopaminergic fibers following MPP+ exposure. After termination of MPP+ treatment, cultures that were not exposed to GDNF prior to MPP+ treatment showed a steady, significant decrease in uptake capacity (five days after MPP+, DA uptake was 14.0±0.1% of control, compared to 22.6±1.9% of control at 24h after MPP+, $p<0.05$). The reduction of DA uptake capacity observed after five days was prevented if 1 ng/mL GDNF was added following MPP+ treatment. The reduction in uptake capacity by dopaminergic cells was reflected by a loss in the number of TH+ cells following MPP+ treatment. However, GDNF treatment following MPP+ was ineffective at preventing this reduction (Hou et al., 1996). In contrast, cultures treated with GDNF before and after MPP+ treatment showed steady recovery in uptake capacity (between 20% of control 24 h after MPP+ to 57% control after five days, $p<0.001$, Figure 3). The recovery took place even if GDNF treatment ceased after exposure to MPP+, though not to the same degree (Hou et al., 1996). These experiments indicate that GDNF is able to prevent progressive cell death in MPP+ treated cultures and that GDNF may also promote fiber regrowth following an MPP+ insult and are in agreement with the ability of GDNF to prevent MPTP toxicity *in vivo* (Tomac, et al., 1995).

3. OXIDATIVE STRESS, NEURONAL DEGENERATION AND NEUROPROTECTION IN MESENCEPHALIC CULTURES

Oxidative stress and free radical species have been linked to a number of neurodegenerative diseases, including Huntington's disease, amyotrophic lateral sclerosis and PD (Coyle and Puttfarcken, 1993; Olanow, 1993). In support of this hypothesis, post-mortem studies of Parkinson brains show reduced levels of glutathione (GSH; DiMonte et al., 1992; Sian et al., 1994), elevated iron (Riederer et al., 1989; Sofic et al., 1989; Good et al., 1992), increased lipid peroxides (Dexter et al., 1989) and decreased complex I activity (Schapira et al., 1990; Mizuno et al., 1995). In the mammalian brain, GSH decomposes H_2O_2 and plays an important role in the detoxification of quinones (Reed, 1990; Meister and Anderson, 1983). One source of oxidative stress in the Parkinson brain may be the auto-oxidation of dopamine or L-DOPA. This reaction generates a wide variety of toxic species, including hydrogen peroxide and quinones (Graham, 1978; Cohen, 1985). Mesencephalic cell cultures depleted of GSH can serve as a model for a central nervous system that is vulnerable to oxidative stress.

Cultures of mesencephalic neurons exposed to 200 μM L-DOPA for 48h show a 30% reduction in the number of TH+ cells, and a concomitant 58% loss of [^{3}H]-DA uptake capacity (Mytilineou et al., 1993). In addition, L-DOPA treatment produced a significant increase in the level of GSH. The rise in GSH levels by L-DOPA was blocked by ascorbic acid, suggesting that GSH levels increased in response to an oxidative signal (Mytilineou et al., 1993). GSH depletion sensitizes cells to oxidative stress (Biaglow et al., 1986; Ling et al.,1990). L-buthionine sulfoximine (L-BSO) causes GSH depletion by blocking γ-glutamylcysteine synthetase (Slivka, et al., 1988), the rate-limiting enzyme required for glutathione synthesis (see Meister, 1991 for review). Interestingly, when cells were treated with 6 μM L-BSO and 200 μM L-DOPA, L-DOPA prevented cell death by L-BSO (Mytilineou et al., 1993). These findings suggest that L-DOPA itself may have anti-oxidant properties in mesencephalic cultures. Recent evidence shows that both L-DOPA and deprenyl protect dopaminergic neurons against L-BSO induced toxicity in a concentration-dependent manner (Kokotos Leonardi et al., 1996; Mytilineou et al., 1998). The protective actions of L-DOPA and deprenyl are additive with ascorbic acid, suggesting that both these agents may act as anti-oxidants in dopamine neurons (Kokotos Leonardi et al., 1996).

6-Hydroxydopamine (6-OHDA) has been used as a model of oxidative stress to catecholamine neurons. Damage from 6-OHDA treatment is thought to be a result of hydrogen peroxide and oxygen free radical formation (Heikkila and Cohen, 1972; Sachs and Jonsson, 1975; Graham 1978; Kumar et al., 1995). Several growth factors have been shown to protect neurons against oxidative damage (see Mattson et al., 1993 for review). Glial cell proliferation has been shown to play a role in the neuroprotective mechanisms of some growth factors (Knusel et al., 1990; Engele and Bohn, 1991; Park and Mytilineou, 1992) and GSH, a major cellular anti-oxidant, is predominantly concentrated within glia (Slivka et al., 1987; Raps et al., 1989). To investigate the role of glia in neuroprotection against oxidative stress we compared the actions of GDNF and bFGF, which respectively

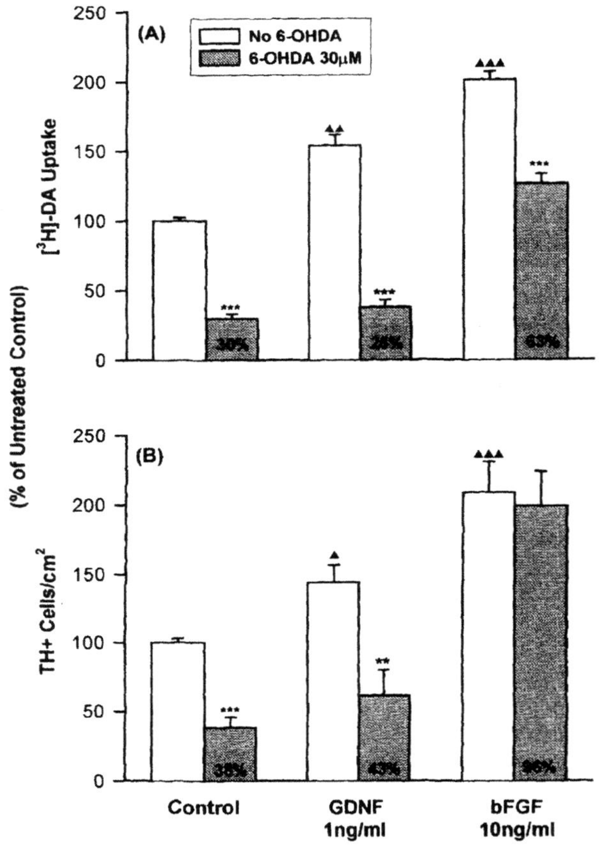

Figure 4. Effect of GDNF and bFGF on 6-OHDA toxicity. Mesencephalic cultures were maintained in the presence of GDNF (1 ng/ml) or bFGF (10 ng/ml). On DIV 5 they were treated with 30 μM 6-OHDA for 45 min. [^{3}H]DA uptake (A) and TH positive neurons (B) were measured on DIV 6. Data are mean values ± SEM expressed as percent of untreated controls. The numbers on the bars show the uptake values expressed as percent of the non-6-OHDA control in each group. **$p<0.01$, ***$p<0.001$ vs. non-6-OHDA control; ▲$p<0.05$, ▲▲$p<0.01$, ▲▲▲$p<0.001$ vs. non-growth factor control (from Hou et al., 1997).

act independently or through glial cell proliferation, against 6-OHDA toxicity in cultured mesencephalic neurons (Hou et al., 1997).

Mesencephalic cultures maintained in a chemically defined medium were treated with either GDNF (1 ng/mL) or bFGF (10ng/mL) beginning on the initial day of plating (DIV 0). These concentrations were selected for their maximal neurotrophic effects on mesencephalic neurons (Hou et al., 1996; Park and Mytilineou, 1992). On DIV5, cultures were exposed to 30 μM 6-OHDA for 45 minutes at 37°C. Following a 24h drug washout period in fresh medium, the survival of dopaminergic neurons was determined by measuring high affinity [^{3}H]-DA uptake and TH-immunoreactivity. GDNF did not protect dopaminergic neurons against 6-OHDA. bFGF, on the other hand, showed a strong protective effect. 6-OHDA decreased DA uptake capacity by 37% in bFGF-treated cultures, compared to 70% in controls ($p<0.001$; Hou et al., 1997 Figure 4).

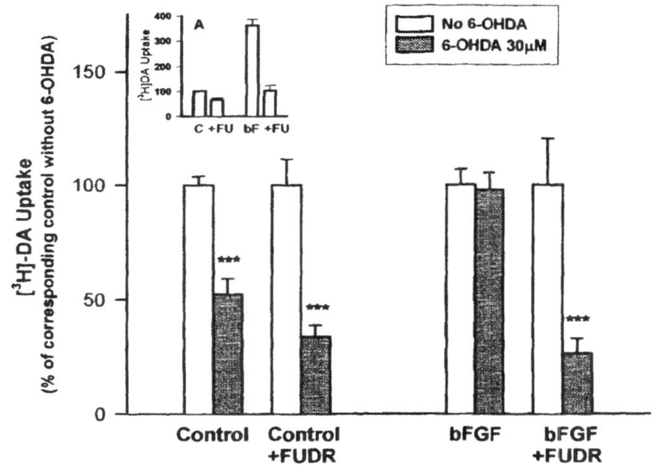

Figure 5. Effect of inhibition of glial proliferation on the bFGF-induced protection from 6-OHDA toxicity. Mesencephalic cultures were maintained in the absence or presence of bFGF (10 ng/ml). The antimitotic agent FμDR was added 24 h after plating. Cultures were treated on DIV 7 with 6-OHDA (30 μM) for 45 min. [^{3}H]DA uptake was assayed on DIV 8. Data are mean values ± SEM expressed as percent of corresponding controls. Insert A depicts uptake values of non-6-OHDA treated groups expressed as % of untreated controls. ***p<0.001 compared to corresponding non-6-OHDA treated control in each group (from Hou et al., 1997).

To determine the role of proliferating glial cells in the protective effect of bFGF against 6-OHDA, cultures were treated with the anti-mitotic inhibitor, FuDR on DIV1. When 6-OHDA was added to FuDR treated cultures on DIV7, the neurotrophic effect of bFGF and protection against 6-OHDA was completely abolished (Figure5).

The relationship between the protective effect of bFGF and GSH content in cultures was determined by measuring GSH levels following treatment with GDNF and bFGF. GDNF did not increase GSH levels, but bFGF doubled GSH levels compared to controls on DIV 9 (Figure 6, Hou et al., 1997). Addition of FuDR prevented the increase in GSH by bFGF, indicating that glial cell proliferation was responsible for the increased GSH levels following bFGF treatment. GSH levels were significantly increased 4h after 6-OHDA

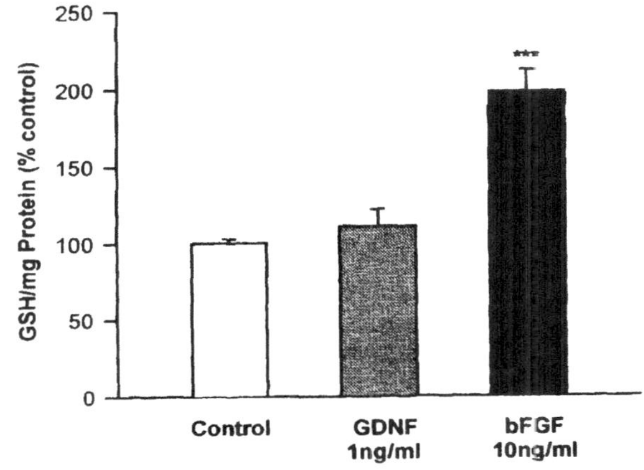

Figure 6. Effect of GDNF and bFGF on GSH levels in mesencephalic cultures. Cultures were maintained in the presence of GDNF (1 ng/ml) or bFGF (10 ng/ml). GSH was assayed on DIV 9. Data are mean values " SEM expressed as percent of untreated controls. ***p<0.001 (Hou et al., 1997).

treatment in cultures treated with bFGF; the up-regulation of GSH in the bFGF group was completely blocked by FuDR. The inhibition of GSH synthesis by L-BSO also reduced the protective effect of bFGF against 6-OHDA (Hou et al., 1997). These findings together provide evidence that a neuronal-glial interaction that stimulates GSH production is a crucial requirement in the neuroprotective mechanism of bFGF against 6-OHDA mediated oxidative stress.

4. EXCITOTOXICITY STUDIES IN FETAL RAT MESENCEPHALIC CULTURES

In situ hybridization and immunohistochemical studies in squirrel monkeys demonstrate a greater extent of labeling for NMDAR1 and GluR2 subunits in dopaminergic neurons in the substantia nigra pars compacta than those in the ventral tegmental area (Paquet, et al., 1997). NMDA and glutamate receptors have also been localized in rat midbrain slices (Mereu et al., 1991). A greater amount of EAA receptors in the substantia nigra may underlie the vulnerability of these neurons in Parkinson's disease.

Daily medium change in mesencephalic cultures is thought to cause damage to dopamine neurons by an accumulation of extracellular glutamate (Driscoll et al, 1991). In support of this hypothesis, damage to neurons resulting from daily changes of culture media can be blocked by NMDA receptor antagonists (Driscoll et al., 1993). Furthermore, *in vitro* studies of mesencephalic dopaminergic neurons show excitotoxic damage by glutamate receptor agonists (Casper and Blum, 1995).

L-deprenyl is a selective, irreversible inhibitor of MAO-B (Knoll 1978). Because of its inhibitory actions on the catabolism of dopamine, deprenyl has been used as an adjunct to L-DOPA in the treatment of Parkinson's disease, where its use provides modest benefits to patients (Birkmeyer and Riederer, 1984; Golbe et al., 1988). Recent studies have shown that deprenyl may delay the progression of PD (Olanow, et al., 1995).

Deprenyl enhances the survival of cultured dopaminergic neurons (Roy and Beddard, 1993). However, its protective properties are not limited solely to dopamine neurons (Salo and Tatton, 1992; Tatton et al., 1994). Several lines of evidence suggest that deprenyl exerts its neuroprotective effects independent of MAO-B inhibition. For example, deprenyl protects mesencephalic cultures against MPP+, the metabolite of MPTP that is produced following MAO-B catabolism of MPTP (Mytilineou et al., 1985). Deprenyl also protects against nigrostriatal MPTP lesions in mice 72h after MPTP, a time point where MPTP should be oxidized and cleared from the brain (Tatton and Greenwood, 1991). The protective effects of deprenyl were therefore investigated in mesencephalic cultures subjected to excitotoxic damage by daily medium change.

Mesencephalic cultures were exposed to 0.5, 5, or 50 μM deprenyl concurrent with daily medium change between DIV 5 and DIV 9. Deprenyl significantly increased the number of TH+ neurons in cultures subjected to daily medium change in a concentration dependent manner (approximately 1.5, 2 and 2.5 fold increases, respectively; Figure 7; Mytilineou et al, 1997). The increase in TH+ neurons observed in deprenyl treated cultures was reflected by increases in the uptake of [^{3}H]-DA. However, significant increases in [^{3}H]-DA uptake were only observed with 5 and 50 μM deprenyl.

To test the possibility that deprenyl protects cultured mesencephalic DA neurons from the excitotoxic effects of daily medium changes, the effects of 50 μM deprenyl were compared to those of 10 μM MK-801, an NMDA receptor antagonist. Treatment of cultures exposed to daily medium changes with MK-801 increased their uptake capacity ap-

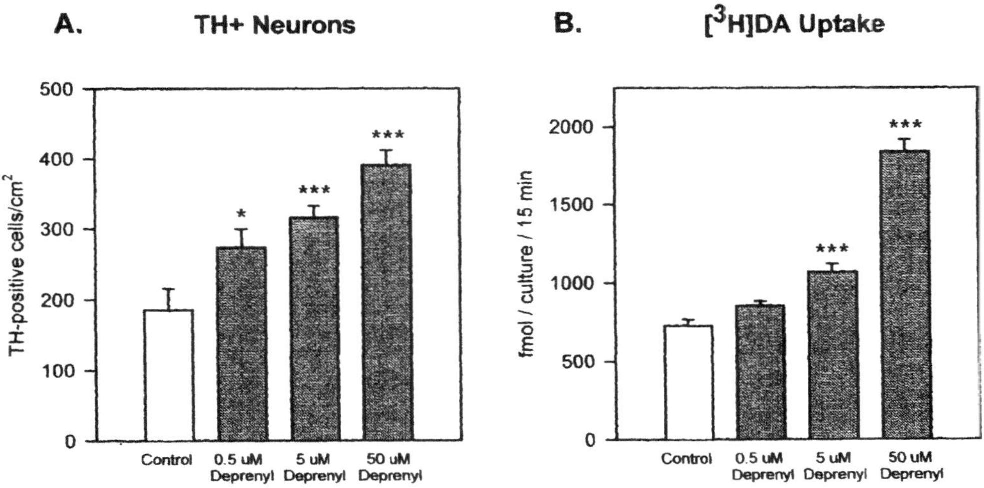

Figure 7. Effect of increasing concentrations of deprenyl on the number of TH-positive neurons (A) and [^{3}H]DA uptake (B) in mesencephalic cultures following four complete changes of feeding medium. Deprenyl was added only during complete medium changes which began on DIV 5. *$p<0.05$, ***$p<0.001$ compared to controls (from Mytilineou et al., 1997).

proximately three-fold (Mytilineou et al, 1997). Deprenyl was not as potent as MK-801; cultures exposed to medium changes in the presence of 50 μM deprenyl had [^{3}H]-DA uptake values twice those of their untreated controls (Mytilineou et al, 1997). Both deprenyl and MK-801 increased the number of TH+ neurons compared to untreated, changed cultures; 50μM deprenyl completely protected DA neurons against medium changes (Mytilineou et al., 1997).

To further examine whether deprenyl protects cultures against NMDA receptor stimulation, cultures were treated with either deprenyl or MK-801 on DIV 9 and then exposed to 200 μM NMDA on DIV 12. TH immunocytochemistry and [^{3}H]-DA uptake were performed 24h later. Treatment with NMDA lowered the [^{3}H]-DA uptake capacity of mesencephalic cultures by 44%, and decreased the number of TH+ cells by 25%. These decreases were prevented by treatment with deprenyl or MK-801 (Mytilineou et al., 1997). The possibility that deprenyl may exert its protective effects by competing with NMDA for binding to glutamatergic receptors was examined. Deprenyl did not compete with [^{3}H]-CGP-39653 for binding to the NMDA receptor (Sills et al., 1991), suggesting that its protective actions are independent of the NMDA receptor (Mytilineou et al., 1997). Interestingly, desmethylselegeline (DMS), a metabolite of deprenyl formed by the cyctochrome P-450 system (Heinonen, et al., 1994) also protected cultured dopaminergic neurons against excitotoxicity resulting from a 30 min treatment of 200 μM NMDA to cultures on DIV 12 (Mytilineou et al, 1997).

The protective effects of deprenyl against the excitotoxic effects of media changes do not appear to be related to its MAO inhibitory properties. Pargyline, a potent and nonselective MAO inhibitor did not protect dopaminergic neurons against media changes. [^{3}H]-DA uptake in cultures exposed to pargyline at 0.5, 5 and 50 μM during media changes did not differ from controls (Mytilineou, et al., 1997). These findings are consistent with observations reported by Tatton et al, (1994) which provide evidence for tran-

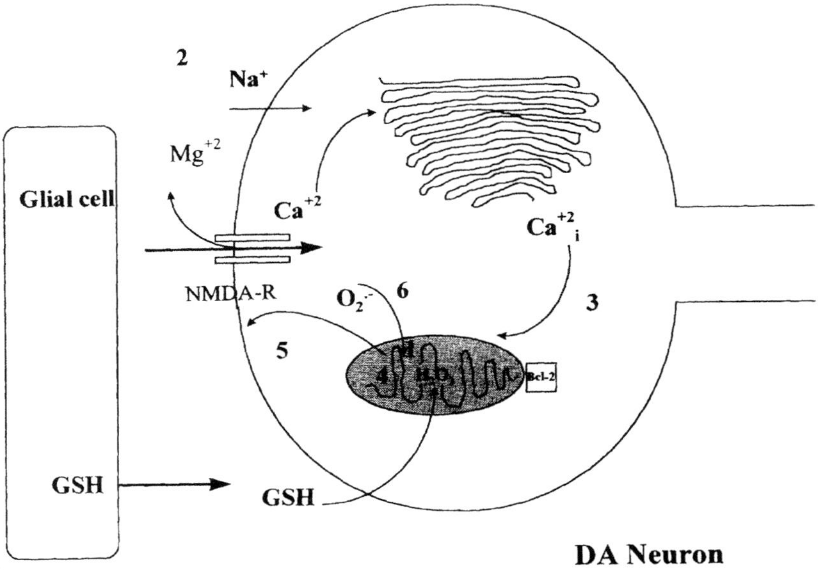

Figure 8. Concerted model of possible mechanisms of neuronal degeneration and neuroprotection in PD. Impaired complex I activity (1) renders dopaminergic neurons more vulnerable to NMDA excitotoxicity by facilitating cell depolarization. Depolarization leads to removal of the Mg^{+2} blockade at the NMDA receptor (2). Stimulation of NMDA receptors results in a massive influx of Ca^{+2}, which in turn liberates Ca^{+2} from intracellular stores. The massive disruption of intracellular Ca^{+2} homeostasis leads to destruction of the mitochondrial potential which leads to free radical production and oxidative stress (4). Free radicals destroy the cell membrane by lipid peroxidation (5) and the resultant oxidative stress also further inhibits complex I activity (6). GSH, produced in the glia, acts as an anti-oxidant and can enhance cell survival after oxidative stress; growth factors that stimulate glial proliferation enhance DA cell survival following oxidative damage, blocking 4, 5, and 6. Deprenyl also seems to have anti-oxidant properties. Bcl-2, a major anti-apoptotic protein located on the mitochondrion is up-regulated by growth factors and blocks the opening of the mitochondrial permeability transition pore, also preventing 4, 5, and 6. Glutamate release blockers and NMDA antagonists also protect against cell death by interfering with 2.

scriptional events involving *de novo* protein synthesis are associated with the benefits of deprenyl.

Deprenyl may protect against NMDA-mediated excitotoxicity by preventing free radical formation arising from changes in intracellular Ca^{+2}. These changes may lead to the accumulation of Ca^{+2} in the mitochondria (Lafon-Cazal, et al., 1993), as well as the generation of free radicals via stimulation of phospholipase A_2, xanthine oxidase, or nitric oxide synthase (McCord, et al, 1985 ; Dumuis et al, 1988; Dawson et al., 1992). Deprenyl suppresses the formation of free radicals (Wu et al., 1993; Chieuh et al., 1994) and upregulates the activities of catalase and superoxide dismutase in the brain (Carillo, 1994; Thiffault, 1995).

NMDA-receptor mediated cell death can occur either by apoptotic or necrotic pathways; these effects are dependent upon NMDA concentration (Bonfoco et al., 1995; Portera-Caillau, et al., 1995; van Lookeren Campagne et al., 1995). Apoptosis has been shown to be associated with a change in the ratio of the oncoproteins Bax and Bcl-2 (Korsmeyer, et al., 1995). Overexpression of bcl-2 can block apoptosis and protect against glutamate-mediated neurotoxicity (Behl et al., 1993). Deprenyl reduces apoptosis and produces concomitant increases in bcl-2 while reducing Bax in PC12 cells (Tatton et al., 1996). These effects appear to be due to transcriptional events (Tatton et al., 1994; 1996). Therefore, deprenyl may protect against cell death in cultured dopaminergic neurons by interfering with apoptosis.

5. POSSIBLE MECHANISMS OF NEURONAL DEGENERATION AND NEUROPROTECTION IN PARKINSON'S DISEASE

Several observations have been made regarding the functional characteristics of nigral dopaminergic neurons *in vitro* and *in vivo* that may explain their susceptibility to degeneration in Parkinson's disease. Glutamatergic receptors are present on nigral dopaminergic neurons which make these cells vulnerable to excitotoxic damage (Paquet et al., 1997; Mereu et al., 1991). In addition, mesencephalic dopamine neurons apparently have a constitutive deficiency of mitochondrial complex I (Marey-Semper et al., 1993); this deficiency has also been observed in post-mortem studies of Parkinson brains (Schapira et al., 1990; 1992).

Inhibition of the mitochondrial electron transport chain by several agents results in secondary excitotoxicity and free radical formation (Schulz et al, 1996). Such agents include MPP+, a complex I inhibitor (Storey, et al, 1978), 3-nitropropionic acid (3-NP) and malonate (Beal et al., 1993; Beal et al., 1993a); the latter two agents are irreversible and reversible inhibitors of succinate dehydrogenase (complex II).

A defect in energy metabolism such as on seen in Parkinson's disease brains may lead to a partial neuronal depolarization, removal of the Mg^{+2} blockade at the NMDA receptor and persistent receptor activation followed by free radical formation (Schulz et al., 1996). Several lines of evidence support this hypothesis. In fetal mesencephalic cultures, the synergistic effects of NMDA receptor stimulation and complex I inhibition by rotenone specifically target dopamine neurons (Marey-Semper et al., 1995). The glutamate release blockers, lamotrigine and BW1003C87, significantly attenuate lesions produced by intrastriatal injections of MPP+ (Schulz et al., 1996) and ARL-15896, a NMDA receptor antagonist, reduces lesions produced from intrastriatal malonate lesions (Greene et al., 1996). Coenzyme Q_{10} (ubiquinone) and nicotinamide, putative enhancers of mitochondrial function, block neurodegeneration resulting from striatal malonate lesions (Beal et al., 1994). In addition, Coenzyme Q_{10} and MK-801 have additive neuroprotective effects against intrastriatal malonate injections (Schulz, et al., 1996). Thus deficiencies in energy metabolism seem to result in impaired intracellular Ca^{+2} homeostasis. This hypothesis was confirmed by Sheehan et al., (1997) in cybrids made from platelet mitochondria of control and PD patients. Cybrids from PD patients had 26% less complex I activity, and showed a 53% slower recovery in carbachol-induced increases in intracellular Ca^{+2} (Sheehan et al., 1997).

It is interesting to note that prolonged elevated levels of intracellular Ca^{+2} arising from either primary or secondary excitotoxicity may further disrupt mitochondrial function by overloading the mitochondrial Ca^{+2} uptake uniporter. This would ultimately lead to opening of the permeability transition pore, followed by collapse of the proton gradient, generation of reactive oxygen species and ultimately, cell death by apoptosis (McCormack and Denton, 1993, Zamzami et al., 1995; Susin et al., 1996). Apoptotic cell death is also concomitant with reactive oxygen species production (Susin et al., 1996). Cell death following excitotoxicity can occur via necrosis or apoptosis, depending upon the severity of the insult (Portera-Caillau et al., 1995; Didier et al., 1996). An agent that interferes with apoptosis following excitotoxicity or oxidative stress may afford protection to degenerating dopaminergic neurons in Parkinson's disease. Examples include nerve growth factor, which protects against apoptosis and oxidative stress in PC-12 cells, (Tong and Perez-Polo, 1996; Kirschner et al, 1996) and brain derived neurotrophic factor (BDNF), neurotrophin 5 (NT-5) and bFGF (Hou et al., 1997), which protect dopaminergic neurons.

Bcl-2 is the best characterized anti-apoptotic protein to date, its apoptosis-inducing counterpart is Bax (Korsmeyer, et al., 1995). Bcl-2 interferes with the opening of the permeability transition pore in T cell hybridoma cells (Susin et al., 1996), and directs the protein kinase Raf-1 to mitochondria, blocking the apoptotic path (Wang et al., 1996).

Neuroprotective agents protect against cell death by interfering with various pathways that ultimately result in death. The neuroprotective actions of deprenyl against NMDA-receptor mediated cell death in mesencephalic cultures are independent of its MAO-B inhibitory properties and competition with the endogenous ligand for the NMDA receptor (Mytilineou et al., 1997). Previous studies have shown a reduction in Bax and an increase in Bcl-2 by deprenyl in differentiated PC12 cells (Tatton, et al., 1996). The glutamate analogue, kainic acid, induces apoptosis in mouse brain and is associated with an up-regulation of bax and a down-regulation of bcl-2 (Gillardon et al., 1995). Therefore, deprenyl may prevent excitotoxic cell death in dopaminergic neurons *in vitro* by an anti-apoptotic mechanism (Mytilineou et al., 1997). Interestingly, deprenyl also induces increased expression of ciliary neurotrophic factor (CNTF) and bFGF in reactive astrocytes (Biagini et al., 1994; Seniuk et al, 1994). Growth factors have also been shown to increase levels of anti-oxidant enzymes and to modulate intracellular Ca^{+2}, which may attenuate glutamate-induced neurotoxicity (Spina et al., 1992; Mattson et al., 1995). Brain derived neurotrophic factor (BDNF) requires bcl-2 in its anti-apoptotic mechanism in chick ciliary ganglia, while CNTF does not (Allsopp et al., 1995). This suggests that there are several independent survival mechanisms within cells to protect them against apoptosis.

Decreased levels of reduced glutathione potentiate the MPP+-induced neurotoxicity (Wullner et al,. 1996). Conversely, dopaminergic neuronal cell death resulting from oxidative stress can be blocked by increases in GSH. Hou et al (1997) observed that growth factors that increase glial cell proliferation also increase GSH levels and protect against 6-OHDA mediated toxicity *in vitro*. This protection was blocked when cultures were depleted of GSH (Hou et al., 1997). Park and Mytilineou (1992) observed increased dopaminergic cell survival following treatment with the growth factors bFGF and EGF. Both these agents had their protective actions blocked when glial cell proliferation was arrested in the cultures by FuDR treatment. These findings present evidence for the requirement of a functional neuronal-glial interaction in order for dopaminergic neurons to be protected against oxidative stress.

In summary, there are several processes that may underlie Parkinson's disease; these may act individually, or, more likely, as a concerted mechanism to produce dopaminergic cell death (see schematic). At the outset, impaired complex I activity renders cells more vulnerable to EAA input and consequent disruptions in intracellular Ca^{+2} homeostasis. The mass influx of Ca^{+2} ultimately disrupts the mitochondrial potential, leading to oxidative stress, and apoptotic cell death. This process is further exacerbated by decreased levels of GSH in Parkinson brains. Rescue of dopaminergic neurons can take place by increasing GSH levels via glial proliferation, apoptotic blockade by growth factors or a blockade of intracellular Ca^{+2} efflux. Agents in the rescue of dopaminergic neurons include, but are not limited to, deprenyl, growth factors and NMDA antagonists as well as glutamate release blockers. Cell culture models of dopamine neuronal degeneration can assist in the understanding of these processes and also in the development of neuroprotective agents.

REFERENCES

Allsopp, T.E., Kiselev, S., Wyatt, S. and Davies, A. (1995) Role of Bcl-2 in the brain derived neurotrophic factor survival response. Eur. J. Neurosci. 7:1266–1272.

Beal, M.F., Henshaw, D.R., Jenkins, B.G., Rosen, B.R. and Schulz, J.B. (1994) Coenzyme Q_{10} and nicotinamide block striatal lesions produced by mitochondrial toxin malonate. Ann. Neurol. 36:882–889.

Beal, M.F., Brouillet, E., Jenkins, B., Henshaw, R., Rosen, B. and Hyman, B.T. (1993) Age dependent striatal excitotoxic lesions produced by the endogenous mitochondrial inhibitor, malonate. J. Neurochem. 61, 1147–150.

Beal, M.F., Brouillet, E., Jenkins, B.G., Ferrante, R.J., Kowall, N.W., Miller, J.M., Storey, E., Srivasta, R., Rosen, B.R. and Hyman, B.T. (1993a) Neurochemical and histologic characterization of excitotoxic lesions produced by the mitochondrial toxin, 3-nitropropionic acid. J. Neurosci. 13:4181–491.

Beck, K.D., Knusel, B., Winslow, J.W., Rosenthal, A., Burton, L.E., Nikolics, K. and Heft, F. (1992) Pretreatment of dopaminergic neurons in culture with brain-derived neurotrophic factor attenuates toxicity of 1-methyl-4-phenylpyridinium. Neurodegeneration 1:27–36.

Behl, C., Hovey, L., Krajewski, S., Schubert, D., and Reed, J.C. (1993) Bcl-2 prevents killing of neuronal cells by glutamate but not by amyloid beta protein. Biochim. Biophys. Res. Commun. 97:949–956.

Biagini, G., Frasoldati, A., Fuxe, K. and Agnati, L.F. (1994) The concept of astrocyte-kinetic drug in the treatment of neurodegenerative disease: evidence for l-deprenyl-induced activation of reactive astrocytes. Neurochem. Intl. 25:17–22.

Biaglow, J.E., Varnes, M.E., Tuttle, S.W., Olenick, N.L. and Glazier, K. (1986) The effect of L-buthionine sulfoximine on the aerobic radiation response of A549 human lung carcinoma cells, Int. J. Radiat. Oncol. Biol. Phys. 12:1139–1142.

Birkmeyer, W., and Riederer, P. (1984) Deprenyl prolongs the therapeutic efficacy of combined L-DOPA in Parkinson's disease. Adv. Neurol. 40:475–481.

Bonfoco, E., Krainc, D., Ancarona, M., Nicotera, P. and Lipton, S.A. (1995) Apoptosis and necrosis: two distinct events induced, respectively, by mild and intense insults with N-methyl-d-aspartate or nitric oxide/superoxide in cortical cell cultures. Proc. Natl. Acad. Sci USA 92:762–7166.

Burns, R.S., Chiueh, C.C., Markey, S.P., Ebert, M.H., Jacobowitz, D.M. and Kopin, I.J. (1983) A primate model of parkinsonism: Selective destruction of dopaminergic neurons in the pars compacta of the substantia nigra by N-methyl-4-phenyl-1,2,3,6-tetrahydropyridine. Proc. Natl. Acad. Sci USA 80:4546–4550.

Carillo, M.-C., Kitani, K., Kanai, S., Sato, Y., Miyasaka, K. and Ivy, G.O. (1994) L-deprenyl increases activities of superoxide dismutase (SOD) and catalase in certain brain regions of old male mice. Life Sci. 54:975–981.

Carillo, M.-C., Kanai, S., Sato, Y., Ivy, G.O., and Kitani, K. (1992) Sequential changes in activities of superoxide dismutase and catalase in brain regions and liver during deprenyl infusion in male rats. Biochem. Pharmacol. 44:2185–2189.

Casper, D., and Blum, M. (1995) Epidermal growth factor and basic fibroblast growth factor protect dopaminergic neurons from glutamate toxicity in culture. J. Neurochem. 65:016–1026.

Casper, D. Mytilineou, C. and Blum, M. (1991) EGF enhances the survival of dopamine neurons in rat embryonic mesencephalon primary culture. J. Neurosci. Res. 30:372–381.

Castagnoli, N., Chiba, K. and Trevor, A.J. (1985) Potential bioactivation pathways for the neurotoxin 1-methyl-4-phenyl-1,2,3,6-tetrahydropyridine (MPTP) Life Sci 36:25–230.

Chiba, K., Trevor, A. and Castagnoli, N. Jr. (1984) Metabolism of the neurotoxic tertiary amine, MPTP, by brain monoamine oxidase. Biochem.Biophys.Res.Commun. 120:574–578.

Chiba, K., Trevor, A.J. and Castagnoli, N. (1985) Active uptake of MPP+, a metabolite of MPTP, by brain synaptosomes. Biochem. Biophys. Res. Commun. 128:1228–1232.

Chiueh, C.C., Huang, S.J. and Murphy, D.L. (1994) Suppression of hydroxyl radical formation by MAO inhibitors: a novel possible neuroprotective mechanism in dopaminergic neurotoxicity. J. Neural Transm. Suppl. 41:189–196.

Choi, D.W., Maulucci-Gedde, M. and Kriegstein, A.R. (1987) Glutamate neurotoxicity in cortical cell cultures. J. Neurosci. 7:357–368.

Choi, D.W. (1990) Glutamate neurotoxicity: a three stage process. In *Neurotoxicity of Excitatory Amino Acids* (A. Guidotti, ed) pp. Raven Press, New York

Cohen, G. (1985) Oxidative stress in the nervous system, in *Oxidative Stress* (Sies, H., ed.) pp. 383–401. Academic Press, London.

Corboy, D.L., Wagner, M.L. and Sage, J.I. (1995). Apomorphine for motor fluctuations and freezing in Parkinson's disease. Ann. Pharmacother. 29(3), 282–288

Coyle, J.T. and Putfarcken, P. (1993) Oxidative stress, glutamate and neurodegenerative disorders. Science 262:689–694.

Danias, P., Nicklas, W.J., Ofori, S., Shen, J. and Mytilineou, C. (1989) Mesencephalic dopamine neurons become less sensitive to 1-methyl-4-phenyl-1,2,3,6-tetrahydropyridine toxicity during development in vitro. J. Neurochem. 53:1149–1155.

Dawson, T.M., Dawson, V.L. and Snyder, S.H. (1992) A novel neuronal messenger molecule in brain: the free radical, nitric oxide. Ann. Neurol. 32:297–311.

Dexter, D.T., Wells, F.R., Lees, A.J., Agid, F., Agid, Y., Jenner, P. and Marsden, C.D. (1989) Increased nigral iron content and alterations in other metal ions occurring in brain in Parkinson's disease. J.Neurochem. 52:1830–1836.

Dexter, D.T., Holley, A.E., Flitter, W.D., Slater, T.F., Wells, F.R., Daniel, S.E. and Lees, A.J. (1994) Increased levels of lipid hydroperoxides in the parkinsonian substantia nigra: an HPLC and ESR study. Mov. Disord. 9: 92–97.

Didier, M., Bursztajn, S., Adamec, E., Passani, L., Nixon, R.A., Coyle, J.T., Wei, J.Y., and Berman, S.A. (1996) DNA strand breaks induced by sustained glutamate excitotoxicity in primary neuronal cultures. J. Neurosci. 16:2238–2250.

Di Monte, D.A., Chan, P. and Sandy, M.S. (1992) Glutathione in Parkinson's disease: a link between oxidative stress and mitochondrial damage? Ann. Neurol. 32 (Suppl) S111-S115.

Driscoll, B.F., Deibler, G.E., Law, M.J. and Crane, A.M. (1993) Damage to neurons in culture following medium change: a role of glutamine and extracellular generation of glutamate. J. Neurochem. 61: 1795–1800.

Driscoll, B.F., Law, M.J. and Crane, A.M. (1991) Cell damage associated with changing the medium of mesencephalic cultures in serum-free medium is mediated by N-methyl-d-aspartate receptors. J. Neurochem. 56:1201–1206.

Dumuis, A., Sebben., M., Haynes, L., Pin, J.-P., and Bockaert, J. (1988) NMDA receptors activate the arachidonic acid cascade in striatal neurons. Nature 336:68–70.

Engele, J. and Bohn, M.C. (1991) The neurotrophic effects of fibroblast growth factors on dopaminergic neurons in vitro are mediated by mesencephalic glia J. Neurosci 11:3070- 3078. Published erratum appears in J. Neurosci 1992 Mar; 12:685.

Ferrari, G., Minozzi, M.-C., Toffano, G., Leon, A. and Skaper, S.D. (1989) Basic fibroblast growth factor promotes the survival and development of mesencephalic neurons in culture. Dev. Biol. 133:140–147.

Fransden, A., and Schousboe,A. (1993) Excitatory amino acid-mediated cytotoxicity and calcium homeostasis in cultured neurons. J. Neurochem. 60:1201–1211.

Friedman, L. and Mytilineou, C. (1987) The toxicity of MPTP to dopamine neurons in culture is reduced at high concentrations. Neurosci. Lett. 79, 65–72.

Gillardon, F., Wickert, H. and Zimmerman, M. (1995) Up-regulation of bax and down-regulation of bcl-2 is associated with kainate-induced apoptosis in mouse brain. Neurosci. Lett. 192:85–88.

Glinka, Y., Tipton, K.F. and Youdim M.B. (1996) Nature of inhibition of mitochondrial respiratory complex I by 6-hydroxydopamine. J. Neurochem. 66, 2004–2010.

Golbe, L.I., Lieberman, A.N., Muentner, M.D., Ahlskog, J.E., Gopinathan, G., Neophytides, A. and Duvoisin, R.C. (1988) Deprenyl in the treatment of symptom fluctuations in advanced Parkinson's disease. Clin. Neuropharmacol. 11:45–55.

Good, P.F., Olanow, C.W., and Perl, P.D. (1992) Neuromelanin-containing neurons of the substantia nigra accumulate iron and aluminum in Parkinson's disease: a LAMMA study. Brain Res. 593: 343–346.

Graham, D.G. (1978) Oxidative pathways for catecholamines in the genesis of neuromelanin and cytotoxic quinones. Mol. Pharmacol 14:633–643.

Greene, J.G., Porter, R.H., and Greenmayer, J.T. (1996) ARL-15896, a novel N-methyl-d-aspartate receptor ion channel antagonist: neuroprotection against mitochondrial metabolic toxicity and regional pharmacology. Exp. Neurol. 137, 66–72.

Heikkila, R.H. and Cohen, G. (1972) Further studies on the generation of hydrogen peroxide by 6-hydroxydopamine: Potentiation by ascorbic acid. Mol. Pharmacol. 8:241–248.

Heikkila, R.E., Hess, A. and Duvoisin, R.C. (1984) Dopaminergic neurotoxicity of 1-methyl-4-phenyl-1,2,5,6-tetrahydropyridine in mice. Science 224:1451–1453.

Heinonen, E.H., Markku, A.I. and Lammintausta, R.A.S., (1994) Pharmacokinetic aspects of l-deprenyl (selegiline) and its metabolites. Clin. Pharmacol. Ther. 56:742–749.

Hou, J.-G. G. , Cohen, G. and Mytlineou, C. (1997) Basic fibroblast growth factor stimulation of glial cells protects dopamine neurons from 6-hydroxydopamine toxicity: involvement of the glutathione system. Brain Res. J. Neurochem. 69: 76-83.

Hou, J.-G. G., Lin, L.-F. H., and Mytilineou, C. (1996) Glial cell line-derived neurotrophic factor exerts neurotrophic effects on dopaminergic neurons in vitro and promotes their survival and regrowth after damage by 1-methyl-4-phenylpyridium. J. Neurochem. 66:74–82.

Hyman, C., Juhasz, M., Jackson, C., Radziejewski, C. and Lindsay, R.M. (1991) BDNF is a neurotrophic factor for dopaminergic neurons of the substantia nigra. Nature 350:230- 232.

Hynes, M.A., Poulsen, K., Armanini, M., Berkmeier, L., Phillips, H., and Rosenthal., A. (1994) Neurotrophin 4/5 is a survival factor for embryonic midbrain dopaminergic neurons in enriched cultures. J. Neurosci. Res. 37:144–154.

Javitch, J. A. and Snyder, S.H. (1985) Uptake of MPP(+) by dopamine neurons explains selectivity of parkinsonism-inducing neurotoxin, MPTP. Eur. J. Pharmacol. 106:445- 456.

Javitch, J.A., D'Amato, R.J., Sttrittmater, S.M. and Snyder, S.H. (1985) Parkinsonism-inducing neurotoxin, N-methyl-4-phenyl-1,2,3,6-tetrahydripyridine: Uptake of the metabolite N-methyl-4-phenylpyridine by dopamine neurons explains selective toxicity. Proc. Natl. Acad. Sci. U.S.A., 82:2173–2177.

Jenner, P. (1993) Presymptomatic detection of Parkinson's disease. (1993) J. Neural Trans. Suppl. 40, 23–36.

Kato, P., Puttfarcken, P.S., Lyons, W.E., and Coyle, J.T. (1991) Developmental time course and ionic dependence of kainate-mediated toxicity in rat cerebella granule cell cultures. J. Pharmacol. Exp. Ther. 256:402–411.

Kirschner, P.B., Jenkins, B.G., Schulz, J.B., Finkelstein, S.P., Matthews, R.T., Rosen, B.R. and Beal, M.F. (1996) NGF, BDNF and NT-5, but not NT-3 protect against MPP+ toxicity and oxidative stress in neonatal animals. Brain Res. 713:178–185.

Knoll, J. (1978) The possible mechanisms of action of (-) deprenyl in Parkinson's disease. J. Neural Transm. 43:177–198.

Knusel, B., Michel, P.P., Schwaber, J.S. and Hefti, F. (1990) Selective and non-selective stimulation of central cholinergic and dopaminergic development in vitro by nerve growth factor, basic fibroblast growth factor, epidermal growth factor, insulin and the insulin-like growth factors I and II. J. Neurosci 10:558–570.

Kokotos Leonardi , E.T. , Cheng, B., Radcliffe, P.M. Olanow, C.W. Cohen G. and Mytilineou, C. (1996) L-Dopa and L-deprenyl protect against L-BSO induced cell death in cultures mesencephalic neurons. Soc. Neurosci. Abst. 22:1718.

Korsmeyer, S.J., Yun, S.M., Oltvai, A.N., Veis Novack, D.J. and Linette, G.P. (1995) Reactive oxygen species on the regulation of cell death by the bcl-2 gene family. Biochim. Biophys. Acta 1271:63–66.

Krebs, M.-O., Trovero, F., Desban, M., Gauchy, C., Glowinski, J. and Kemel, M.-L. (1991) Distinct presynaptic regulation of dopamine release through NMDA receptors in striosome- and matrix-enriched areas of the rat striatum. J. Neurosci. 11:1256–1262.

Kumar, R., Agarwal, A.K., and Seth, P.K. (1995) Free-radical generated neurotoxicity of 6-hydroxydopamine. J. Neurochem. 64:1703–1707.

Lafon-Cazal, M., Pietri, S., Culcasi, M. and Bockaert, J. (1993) NMDA-dependent superoxide protection and neurotoxicity. Nature 364:535–537.

Langston, J.W., Ballard, P., Tetrud, J.W., and Irwin, I. (1983) Chronic parkinsonism in humans due to a product of meperidine-analog synthesis. Science 219:979–980.

Langston, J.W., Forno, L.S., Rebert, C.S., and Irwin, I. (1984) Selective nigral toxicity after systemic administration of 1-methyl-4-phenyl-1,2,3,6-tetrahydropyridine (MPTP) in the squirrel monkey. Brain Res. 292: 390–394.

Levitt, P., Pintar, J.E. and Breakefield, X.O. (1982) Immunocytochemical demonstration of monoamine oxidase B in brain astrocytes and serotonergic neurons. Proc. Natl. Acad. Sci, USA 115:1617–1624.

Lin, L.-F.H., Zhang, T.J., Collins, F., and Armes, L.G. (1994) Purification and initial characterization of B49 glial cell-line derived neurotrophic factor. J. Neurochem. 63:758–768.

Lin, L.-F. H., Doherty, D.H., Lile, J.D., Bektesh, S. and Collins, F. (1993) GDNF: A glial cell line derived neurotrophic factor for midbrain dopaminergic neurons. Science 260:1130–1132.

Ling, C.C. Wong, R.S.L., and Basas, R.D. (1990) Glutathione depletion and cytotoxicity of buthionine sulphoximine and SR2508 in rodent and human cells. Int. J. Radiat. Oncol. Biol. Phys. 18:325–330.

Marey-Semper, I., Gelman, M. and Levi-Strauss, M. (1995) A selective toxicity toward cultured mesencephalic dopaminergic neurons is induced by the synergistic effects of energetic metabolism impairment and NMDA receptor activation. J. Neurosci. 15:5912–5918.

Marey-Semper, I., Gelman, M. and Levi-Strauss, M. (1993) The high sensitivity to rotenone of striatal dopamine uptake suggests the existence of a constitutive metabolic deficiency in dopaminergic neurons from the substantia nigra. Eur. J. Neurosci. 5:1029–1034.

Mattson, M.P., Lovell,, M.A., Furukawa, K. and Markesbery, W.R. (1995) Neurotrophic factors attenuate glutamate-induced accumulation of peroxides, elevation of intracellular Ca^{+2} concentration , and neurotoxicity and increase antioxidant enzyme activities in hippocampal neurons. J. Neurochem. 65:1740–1751.

Mattson, M.P., Cheng, B. and Smith-Swintosky, V.L. (1993) Growth factor-mediated protection from excitotoxicity and disturbances in calcium and free radical metabolism. Sem. Neurosci. 5:295–307.
McCord, J. (1985) Oxygen-derived free radicals in post-ischemic tissue injury. N. Engl. J. Med. 312: 159–163.
McCormack, J.G. and Denton, R.M. (1993) Mitochondrial Ca^{+2} transport and the role of intramitochondrial Ca^{+2} in the regulation of energy metabolism. Dev. Neurosci. 15:165- 173.
McGeer, P.L., Itagaki, S., Akiyama, H. and McGeer, E.H. (1988) Rate of cell death in parkinsonism indicates active neuropathological processes. Ann. Neurol. 24, 574–576.
Meister, A. (1991) Glutathione deficiency produced by inhibition of its synthesis, and its reversal; applications in research and therapy. Pharmacol. Ther. 51(2):155–194.
Meister, A. and Anderson, M.E. (1983) Glutathione. Ann. Rev. Biochem. 52:711–760.
Mereu, G., Costa, E., Armstrong, D.M. and Vicini, S. (1991) Glutamate receptor subtypes mediate excitatory synaptic currents of dopamine neurons in midbrain slices. J. Neurosci. 11:1359–1366.
Michel, P.P., Dandapani, B.K., Knusel, B., Sanchez-Ramos, J. and Hefti, F. (1990) Toxicity of -methyl-4-phenylpyridinium for rat dopaminergic neurons in culture: selectivity and irreversibility. J. Neurochem. 54:1102–1109.
Mizuno, Y., Ikebe, S., Hattori, N., Nakagawa-Hattori, Y., Mochizuki, H., Tanaka, M. and Ozawa, T. (1995) Role of mitochondria in the etiology and pathogenesis of Parkinson's disease. Biochem. Biophys. Acta 1271:265–274.
Mytilineou, C., Kokotos Leonardi, E. Radcliffe, P.. Heinonen, E.H., Han, S.-K. Cohen, G., and Olanow, C.W. (1998) Deprenyl and desmethylselegeline protect mesenphalic neurons from toxicity induced by glutathione depletion. J. Pharm. Exp. Ther. 284: 700-706.
Mytilineou, C., Radcliffe, P., Kokotos Leonardi, E., Werner, P. and Olanow, C.W. (1997) L- deprenyl protects mesencephalic dopamine neurons from glutamate receptor-mediated toxicity in vitro. J. Neurochem. 68:33–39.
Mytilineou, C., Radcliffe, P.M. and Olanow, C.W. (1997a) L-(-)-desmethylselegeline, a metabolite of selegiline [L-(-)-deprenyl], protects mesencephalic dopamine neurons from excitotoxicity in vitro. J. Neurochem. 68:434–436.
Mytilineou, C., Werner, P., Molinari, S., DiRocco, A., Cohen, G. and Yahr, M.D. (1994) Impaired oxidative decarboxylation of pyruvate in fibroblasts from patients with Parkinson's disease. J. Neural Transm. [P-D Sect] 8:223–228.
Mytilineou, C., Han, S.-K. and Cohen, G. Toxic and protective effects of L-DOPA on mesencephalic cell cultures. J. Neurochem. 61:1470–1478.
Mytilineou, C. and Friedman, L.K. (1988) Studies in the metabolism of 1-methyl-4-phenyl- 1,2,3,6-tetrahydropyridine in cultures of embryonic rat mesencephalon. J. Neurochem. 51:750–755.
Mytilineou, C., Friedman, L.K. and Danias, P. (1988) Studies on the toxicity of MPTP to dopamine neurons in tissue and cell cultures. *Progress in Parkinson Research* F. Hefti and W.J. Weiner, eds. (Plenum, New York). pp. 127–136.
Mytilineou, C., Cohen, G. and Heikkila, R.E. (1985) 1-methyl-4-phenylpyrindine (MPP+) is toxic to mesencephalic dopamine neurons in culture. Neurosci Lett 57:19–24.
Mytilineou, C. and Cohen, G. (1984) 1-Methyl-1,2,3,6-tetrahydopyridine destroys dopamine neurons in explants of rat embryo mesencephalon. Science 225:529–531
Nicklas, W.J., Vyas, I. and Heikkila, R.E. (1985) Inhibition of NADH-linked oxidation in brain mitochondria by 1-methyl-4-phenylpyrindine, a metabolite of the neurotoxin, 1-methyl-4-phenyl-1,2,3,6-tetrahydropyridine. Life Sci 36:2503–2508.
Nicklas, W.J., Youngster, S.K., Kindt, M.V. and Heikkila, R.E. (1987) MPTP, MPP+ and mitochondrial function. Life Sci 40:721–729.
Olanow, C.W., Hauser, R.A., Gauger, L., Mapapira, T., Koller, W., and Hubble, J. (1995) The effect of deprenyl and levodopa on the progression of Parkinson's disease. Ann. Neurol. 38:771–777.
Olanow, C.W. (1993) A radical hypothesis for neurodegeneration, TINS 16:439–444.
Otto, D. and Unsicker, K. (1990) basic FGF reverses chemical and morphological deficits in the nigrostriatal system of MPTP-treated mice. J. Neurosci. 10:1912–1921.
Otto, D. and Unsicker, K. (1993) FGF-2 mediated protection of cultures mesencephalic dopaminergic neurons against MPTP and MPP+: specificity and impact of culture conditions, non-dopaminergic neurons and astroglial cells. J. Neurosci. Res. 34: 382–393.
Owen, A.D., Schapira, A.H., Jenner, P. and Marsden, C.D. (1996) Oxidative stress and Parkinson's disease. Ann. N.Y. Acad. Sci. 786:217–223.
Paquet, M., Tremblay, M., Soghomonian, J.-J., and Smith, Y. (1997) AMPA and NMDA glutamate receptor subunits in midbrain dopaminergic neurons in the squirrel monkey: an immunohistochemical and in situ hybridization study. J.Neurosci 17(4):1377–1396.

Park, T.H. and Mytilineou, C. (1992) Protection from 1-methyl-4-phenylpyridinium (MPP+) toxicity and stimulation of regrowth of MPP+-damages dopaminergic fibers by treatment of mesencephalic neurons with EGF and basic FGF. Brain Res. 599:83–97.

Parker, W.D., Boyson, S.J. and Parks, J.K. (1989) Abnormalities of the electron transport chain in idiopathic Parkinson's disease. Ann. Neurol. 26, 719–723.

Polymeropoulos, M.H., Higgins, J.J., Golbe, L.I., Johnson, W.G., Id, S.E., Di-Iorio, G., Sanges, G., Stenroos, W.S., Pho, L.T., Schaffer, A.A., Lazzarini, A.M., Nussbaum, R.L., and Duvoisin, R.C. (1996) Mapping of a gene for Parkinson's disease to chromosome 4q1-q23. Science 274:1197–1199.

Portera-Cailliau, C., Hedreen. J.C., Price, D.L., and Koliatsos, V.E. (1995) evidence for apoptotic cell death in Huntington disease and excitotoxic animal models. J. Neurosci. 15:3775–3787.

Rabey, J.M. (1995) Second generation of dopamine agonists: pros and cons. J. Neural Trans. Suppl. 45, 213–24.

Raps, S.P., Lai, J.C., Hertz, L. and Cooper, A.J. (1989) Glutathione is present in high concentrations in cultures astrocytes but not in cultured neurons. Brain Res. 493:398–401.

Reed, D.J. (1990) Glutathione: Toxicological implications. Ann. Rev. Pharmacol. Toxicol. 160:55–62.

Riederer, P., Sofic, E., Rausch, W.D., Schmidt, B., Reynolds, G.P., Jellinger, K. and Youdim, M.B. (1989) Transition metals, ferritin, glutathione and ascorbic acid in Parkinsonian brains. J. Neurochem. 52:515–520.

Roy, E. and Bedard, P.J. (1993) L-deprenyl increases survival of rat foetal nigral neurones in culture. Neuroreport 4:1183–1186.

Sachs, C. and Jonsson, G. (1975) Mechanisms of action of 6-hydroxydopamine. Biochem. Pharmacol. 24:1–8.

Saggu, H., Cooksey, J., Dexter, D., Wells, F.R., Lees, A., Jenner, P., and Marsden, C.D. (1989) A selective increase in particulate superoxide dismutase activity in parkinsonian substantia nigra. J. Neurochem. 53: 692–697.

Salo, P.T. and Tatton, W.G. (1992) L-deprenyl reduced the death of motoneurons caused by axotomy. J. Neurosci. Res. 31:394–400.

Sanchez-Ramos, J., Barrett, J.N., Goldstein, M., Weiner, W.J., and Hefti, F. (1986) 1-methyl-4-phenylpyridinium but not 1-methyl-4-phenyl-1,2,3,6-tetrahydropyridine (MPTP) selectively destroys dopaminergic neurons in cultures of dissociated rat mesencephalic neurons. Neurosci. Lett., 72:215–220.

Schapira, A.H., Cooper, J.M., Dexter, D., Clark, J.B., and Marsden, C.D. (1990) Mitochondrial complex I deficiency in Parkinson' disease. J. Neurochem. 54:823–827.

Schapira, A.H., Mann, C.V.M., Cooper, J.M., Krige, D., Jenner, P.J. and Marsden, P.J. (1992) Mitochondrial function in Parkinson's disease. Ann. Neurol. 32, S116-S124.

Schulz, J.B., Matthews, R.T., Henshaw, D.R., and Beal, M.F. (1996) Neuroprotective strategies for the treatment of lesions produced by mitochondrial toxins: implications for neurodegenerative diseases. Neuroscience 71:1043–1048.

Seniuk, N.A., Henderson, J.T., Tatton, W.G. and Roder, J.C. (1994) Increased CNTF gene expression in process-bearing astrocytes following injury is augmented by R(-)-deprenyl. J. Neurosci. Res. 37:278–286.

Sheehan, J.P., Swerdlow, R.H., Parker, W.D., Miller,S.W., Davis, R.E. and Tuttle, J.B. (1997) Altered calcium homeostasis in cells transformed by mitochondria from individuals with Parkinson's disease. J. Neurochem. 68:1221–1223.

Shoffner, J.M., Watts, R.L., Juncos, J.L., Torroni, A. and Wallace, D.C. (1991) Mitochondrial oxidative phosphorylation defects in Parkinson's disease. Ann. Neurol. 30:332–339.

Sian, J., Dexter, D.T., Lees, A.J., Daniel, S., Agid, Y., Javoy-Agid, F., Jenner, P. and Marsden, C.D. (1994) Alterations in glutathione levels in Parkinson's disease and other neurodegenerative disorders affecting basal ganglia. Ann. Neurol. 36:348–355.

Sills, M.A., Fagg, G., Pozza,M., Angst, C., Brundish, D.E., Hurt, S.D., Wilusz, E.J. and Williams, M. (1991) [^{3}H]-CGP 39653: a new N-methyl-d-aspartate antagonist radioligand with low nanomolar affinity in rat brain. Eur.J. Pharmacol. 192:19–24.

Slivka, A., Spina, M.B., Calvin, H.I. and Cohen, G. (1988) Depletion of glutathione in preweanling mice by L-buthionine sulfoximine. J. Neurochem. 50:1391–1393.

Slivka, A., Mytilineou, C. and Cohen, G. (1987) Histochemical evaluation of glutathione in brain. Brain Res. 409:275–284.

Sofic, E., Paulus, W., Jellinger, K., Riederer, P. and Youdim, M.B.H. (1991) Selective increase of iron in substantia nigra zona compacta of Parkinson's brains. J. Neurochem. 56:978–982.

Spina, M.B., Squinto, S.P., Miller, J., Lindsay, R.M. and Hyman, C. (1992) Brain-derived neurotrophic factor protects dopamine neurons against 6-hydroxydopamine and N-methyl-4-pyridinium ion toxicity: involvement of the glutathione system. J. Neurochem. 59:99–106.

Storey, E., Hyman, B.T., Jenkins, B.T., Brouillet, E., Miller, J.M., Rosen, B.R. and Beal, M.F. (1992) MPP produces excitotoxic lesions in rat striatum due to impairment of oxidative metabolism. J. Neurochem. 58:1975–1978.

Susin, S.A., Zamzami, N., and Kroemer, G. (1996) The cell biology of apoptosis: evidence for the implication of mitochondria. Apoptosis 1:231–242.

Tatton, W.G., Ju, W.Y.H. Holland, D.P., Tai, C., and Kwan, M. (1994) L-deprenyl reduces PC12 cell apoptosis by inducing new protein synthesis. J. Neurochem. 63:1572–575.

Tatton, W.G. and Greenwood, C.E. (1991) Rescue of dying neurons: a new action for L-deprenyl in MPTP parkinsonism. J. Neurosci. Res. 30:666–672.

Tatton, W.G., Ju, W.Y.H., Wadia, J. and Tatton, N.A. (1996) Reduction of neuronal apoptosis by small molecules: promise for new approaches to neurological injury, in *Neurodegeneration and Neuroprotection in Parkinson's Disease* (Olanow, C.W., Jenner, P., and Youdim, M., eds) pp. 209–220. Academic Press, London.

Thiffault, C., Aumont, N., Quirion, R. and Poirier, J. (1995) Effect of PTP and l-deprenyl on anti-oxidant enzymes and lipid peroxidation levels in mouse brain. J. Neurochem. 65:2725–2733.

Tomac, A., Lindqvist, E., Lin, L.-F.H., Ogren, S.O., Young, D., Hoffer, B.J. and Olson, L. (1995) Protection and repair of the nigrostriatal dopaminergic system by GDNF in vivo. Nature 373: 335–339.

Tong, L. and Perez-Polo, J.R. (1996) Effect of nerve growth factor on AP-1, NF-kappa B and Oct DNA binding activity in apoptotic PC12 cells: extrinsic and intrinsic elements. J. Neurosci. Res. 45:1–12.

van Lookeren Campagne, M., Lucassen, P.J., Vermeulen, J.P. and Balazs, R. (1995) NMDA and kainate induce internucleosomal cleavage associated with both apoptotic and necrotic cell death in the neonatal rat brain. Eur. J. Neurosci. 7:1627–1640.

Wang, H.-G., Rapp, U.R., Reed, J.C. (1996) Bcl-2 targets the protein kinase Raf-1 to mitochondria. Cell 87:629–638.

Wu, R.M., Chiueh, C.C., Pert, A. and Murphy, D.L. (1993) Apparent antioxidant effect of L-deprenyl on hydroxyl radical formation and nigral injury elicited by MPP+ in vivo. Eur. J. Pharmacol. 243:241–247.

Wullner, U., Lochsmann, P.-A., Schulz, J.B., Schmid, A., Dringen, R., Eblen, F., Turski, L., and Klockgether, T. (1996) Glutathione depletion potentiates MPTP and MPP+ toxicity in nigral dopaminergic neurones. NeuroReport 7:921–927

Zamzami, N., Marchetti, P., Castedo, M., et al. (1995) Sequential reduction of mitochondrial transmembrane potential and generation of reactive oxygen species in programmed cell death. J. Exp. Med. 182:367–377.

Zang, L.Y. and Misra, H.P. (1993) Generation of reactive oxygen species during the monoamine oxidase-catalyzed oxidation of the neurotoxicant, 1-methyl-4-phenyl-1,2,3,6-tatrahydropyridine. J. Biol. Chem. 268:16504–16512.

13

AGING AND DEMENTIA OF THE ALZHEIMER TYPE IN PERSONS WITH MENTAL RETARDATION

Henryk M. Wisniewski and Wayne Silverman

Departments of Pathological Neurobiology and Psychology
New York State Institute for Basic Research in Developmental Disabilities
Staten Island, New York 10314

Aging during the senior years is part of the normal process of life span development. We are born, mature, and if we are fortunate, we grow old before we die. Unfortunately, the process of growing old is associated with declining status in many respects, physical and cognitive. These declines can be related to the concept of reserve: the greater the reserve the longer the organism will be able to avoid any functional deficit. Disease processes, by and large, appear to diminish our reserves. Therefore, in persons suffering from various diseases, age-associated functional impairments are observed earlier and the life span may be shortened.

As we all know, many aspects of development throughout our life span are influenced by genes. There are animal species with very short and very long life spans. There appear to be multiple genes that control our aging and life span, and already "aging genes" which appear to have a great effect on the occurrence of old-age-associated diseases have been identified that affect target organs in such a way that function declines, leading to diseases and death. The best examples of genes controlling the aging processes are associated with progeric syndromes. In general, these syndromes (Progeria, Cockayne syndrome, Down syndrome (DS)) are characterized by phenotypic expression of premature aging (Brown and Wisniewski, 1983). Some of the syndromes are associated with mental retardation such as Cockayne and Down syndromes.

Of the syndromes associated with genetically controlled premature aging, the most studied and best characterized is DS. Actually, DS is not related to a specific gene-defect, but it is rather a chromosomal abnormality that may produce gene dose effects on the developing organism. In fact, DS illustrates that not only a defect in the gene structure, but also an excess of a gene product or multiple gene products, such as is produced in trisomy 21, can lead to pathological development.

The great interest in studying DS resulted from the fact that it produces a very distinctive phenotype, that it is the most common syndrome associated with mental retarda-

Molecular and Cellular Mechanisms of Neuronal Plasticity, edited by
Ehrlich. Plenum Press, New York, 1998.

tion, and that it has been recognized as a chromosomal pathology for many years. Great interest in studying DS also has been stimulated by the observation that DS is characterized by the early development of Alzheimer disease (AD) type neuropathology. One of the neuropathological hallmarks of AD, the accumulation of amyloid-ß plaques, has been shown to be present in individuals with DS as young as 14 years-of- age. By the time they are thirty years old, virtually all individuals with DS have amyloid-ß plaques formed within their brains (Wisniewski et al., 1994). Because these lesions are characteristic of AD, it was assumed that any cognitive or behavioral abnormalities that develop in adults with DS after they reach thirty years-of-age must be due to the development of AD. However, it has also been known that in spite of the presence of AD pathology in these persons only a small number of individuals with DS before age 50 develop signs and symptoms of AD that have been clinically recognized (Zigman et al., 1996).

Numerous hypotheses have been forwarded to explain this discrepancy, including the difficulty in measuring negative change in people with mental retardation (Ellis et al., 1974); the possibility that there may be a higher threshold, as measured by neuropathological markers, that must be exceeded before Alzheimer's dementia is expressed in adults with DS (Wisniewski and Rabe, 1986), and the possibility that the specific DS gene genotype (i.e., free trisomy 21, mosaic, translocation) may be related to individual differences in the age of onset of dementia (Wisniewski et al., 1987).

Two more recent hypotheses have been proposed to explain the complex relationship between Alzheimer's-type pathology and dementia in adults with DS. The first hypothesis has been explored in a number of studies by Wisniewski and his colleagues (Wisniewski et al., 1994a,b, 1996). They have investigated the process of plaque development in the brain, identifying two distinct subtypes of plaques. These are plaques associated with fibrils (fibrillized, neuritic, or malignant plaques) and plaques that do not contain fibrils (nonfibrillized, benign, or diffuse). The neuritic plaques are associated with disruption of neural processes, probably resulting in functional impairment, whereas diffuse plaques appear to be benign age-associated changes having little or no functional consequences. The stains that have been traditionally used at autopsy to identify plaques in the brains of adults with Down syndrome, which predominantly rely on silver impregnation, do not clearly differentiate these two types of plaques (Wisniewski et al., 1989). In contrast, fluorescent staining with thoflavine-S is selectively sensitive to neuritic plaques. Using this stain in combination with others, Wisniewski et al. (1994b) have demonstrated that adults with DS younger than 50 years-of-age have numerous diffuse (i.e., benign) plaques, but few neuritic (i.e., malignant) plaques. Alternatively, when adults with DS are older than 50 years, neuritic plaques become more predominant. The observation that neuritic plaques are not usually found in large numbers until adults with DS are over age 50 may explain why individuals with DS who are between 30 and 50 years-of-age do not exhibit a high prevalence of Alzheimer's dementia. This may explain the 10- to 20-year lag between the onset of neuropathology and the onset of Alzheimer's-type dementia.

A second possible reason for the discrepancy between the neuropathology and dementia of the Alzheimer's type may be related to differential patterns in the development of neurofibrillary tangles (NFT). Wegiel et al. (1996) have shown that the development of a sufficient number of neurofibrillary tangles to affect clinical function in adults with DS does not appear to occur until these people reach approximately 60 years-of-age (i.e., 10 to 20 years after the development of plaque pathology).

From various studies it is known that the quantity of NFT correlates more strongly with dementia than the quantity of amyloid beta (Aß) plaques (Barcikowska et al., 1989). Based on these new neuropathological studies showing that the brain damaging lesions of

AD in people with DS develop after 50 years-of-age and that there are individual differences in the severity of the neuropathology, it becomes clear why few people with DS younger than 50 years-of-age develop the signs and symptoms of dementia. Their brain processes have not in fact been disrupted.

It is also well established that with the exception of a few cases with point mutation in APP, AD is an old-age-associated disease. The older we get the higher the risk of developing AD. As indicated above, DS is characterized by premature aging. It is possible, therefore, that other as yet unspecified aging genes that may be on chromosome 21 could be as important in making adults with DS vulnerable to many age-associated changes, including AD, some 20 to 30 years prematurely. This reasoning leads us to predict that the age-specific risk for AD among adults with DS should look remarkably similar to that of the general population, but with an earlier age of onset. Studies of individuals without mental retardation with DS (Popovitch et al., 1990) also have shown that these individuals may be at increased age-specific risk for AD-type neuropathology than are adults without mental retardation. If further research supports this finding, the group of adults with mental retardation at risk for this dementia may be larger than was previously expected. In addition, as a function of their pre-existing mental retardation, the costs of diagnosis for this group may be substantially higher because of their atypical presentation of symptoms. It will be important to prepare the necessary resources and supports for the group, which may add substantially to total health care costs.

REFERENCES

Barcikowska, M., Wisniewski, H.M., Bancher, C., and Grundke-Iqbal, I. (1989) About the presence of paired helical filaments in dystrophic neurites participating in the plaque formation. *Acta Neuropathol.* **78**, 225–231.

Brown, W.T., and Wisniewski, H.M. (1983) Genetics of human aging. Review of biological research. *Aging* **1**, 81–99.

Ellis, W.G., McCulloch, J.R., and Corley, C.L. (1974) Presenile dementia in Down's syndrome. *Neurology* **24**, 101–106.

Popovitch, E.R., Wisniewski, H.M., Barcikowska, M., Silverman, W., Bancher, C., Sersen, E., and Wen, G.Y. (1990) Alzheimer neuropathology in non-Down's syndrome mentally retarded adults. *Acta Neuropathol.* **80**, 362–367.

Wegiel, J., Wisniewski, H.M., Dziewiatkowski, J., Popovitch, E.R., and Tarnawski, M. (1996) Differential susceptibility to neurofibrillary pathology among patients with Down syndrome. *Dementia* **7**, 135–141.

Wisniewski, H.M., and Rabe, A. (1986) Discrepancy between Alzheimer-type neuropathology and dementia in persons with Down's syndrome. *Annals N.Y. Acad. Sci.* **477**, 247–260.

Wisniewski, H.M., Silverman, W., and Wegiel, J. (1994a) Aging, Alzheimer disease and mental retardation. *J. Intellect. Disabil. Res.* **38**, 233–239.

Wisniewski, H.M., Wegiel, J., Kotula, L. (1996) Some neuropathological aspects of Alzheimer's disease and its relevance to other disciplines. *Neuropathol. Appl. Neurobiol.* **22**, 3–11.

Wisniewski, H.M., Wegiel, J., and Popovitch, E.R. (1994b) Age-associated development of diffuse and thioflavine-S-positive plaques in Down syndrome. *Dev. Brain Dysfunct.* **7**, 330–339.

Wisniewski, H.M., Wen, G.Y., Grundke-Iqbal, I., Iqbal, K., Mehta, P.D., Merz, P.A., Miller, D., Currie, J., Bobin, S.A., Robakis, N.K., Rabe, A., and Brown, W.T. (1987) Pathological, biochemical, and genetic aspects of Alzheimer's disease in aged people and Down syndrome, in *Contributions of Chemistry to Health* (Machleidt, H., ed), pp. 297–315. Proceedings of the Fifth CHEMRAWN Conference, Heidelberg, Germany.

Wisniewski, H.M., Wen, G.Y., and Kim, K.S. (1989) Comparison of four staining methods on the detection of neuritic plaques. *Acta Neuropathol.* **78**, 225–231.

Zigman, W., Silverman, W., and Wisniewski, H.M. (1996) Aging and Alzheimer's disease in Down syndrome: Clinical and pathological changes. *Mental Retardation & Developmental Dis. Res. Reviews* **2**, 73–79.

INDEX

GPSR Compliance

The European Union's (EU) General Product Safety Regulation (GPSR) is a set of rules that requires consumer products to be safe and our obligations to ensure this.

If you have any concerns about our products, you can contact us on ProductSafety@springernature.com

In case Publisher is established outside the EU, the EU authorized representative is:

Springer Nature Customer Service Center GmbH
Europaplatz 3
69115 Heidelberg, Germany

Batch number: 10755292

Printed by Printforce, the Netherlands